Natural Materials for Food Packaging Application

Natural Materials for Food Packaging Application

Edited by Jyotishkumar Parameswaranpillai, Aswathy Jayakumar, E. K. Radhakrishnan, Suchart Siengchin, and Sabarish Radoor

WILEY-VCH

Editors

Dr. Jyotishkumar Parameswaranpillai
Alliance University
Chandapura-Anekal Main Road,
Bengaluru 562106,
Karnataka, India

Dr. Aswathy Jayakumar
King Mongkut's University of Technology
North Bangkok
1518 Pracharaj 1
Wongsawang Road, Bangsue
Bangkok, 10800
Thailand

Dr. E. K. Radhakrishnan
Mahatma Gandhi University
Priyadarshini Hills P. O.
Kottayam,
Kerala, 686560
India

Prof. Suchart Siengchin
King Mongkut's University of Technology
1518 Pracharaj 1
Wongsawang Road, Bangsue
Bangkok, 10800
Thailand

Dr. Sabarish Radoor
King Mongkut's University of Technology
1518 Pracharaj 1
Wongsawang Road, Bangsue
Bangkok, 10800
Thailand

All books published by **WILEY-VCH** are carefully produced. Nevertheless, authors, editors, and publisher do not warrant the information contained in these books, including this book, to be free of errors. Readers are advised to keep in mind that statements, data, illustrations, procedural details or other items may inadvertently be inaccurate.

Library of Congress Card No.: applied for

British Library Cataloguing-in-Publication Data:
A catalogue record for this book is available from the British Library.

Bibliographic information published by the Deutsche Nationalbibliothek
The Deutsche Nationalbibliothek lists this publication in the Deutsche Nationalbibliografie; detailed bibliographic data are available on the Internet at <http://dnb.d-nb.de>.

© 2023 WILEY-VCH GmbH, Boschstr. 12, 69469 Weinheim, Germany

All rights reserved (including those of translation into other languages). No part of this book may be reproduced in any form – by photoprinting, microfilm, or any other means – nor transmitted or translated into a machine language without written permission from the publishers. Registered names, trademarks, etc. used in this book, even when not specifically marked as such, are not to be considered unprotected by law.

Print ISBN 978-3-527-35040-7
ePDF ISBN 978-3-527-83732-8
ePub ISBN 978-3-527-83731-1
oBook ISBN 978-3-527-83730-4

Cover Image: © Peshkova/Shutterstock
Typesetting Straive, Chennai, India

Contents

Preface *xiii*
About the Editors *xv*

1 Introduction to Natural Materials for Food Packaging *1*
Manickam Ramesh, Lakshminarasimhan Rajeshkumar, Venkateswaran Bhuvaneswari, and Devarajan Balaji
1.1 Introduction *1*
1.2 Natural Biodegradable Polymers *4*
1.2.1 Starch-Based Natural Materials *4*
1.2.2 Poly-Lactic Acid-Based Natural Materials *5*
1.2.3 Poly-Caprolactone (PCL)-Based Natural Materials *5*
1.2.4 Poly-Hydroxy Alkanoate-Based Natural Materials *6*
1.2.5 Polyglycolide-Based Natural Materials *6*
1.2.6 Polycarbonate-Based Natural Materials *7*
1.2.7 Soy-Based Bio-degradable Polymers *7*
1.2.8 Polyurethanes *7*
1.2.9 Polyanhydrides *7*
1.3 Biodegradable Polymer Blends and Composites *8*
1.3.1 Polylactic Acid and Polyethylene Blends *8*
1.3.2 PLA and Acrylobutadiene Styrene (ABS) Blends *8*
1.3.3 PCL and Polyethylene Blends *8*
1.3.4 PCL and Polyvinyl Chloride Blends *9*
1.3.5 TPS and Polypropylene Blends *9*
1.3.6 TPS/PE Blends *9*
1.3.7 Poly(Butylene Succinate) Blends *10*
1.4 Properties of Natural Materials for Food Packaging *10*
1.4.1 Barrier Properties *10*
1.4.2 Biodegradation Properties *11*
1.4.3 Consequences of Storage Time *12*
1.5 Environmental Impact of Food Packaging Materials *14*
1.6 Conclusion *14*
References *15*

2	**Plant Extracts-Based Food Packaging Films** *23*	
	Aris E. Giannakas	
2.1	Introduction *23*	
2.2	Extraction Methods for Plant Extracts *24*	
2.3	Research Investigation of Bibliographic Data *25*	
2.4	Chitosan Plant Extract-Based Food Packaging Films *27*	
2.5	Starch/Extract-Based Food Packaging Films *30*	
2.6	Cellulose and Cellulosic Derivatives-Based Food Packaging Films Modified with Plant Extract *32*	
2.7	Gelatin and Alginate/Plant Extract-Based Food Packaging Films *34*	
2.8	Composites/Plant Extract-Based Food Packaging Films *35*	
2.8.1	Chitosan Composites/Plant Extract-Based Food Packaging Films *36*	
2.8.2	Starch Composites/Extract-Based Food Packaging Films *38*	
2.8.3	Other Composites Plant Extract-Based Food Packaging Films *39*	
2.9	Conclusion *41*	
	Acknowledgment *41*	
	References *42*	
3	**Essential Oils in Food Packaging Applications** *51*	
	Madhushree Hegde, Akshatha Chandrashekar, Mouna Nataraja, Niranjana Prabhu, Jineesh A. Gopi, and Jyotishkumar Parameswaranpillai	
3.1	Introduction *51*	
3.2	Chemistry and Classification of Essential Oils *52*	
3.3	Essential Oils in Food Packaging Applications *55*	
3.3.1	Effect of Essential Oil on the Mechanical, Barrier, and Other Physical Properties of Food Packaging Materials *55*	
3.3.1.1	Tensile Properties *55*	
3.3.1.2	Barrier Properties *56*	
3.3.1.3	Other Physical Properties *56*	
3.3.2	Antioxidant Properties of Essential Oil Incorporated Food Packaging Materials *58*	
3.3.3	Antibacterial Properties of Essential Oil Incorporated Food Packaging Materials *61*	
3.4	Challenges and Future Trends Associated with the Use of Essential Oil in Food Packaging Applications and Future Trends *65*	
3.5	Conclusions *65*	
	References *66*	
4	**Agro-Waste Residue-Based Food Packaging Films** *75*	
	Rajarathinam Nithya and Arunachalam Thirunavukkarasu	
4.1	Introduction *75*	
4.2	Agro-Waste-Based Biopolymers *76*	
4.2.1	Cellulose *76*	
4.2.2	Hemicellulose *77*	
4.2.3	Lignin *77*	
4.2.4	Starch *78*	
4.2.5	Pectin *79*	

4.3	Edible Coatings and Films – Classification and Properties	*80*
4.4	Conclusion and Future Prospects	*83*
	References	*83*

5   Hydrogel-Based Food Packaging Films *89*
Kunal Singha and Kumar Rohit

5.1	Introduction	*89*
5.2	Hydrogel Nature, Definition	*91*
5.2.1	Hydrogel Types and Features	*91*
5.2.1.1	Classification According to Polymeric Composition	*91*
5.2.1.2	Classification Based on Configuration: Classification is Done Based on the Setting	*91*
5.2.1.3	Classification Based on the Type of Cross-Linking	*91*
5.2.1.4	Classification Based on Physical Appearance	*92*
5.2.1.5	Classification According to Network Electrical Charge	*92*
5.3	Preparation of Hydrogel Film	*92*
5.4	Hydrogel as Food Packaging Material	*92*
5.4.1	Hydrogels Functional Properties	*93*
5.5	Classification of Hydrogel	*93*
5.6	Hydrogels Functional Properties	*93*
5.7	Potential Application of Hydrogel in Food Packaging Systems	*95*
5.7.1	Applications of Hydrogels In Vitro and Food Matrices	*96*
5.7.2	Biodegradable Packaging	*96*
5.7.3	Biodegradability	*97*
5.7.4	Other Potential Applications in the Food Industry	*98*
5.8	Latest Development in the Hydrogel in the Field of Food Packaging	*98*
5.9	Futuristic Uses of Hydrogel in Miscellaneous Process	*99*
5.10	Conclusions	*100*
	References	*101*

6   Natural Fiber-Based Food Packaging Films *105*
G. Rajeshkumar, M. Karthick, A.K. Aseel Ahmed, T. Vikram Raj, V. Abinaya, K. Madhu Mitha, and R. Ronia Richelle

6.1	Introduction	*105*
6.2	Manufacturing of Fiber-Reinforced Biofilms	*107*
6.3	Rice Straw-Based Films	*109*
6.4	Wheat Straw-Based Films	*109*
6.5	Jute-Based Films	*111*
6.6	Pineapple-Based Films	*112*
6.7	Flax-Based Films	*113*
6.8	Kenaf-Based Films	*114*
6.9	Hemp-Based Films	*115*
6.10	Conclusions	*115*
	References	*116*

7	**Natural Clay-Based Food Packaging Films** *121*	
	Ram Kumar Deshmukh, Dakuri Ramakanth, Konala Akhila, and Kirtiraj K. Gaikwad	
7.1	Introduction *121*	
7.2	Clay Materials Classification *127*	
7.2.1	TO or 1:1 Type (One-One Tetra-octahedral Layer) *127*	
7.2.2	TOT or 2:1 Type (One-Octahedral in Between Two Tetrahedral Layers) *128*	
7.2.3	2:1:1 or TOTO Type (Two Tetrahedral with Two Octahedral) *128*	
7.3	Preparation of Natural Clay Nanocomposites *128*	
7.3.1	In situ Polymerization Method *130*	
7.3.2	Solution-Induced Intercalation *130*	
7.3.3	Melt Processing *130*	
7.4	Properties of Natural Clay-Based Nanocomposite Polymer *130*	
7.4.1	Mechanical Properties *131*	
7.4.2	Barrier Properties *132*	
7.4.3	Thermal Stability of Clay-Based Polymer Composites *133*	
7.4.4	Oxygen and Ethylene Scavenging Activity of Nano-Clay Polymer Composite *133*	
7.5	Application of Natural Clay in Food Packaging Film *135*	
7.5.1	Montmorillonite (MMT)-Based Nanocomposite *139*	
7.5.2	Laponite-Reinforced Polymer Nanocomposite *141*	
7.5.3	Sepiolite-Reinforced PNC *141*	
7.5.4	Bentonite-Reinforced Polymer Nanocomposite *142*	
7.5.5	Hectorite-Reinforced Polymer Nanocomposite *143*	
7.5.6	Rectorite-Reinforced Polymer Nanocomposite *144*	
7.5.7	Other Nanoclay Materials-Based Nanocomposites *145*	
7.6	Challenges of Using Clay in Food Packaging Applications *145*	
7.6.1	Migration and Exposure of Nanoclay Materials to Humans and the Environment *146*	
7.6.2	Toxicity of Nanoclay *148*	
7.7	Future Outlook and Conclusion *149*	
	References *150*	
8	**Curcumin-Based Food Packaging Material** *165*	
	Leidy T. Sanchez, Andres F. Cañon-Ibarra, J. Alejandro Arboleda-Murillo, and Cristian C. Villa	
8.1	Structural Characteristics of Curcumin *165*	
8.2	Antimicrobial, Antifungal, and Antioxidant Properties of Curcumin *166*	
8.3	Nanoencapsulation of Curcumin *167*	
8.4	Curcumin-Based Food Packaging *168*	
8.5	Curcumin-Based Nanocomposite Food Packaging *169*	
8.6	Curcumin-Based Active Food Packaging *169*	
8.7	Curcumin-Based Intelligent Food Packaging *170*	
8.8	Perspectives *171*	
	References *171*	

9	**Sustainable Materials from Starch-Based Plastics** *179*
	Asanda Mtibe and Maya J. John
9.1	Introduction *179*
9.1.1	Starch *179*
9.1.2	Preparation of Thermoplastic Starch (TPS) *180*
9.1.3	Plasticization of Starch *180*
9.1.4	Processing of TPS *183*
9.1.5	Properties of TPS *185*
9.1.5.1	Mechanical Properties *185*
9.1.5.2	Thermal Properties *186*
9.1.5.3	Barrier Properties *186*
9.2	TPS-Biopolymer Blends *187*
9.3	TPS-Biopolymer Composites *188*
9.4	Global Producers, Market Volumes, and Applications of Starch-Based Plastics *191*
9.5	Conclusions *193*
	References *193*
10	**Main Marine Biopolymers for Food Packaging Film Applications** *199*
	Jesús Rubén Rodríguez-Núñez, Diana Gabriela Montoya-Anaya, Judith Fortiz-Hernández, Yolanda Freile-Pelegrín, and Tomás Jesús Madera-Santana
10.1	Introduction *199*
10.2	Polysaccharides from Seaweeds *200*
10.2.1	Main Seaweed Polysaccharides *201*
10.2.2	Alginate *202*
10.2.2.1	Properties and Limitations of Alginate *204*
10.2.2.2	Applications of Alginate in Edible Films and Coatings *205*
10.2.3	Agar *205*
10.2.3.1	Applications of Agar in Edible Films and Coatings *210*
10.2.4	Carrageenan *213*
10.2.5	Fucoidan *216*
10.2.6	Ulvan *218*
10.3	Modified Chitosan for Food Film Applications *220*
10.3.1	Chemical Modifications of Chitosan for Food Packaging *220*
10.3.2	Chitosan Blends/Composites for Films and Coating for Food Applications *222*
10.3.3	Nanomaterials of Chitosan for Food Packaging *224*
10.4	Conclusions and Future Trends *226*
	References *227*
11	**Chitosan-Based Food Packaging Films** *241*
	Kunal Singha and Kumar Rohit
11.1	Introduction *241*
11.1.1	A Brief History of Food Packaging Materials Used *241*
11.1.2	Characteristics of Typical Food Packaging Materials *242*
11.1.3	Need for Biodegradable Food Packaging Materials *242*
11.2	Chitin and Chitosan Chemical Structure *243*

11.3	Chitosan as a Potential Biodegradable Food Packaging Material *243*	
11.3.1	Chitosan as Food Packaging Material *244*	
11.3.2	Chitosan Film in Food Packaging and Their Types *244*	
11.3.2.1	Chitosan-Based Films *245*	
11.3.2.2	Flexible Packaging Films *245*	
11.3.3	Chitosan Film in Food Packaging *245*	
11.3.4	Films Embedded with Nanomaterials *245*	
11.3.5	Films Embedded with Clays *246*	
11.3.6	Films Embedded with Polysaccharide Particles, Fibres, and Whiskers *247*	
11.3.7	Films Embedded with Natural Oils and Extracts *247*	
11.4	Future Research Directions and Developments *249*	
11.4.1	Chitin/Chitosan Derivatives and Their Interactions with Microorganisms: A Comprehensive Review and Future Perspectives *249*	
11.4.2	A Future Perspective in Crop Protection: Chitosan and its Oligosaccharides *249*	
11.4.3	Chitosan in Molecularly-Imprinted Polymers: Current and Future Prospects *250*	
11.4.4	Crosstalk Between Chitosan and Cell Signaling Pathways *250*	
11.4.5	Resorbable Chitosan Matrix – As a Promising Biomaterial for the Future *250*	
11.5	Conclusions *251*	
	References *251*	

12 Effect of Natural Materials on Thermal Properties of Food Packaging Film: An Overview *255*
H. M. Prathibhani C. Kumarihami, Nishant Kumar, Pratibha, Anka T. Petkoska, and Neeraj
Abbreviations *255*

12.1	Introduction *256*	
12.2	Biodegradable Films: An Alternative for Food Packaging *257*	
12.2.1	Biodegradable Polymers *258*	
12.3	Thermal Properties of Food Packaging *259*	
12.4	Effects of Natural Materials on the Thermal Stability of Food Packaging *260*	
12.4.1	Effects of Plant Extract *260*	
12.4.2	Effects of Essential Oils *261*	
12.4.3	Effects of Color Agent *262*	
12.4.4	Effects of Nanomaterials *263*	
12.4.5	Effects of Plasticizers *265*	
12.4.6	Effects of Emulsifiers *266*	
12.5	Conclusions *266*	
	References *267*	

13 Mechanical Properties of Natural Material-Based Packaging Films: Current Scenario *275*
Johnsy George, Muhammed Navaf, Aksalamol P. Raju, Ranganathan Kumar, and Kappat V. Sunooj

13.1 Introduction *275*
13.2 Mechanical Properties of Packaging Films *276*
13.2.1 Tensile Strength (TS) *277*
13.2.2 Young's Modulus (Y) *277*
13.2.3 Elongation at Break (EB) *278*
13.2.4 Seal Strength *278*
13.2.5 Tear Resistance *278*
13.2.6 Puncture Resistance *279*
13.2.7 Impact Resistance *279*
13.2.8 Burst Strength *279*
13.3 Mechanical Properties of Natural Polymer-Based Packaging Films *279*
13.3.1 Naturally Occurring Polymers *280*
13.3.1.1 Starch *280*
13.3.1.2 Cellulose *283*
13.3.1.3 Chitosan *284*
13.3.1.4 Alginates *285*
13.3.1.5 Pectin *285*
13.3.1.6 Casein *286*
13.3.1.7 Whey Protein *287*
13.3.1.8 Collagen *287*
13.3.1.9 Gelatin *288*
13.3.1.10 Zein Protein *289*
13.3.1.11 Soy Protein *290*
13.3.1.12 Gluten Protein *291*
13.3.2 Polymers Synthesized from Natural/Bioderived Monomers *292*
13.3.2.1 Polylactic Acid (PLA) *292*
13.3.2.2 Polyethylene Furanoate (PEF) *295*
13.3.2.3 Polybutylene Succinate (PBS) *295*
13.3.2.4 Poly(Butylene Adipate-*co*-Terephthalate) *296*
13.3.2.5 Bio-based Polyethylene *296*
13.3.2.6 Bio-Based Polypropylene (Bio-PP) *296*
13.4 Mechanical Properties of Natural Polymers Synthesized from Microorganisms-Based Packaging Films *296*
13.4.1 Polymer Processed from Microorganisms *296*
13.4.1.1 Polyhydroxyalkanoate (PHA) *296*
13.4.1.2 Bacterial Cellulose *298*
13.4.1.3 Xanthan *299*
13.4.1.4 Pullulan *299*
13.4.1.5 Gellan *300*
13.4.1.6 Levan *300*
13.5 Conclusion *300*
References *301*

14 Effects of Natural Materials on Food Preservation and Storage *313*
Subhanki Padhi and Winny Routray

14.1 Introduction *313*
14.1.1 Major Objective of Food Preservation and Storage *313*
14.1.2 Available Solutions from the Natural Resources and Combination with Technology *314*
14.2 Biomolecules Utilized for Preservation, Their Properties, and Uses *315*
14.2.1 Polysaccharides *315*
14.2.2 Essential Oil *316*
14.2.3 Phenolic Compounds *318*
14.2.4 Aromatic Compounds *319*
14.2.5 Proteins *320*
14.2.6 Bacteriocins *320*
14.2.7 Other Animal-Based Antimicrobials *321*
14.3 Different Extraction Processes Employed for Natural Materials *321*
14.4 Effects of Natural Materials on Different Product Quality and Storage *323*
14.4.1 Drying Methods and Corresponding Properties *323*
14.4.2 Enhancement of Packaging Characteristics *323*
14.4.3 Maintenance of Physiochemical Properties of Raw and Processed Products *324*
14.5 Conclusion *325*
References *326*

15 Marketing, Environmental, and Future Perspectives of Natural Materials in Packaging *333*
Prakash Binu, Sasi Arun Sasi, Velamparambil Gopalakrishnan Gopikrishna, Abdul Shukkur, Balu Balachandran, and Mahesh Mohan

15.1 Introduction *333*
15.2 Biodegradable Food Packaging *334*
15.3 Different Bio-Based Packaging Materials *336*
15.3.1 Bioplastics *336*
15.3.2 Biopolymers *336*
15.4 Nano Food Packaging *338*
15.5 Natural Antimicrobial Agents in Food Packaging *338*
15.6 Edible Films in Food Packaging *339*
15.7 Environment and Food Packaging *341*
15.8 Sustainable Packaging *342*
15.9 Marketing of Natural Materials in Packaging *343*
15.10 Future Perspectives of Natural Materials in Packaging *344*
15.11 Conclusion *345*
References *345*

Index *353*

Preface

For packaging applications, nonbiodegradable polymers such as polyethylene terephthalate, polyethylene, polyvinyl chloride, polypropylene, and polystyrene are widely used due to their low cost, easy processing, and great resistance properties while handling. However, these nonbiodegradable polymers are toxic to the environment. It is important to add that out of all these plastics produced, c. 36% is used for packaging applications, and out of this more than 70% of the plastic is used for food packaging. Finally, most of these plastics (approximately 92%) are either landfilled or dumped in water bodies. Therefore, recent efforts are going on to introduce natural polymers to preserve food products. However, there are many parameters, such as transparency, transport properties, thermo-mechanical stability, and cost, to be considered before finalizing any biopolymer for packaging applications. This book discussed in detail all the advancements, prospects, and limitations of natural polymers in food packaging applications.

In this book, we have 15 chapters, emphasizing the global trend of using natural polymers for food packaging. **Chapter 1**. *Introduction to natural materials for food packaging* highlights the importance of natural biodegradable polymers, such as starch, polylactic acid, polycaprolactone, and poly-hydroxy alkenoates, and their blends in food packaging applications. The properties of natural polymers, such as biodegradation, barrier properties, aging properties, and environmental impact, are also highlighted. **Chapter 2**. *Plant extract-based food packaging films* highlights the importance of plant extract in food packaging. Plant extracts are known for their antioxidant and antimicrobial properties. The authors reviewed critically the effect of plant extract on biopolymer-based packaging films and underlined that the incorporation of plant extract into packaging films is a promising method to improve food quality and to extend the food shelf life. **Chapter 3**. *Essential oils in food packaging applications* discussed the impact of using essential oils on the physical, antimicrobial, and antioxidant properties of biobased polymer films. **Chapter 4**. *Agro-waste residue based food packaging films* discussed the concepts of reduce and reuse. This chapter discussed the properties of cellulose, hemicellulose, lignin, starch, and pectin-based biofilms (isolated from agro-waste). Functional properties in these films can be introduced by the incorporation of antioxidant and antimicrobial agents such as tea polyphenols, AgNPs, mint, rosemary oils. **Chapter 5**.

Hydrogel-based food packaging films discussed the basics, properties, latest developments, and uses of hydrogels in the food industry.

Chapter 6. *Natural fiber based food packaging films* discussed the reinforcement effect of natural fibers (from rice straw, wheat straw, jute fiber, pineapple fiber, flax fiber, kenaf fiber, hemp fiber, etc.) in different biopolymer matrices (starch, chitosan, polyvinyl alcohol, polylactic acid, poly(3-hydroxybutyrate-*co*-3-hydroxyvalerate), etc.) for food packaging films. **Chapter 7.** *Natural clay-based food packaging films* discussed the possibility and limitations of nanoclay-reinforced polymer composites for food packaging applications. **Chapter 8.** *Curcumin-based food packaging material* discussed the growing interest in using curcumin for active and intelligent food packaging. The limitations of using curcumin in food packaging industries are also discussed in the chapter. **Chapter 9.** *Sustainable materials from starch-based plastics* gives an overview of the preparation and properties of thermoplastic starch, thermoplastic starch-based blends, and thermoplastic starch-based composites. The current market volume and applications are also discussed. **Chapter 10.** *Main marine biopolymers for food packaging film applications* summarizes the uses, modifications, and applications of marine polysaccharides for food packaging applications. **Chapter 11.** *Chitosan-based food packaging films* gives an overview of the application of one of the most popular biopolymers "chitosan-based" films for the food packaging industry. Chitosan-based films are best known for their biocompatibility and antibacterial properties. **Chapter 12.** *Effect of natural materials on thermal properties of food packaging film: an overview* gives an overview of how active ingredients such as plant extract, essential oils, color agents, nanomaterials, plasticizers, and emulsifiers influence the thermal properties of the food packaging biopolymers. **Chapter 13.** *Mechanical properties of natural material-based packaging films: current scenario* gives an overview of the mechanical properties of biopolymers based on starch, cellulose, chitosan, alginates, pectin, casein, whey protein, collagen, gelatin, zein protein, soy protein, and gluten protein. The authors also reviewed the mechanical properties of polymers derived from natural materials and microorganisms. **Chapter 14.** *Effects of natural materials on food preservation and storage* gives a comprehensive overview of the impact of natural materials on food preservation and maintenance of quality. **Chapter 15.** *Marketing, environmental, and future perspectives of natural materials in packaging* gives an overview of the importance of natural materials in food packaging from a global perspective.

01-01-2023

Jyotishkumar Parameswaranpillai, India
Aswathy Jayakumar, Thailand
E. K. Radhakrishnan, India
Suchart Siengchin, Thailand
Sabarish Radoor, Thailand

About the Editors

Dr. Jyotishkumar Parameswaranpillai is currently an associate professor at Alliance University, Bangalore. He received his PhD in Chemistry (Polymer Science and Technology) from Mahatma Gandhi University, Kottayam, India, in 2012. He has research experience in various international laboratories such as the Leibniz Institute of Polymer Research Dresden (IPF), Germany; Catholic University of Leuven, Belgium; University of Potsdam, Germany; and King Mongkut's University of Technology North Bangkok (KMUTNB), Thailand. He has more than 240 international publications. He is a frequent invited and keynote speaker and a reviewer for more than 70 international journals, book proposals, and international conferences. He received numerous awards and recognitions including the prestigious INSPIRE Faculty Award 2011, Kerala State Award for the Best Young Scientist 2016, and Best Researcher Award 2019 from King Mongkut's University of Technology North Bangkok. He is named among the world's Top 2% of the most-cited scientists in the Single Year Citation Impact (2020, 2021) by Stanford University. His research interests include polymer coatings, shape memory polymers, antimicrobial polymer films, green composites, nanostructured materials, water purification, polymer blends, and high-performance composites.

Dr. Aswathy Jayakumar is currently working as a Post-doctoral Researcher at Kyung Hee University, Seoul, South Korea. She completed her Post Doctoral Fellowship from King Mongkut's university of Technology, North Bangkok Thailand (2022). She received her Ph.D in Biotechnology from School of Biosciences, Mahatma Gandhi University, Kottayam, India (2021). She has authored more than 60 international publications. She received the Best paper award in Biotechnology 2019 in Kerala Science Congress (Kerala State Award). Her area of research is functional biology of endophytic microorganisms, molecular and genomic studies, bionanocomposites-based food packaging films, carbon quantum dots and their applications.

Dr. E. K. Radhakrishnan, PhD, is an associate professor at the School of Biosciences; Director of Business Innovation and Incubation Center; and Joint Director of the Inter University Centre for Organic Farming and Sustainable Agriculture, Mahatma Gandhi University, Kottayam, India. During his 12 years of research, he has published over 100 research publications, many book chapters, and several review papers. He edited two books with Springer Nature and Elsevier, and six books are in progress. His work has been cited almost 3709 times, and his h-index is 33 and i10-index is 79. To date, he has delivered over 40 invited talks at various national and international conferences and seminars. He has completed several research projects for various funding agencies and has five ongoing projects as PI. His research areas include plant–microbe interactions, microbial natural products, microbial synthesis of metal nanoparticles, and development of polymer-based nanocomposites with antimicrobial effects for food packaging and medical applications. He completed his doctoral degree at the Rajiv Gandhi Centre for Biotechnology, Thiruvananthapuram, Kerala, India, and his postdoctoral studies at the University of Tokyo, Japan.

Prof. Dr.-Ing. habil. Suchart Siengchin is President of King Mongkut's University of Technology North Bangkok (KMUTNB), Thailand. He received his Dipl.-Ing. in Mechanical Engineering from the University of Applied Sciences Giessen/Friedberg, Hessen, Germany; his MSc in Polymer Technology from the University of Applied Sciences Aalen, Baden-Wuerttemberg, Germany; his MSc in Material Science at the Erlangen-Nürnberg University, Bayern, Germany; his Doctor of Philosophy in Engineering (Dr.-Ing.) from the Institute for Composite Materials, University of Kaiserslautern, Rheinland-Pfalz, Germany; and his postdoctoral research from the School of Materials Engineering, Purdue University, United States. In 2016, he completed the Habilitation (Dr.-Ing. habil.) in Mechanical Engineering from Chemnitz University of Technology, Saxony, Germany, and worked as a lecturer in the Mechanical and Process Engineering Department at the Sirindhorn International Thai-German Graduate School of Engineering (TGGS), KMUTNB. He has been full professor at KMUTNB, became the Vice President for Research and Academic Enhancement in 2012, and was elected President of KMUTNB in November 2016. He won the Outstanding Researcher Award in 2010, 2012, and 2013 at KMUTNB and the National Outstanding Researcher Award for 2021 in engineering and industrial research from the National Research Council of Thailand (NRCT). His research interests are in polymer processing and composite materials. He is editor-in-chief of *Applied Science and Engineering Progress*, International Advisory Board of *eXPRESS Polymer Letters* and the *Journal of Production Systems and Manufacturing Science*, and the author of more than 321 peer-reviewed journal articles and edited books and book chapters in more than 139 books. He has participated with presentations in more than 49 international and national conferences with respect to materials science and engineering topics.

Dr. Sabarish Radoor received B.Sc (Polymer Chemistry) from Calicut University, Calicut, Kerala, India in the year 2006, M.Sc (Applied Chemistry) from Calicut University, Calicut, Kerala, India in the year 2008, M.Tech (Industrial Catalysis) from Cochin University of Science and Technology, Kochi in the year 2011 and Ph.D. (Chemistry) from National Institute of Technology, Calicut in the year 2019. He worked as a Post-doctoral Researcher at Production and Material Engineering, Department at The Sirindhorn International Thai-German Graduate School of Engineering (TGGS), KMUTNB. Currently, he works as a Post-doctoral fellow at the Department of Polymer-Nano Science and Technology, Jeonbuk National University, Republic of Korea. He has published over 70 articles including book chapters in high-quality international peer-reviewed journals. His current research areas include wastewater treatment, natural fiber composites, zeolites, and Intelligent food packaging.

1

Introduction to Natural Materials for Food Packaging

Manickam Ramesh[1], *Lakshminarasimhan Rajeshkumar*[2], *Venkateswaran Bhuvaneswari*[2], *and Devarajan Balaji*[2]

[1] *KIT-Kalaignarkarunanidhi Institute of Technology, Department of Mechanical Engineering, Coimbatore, Tamil Nadu 641402, India*
[2] *KPR Institute of Engineering and Technology, Department of Mechanical Engineering, Coimbatore, Tamil Nadu 641407, India*

1.1 Introduction

Food packaging material should also maintain the lifetime of the food by dodging adverse conditions such as spoilage microorganisms, mechanical vibration, shocks, gases, moisture, chemical contamination, bad odor, and exposure to oxygen. Fresh and healthy foods are the everlasting demands of the consumers in the global market after the inception of packaged foods [1, 2]. Bio-based polymers have been the first-choice materials for food packaging applications which not only promote sustainable material development but also overcome environmental concerns causing very less ecological threats. To enhance the shelf life of the food material as well as to uphold its quality as it is from the date of manufacture, plastic-based food packaging materials are used [3]. Muizniece-Brasava et al. [4] stated in their study that as per the recent statistics, an 8% annual increase in production of packaging materials from petroleum-based materials has been seen, but on the other hand, only 5% of those materials were potentially recycled. This resulted in almost million ton plastic packaging materials in landfills each year due to accumulation of non-recycled materials, thus affecting the environment which is the current-day problem on planet earth.

One concern about plastic pollution has motivated the growth of degradable, natural, and green product materials [5, 6]. Combining a biodegradable polymer consequent from renewable sources with a natural fiber filler to create a bio-composite represents a self-sustaining as well as a technically feasible alternative to so-called "commodity" plastic products in the food packaging sector. Three critical factors must be considered. To start, we must reduce our reliance on petroleum-based

---

Natural Materials for Food Packaging Application, First Edition.
Edited by Jyotishkumar Parameswaranpillai, Aswathy Jayakumar,
E. K. Radhakrishnan, Suchart Siengchin, and Sabarish Radoor.
© 2023 WILEY-VCH GmbH. Published 2023 by WILEY-VCH GmbH.

materials while increasing our use of renewable sources to make plastics, thereby reducing the amount of old carbon put into the atmosphere. Second, the use of biopolymers enables the package to be treated similar to an organic biodegradable residue following its use, thereby helping to decrease polymeric trash bound for landfills and incinerators. Eventually, use of natural fabrics as fillers enables the valorization of agricultural residues, thereby reducing the food production cycle's overall impact. Around each other, use of biocomposites composed of biodegradable polymers based on renewable sources and fillers resulting from agricultural fiber garbage as well as other by-products enables more justifiable products by promoting a cradle-to-cradle approach and the life cycle assessment (LCA) [7, 8]. Figure 1.1 shows various materials used in recent days for food packaging.

Green packaging materials made of biodegradable composite are gaining increasing interest in a range of disciplines owing to their distinctive characteristics in comparison to conventional petrochemical-based plastics [9]. Furthermore, they are fully biodegradable and degrade completely including organic material, H_2O, and carbon dioxide. These characteristics may enable their use in diverse applications, including smart nano-food wrapping [10–12], biomembranes for water purification, recycling of waste, as well as drug delivery. Thus, the primary function of packaging material is to enhance the quality and safety of food while extending its life span [13]. Due to their ability to prevent the transmission of humidity, oxidant, and flavors among foodstuffs and their surrounding environment, edible films may be a worthwhile alternative to plastics in a variety of applications [14]. As a result, use of edible coatings for preserving the quality of various foods has grown rapidly [15]. Recently, a variety of biodegradable food packaging materials, including sipping beverages,

Figure 1.1 Various food packaging materials. Source: Sanyang et al. [8]/with permission from Springer Nature.

sheets, silverware, overwrap, as well as lamination films, have been manufactured and distributed through grocery stores [16–18].

Owing to its least cost along with easy accessibility for industries, carbohydrate, a normally sustainable energy fructose polymer, is the most frequently utilized fresh material to produce biodegradable plastics [19]. Other studies have been conducted to determine its great potential in aqua-soluble pouches for storing detergents and insecticides, as well as to determine its utility in washable lining, satchels, and other medical equipment. Starch is composed of two molecules: amylose (a sequential chemical compound with very few branch offices) and amylopectin (a branched chain molecule). When starch is processed, the existence of amylose in significant amounts provides strength to the films. Tensile stress in layers is observed to reduce when amylopectin is a predominant component of the starch. Maize or corn flour is the primary source of starch, accounting for approximately 80% of the global market. Rice starches exhibit a range of characteristics depending on the paddy variety [20, 21], resulting in biodegradable films with a range of characteristics. Rice starches are being used in place of synthetic films to generate biologically decomposable films owing to their low cost, abundant availability in nature, and acceptable mechanical characteristics. However, these rice starch films lack adequate barrier properties against nonpolar compounds, restricting their application. This led to development of rice-based starch films with improved characteristics [22, 23].

Among the various widely viable bioplastics, poly(3-hydroxybutyrate) (PHB) is a particularly interesting member of the hydroxyl alkanoates family for packaging applications. PHB is a plastic material that can be transformed industrially using standard polymer transformation equipment. Additionally, it has a good mechanical result in terms of strength and stiffness that is comparable to or greater than that of some commodities (for example, PP), as well as barrier characteristics (comparable to PET). PHB worsens in composting environments and other surroundings such as saltwater [3]. While PHB is an interesting choice for self-sustaining packaging applications, it does have some drawbacks that limit its widespread use in the fresh produce food packaging sector. PHB has a high intrinsic fragility, which upsurges over time as a result of a second crystal growth and physical aging. Additionally, owing to the high crystallinity, PHB has a narrow handling window, which makes it unsuitable for some prevalent food applications, such as blow molding [24, 25].

Edible coatings are very thin films of material (generally or less than 0.3 mm in thickness) which are used to cover food goods to substitute or strengthen the natural layers. They can be devoured as a product or after further expulsion. As a result, the ingredients used in the composition should adhere to applicable food policies and guidelines. Furthermore, the adhesives and films should not have a detrimental effect on the food product's organoleptic properties. Edible packaging may take the form of a surface-level coating on the meals or constant layers among compartments/ingredients of heterogeneous products (for example, grilled cheese, pastry shop fillers, and toppings) [26, 27]. Additionally, the coating can be given to individual pieces of a larger product that are not being independently wrapped for practical reasons, including nuts, kiwis, fruit, veggies, fresh-cut watermelon, and fruits.

Edible films as well as layering can be used to counter a variety of barriers associated with food marketing. These features can be specified as restraining the migration of moisture, solute, oil, and gas, enhancing structural stability, retentive volatile flavor compounds, and transporting dietary supplements. Additionally, they enhanced the attractive look by reducing physical damage, trying to conceal scar tissue, and enhancing surface glow. For example, citrus fruits have been encased with hot-melt paraffin wax to retard moisture absorption, eatable connective tissue canisters have been used to offer structural integrity to sausages, and fruits have been encased with sealant to shape the way glow and inhibit actual injury [28, 29].

1.2 Natural Biodegradable Polymers

Biodegradable polymers are considered to be the most likely solution for dodging various environmental hiccups like litter, landfills, and waste pollution which originate due to the use of nonbiodegradable polymers. But owing to the cost of processing and the limited range of selection, the utilization of biodegradable plastics is less than expected in various end applications these days [30]. Hence, it was stated in various researches that blend of one renewable and biodegradable polymer with another during the process of preparation reduces the cost and widens the prospects of industrial application. Materials including chitosan, proteins, starch, lignin, and cellulose are some of the prominently known biodegradable elements derived from polysaccharides and natural oil bioresources. Few other materials like PLA, PBH, and PCL are derived from partially biodegradable raw materials and they are also categorized under biodegradable polymers.

Natural oils, which can be obtained from both animal as well as plant sources, are one of the abovementioned potential substitutes for chemical raw materials. Triglycerides are widely used in agricultural chemicals, inks, as well as coatings, according to numerous studies. Many of the above implementations made use of brand-new triglyceride oil polymerization and monomerization methods. Nanocellulose, on the other hand, is a comparatively newly developed type of nanomaterial with superior physical as well as chemical properties that are widely used. The nanocellulose in these materials has the prospect to change the top layer chemistry of the embedded material, making them more flexible, stronger, as well as lighter than conventional nanomaterial [18, 19].

1.2.1 Starch-Based Natural Materials

Starch has been the most extensively as well as frequently utilized biopolymer derived solely from renewable as well as natural resources. Due to starch's low cost, complete biodegradability, and ease of availability, starch-based polymers are mostly in high demand these days. Any biodegradable polymer can be incorporated into thermoplastic starch (TPS) in an attempt to lessen the manufacturing costs of biopolymers. Aside from starch, polysaccharides obtained from plants are the most abundant as well as the renewable class of polysaccharides. Amylopectin and amylose are the two major glucose polymers in starch. Amylopectin seems to be a polymeric chain of D-glucose atomic chains linked together by −1, 4 branched bonds, while amylose is a short −1, 4 connected D-glucose chain made up of atoms with −1, 6 branched bonds. The hydrophilicity, as well as brittleness of starch, make it difficult to use, despite the fact that it is completely biodegradable, low cost, as well as able to generate film-forming components with low oxygen permeability as well as the capacity to be managed easily. As a result, starch-based polymers cannot be used in common applications like food packaging and plastic bag substitutes. Different researchers have used plasticizers like sorbitol, glycerol, and glycol underneath the activity of shear stress as well as heat throughout the extrusion process to transform starch into TPS to resolve the shortcomings mentioned above as well as enhance processing potential as well as flexibility [20–22].

1.2.2 Poly-Lactic Acid-Based Natural Materials

PLA, a biodegradable polymer, is one of the most frequently used biodegradable polymers, along with starch, a significant variant of the aliphatic polyester lactic acid, which is a byproduct from the fermentation of plants like sugar beets as well as corn. This biodegradable polymer, like starch, is cheap as well as plentiful, so it has received a lot of attention from researchers and manufacturers. Additional advantages include its biocompatibility, commercial availability, complete biodegradability, ease of processing as well as high transparency. These are the primary reasons for its widespread use [23]. Lactic acid is typically produced during the petrochemicals or bacteria fermentation. Plastic film (PLA) is created by condensation polymerization of lactic acid or by opening the ring polymerization reaction of lactide monomer enshrined in lactic acid (L or D-lactic acid). Polycondensation of lactic acid is used to produce PLA with a lower molecular weight if necessary. Ring-opening polymerization reaction and also azeotropic polymerization condensation of lactic acid, on the other hand, produce PLA with higher molecular weight and considerate mechanical properties [24].

Several scientists have attempted to make PLA-based biomaterials by reinforcing the PLA matrix along with nanocellulosic substances over the previous decade. The mechanical strength, as well as stiffness of the biomaterials depending on PLA, was increased when the original nanocellulose has been reinforced in the PLA matrix to receive biomaterials depending on PLA [25]. A variety of chemical and physical surface modification techniques, including polymer grafting or derivatization and macromolecule or surfactant coating, have been utilized to improve the compatibility as well as a scattering of nanocellulose fillers within a hydrophobic and nonpolar PLA matrix, thereby improving the interfacial characteristics between the filler as well as matrix and the effectiveness of nanocellulose-embedded PLA composites [26–29].

1.2.3 Poly-Caprolactone (PCL)-Based Natural Materials

This polymer is a thermoplastic polymer that has better biodegradability, lower viscosity, better thermal computing ability, as well as the least melting point in the range of 55–60 °C [30]. This is because the interfacial bond formed by the straightforward mixing of chemically inconsistent nanocellulose and PCL fibers in reinforced PCL biomaterials may have led to the observation that nanocellulose could only be reinforced in small percentages with PCL matrix. This incompatibility, which results in poor interfacial properties, has been found to be remedied by surface-modifying nanocellulose fibers, which improve the reinforcement's compatibility with other components of biomaterials. Adding polymer chain surface transplanting straightforwardly to the nanocellulose fiber surface could improve the nanocellulose fibers' bioavailability in a PCL polymer matrix significantly. As a result, modified nanocellulose is an excellent choice for reinforcing PCL-based biocomposite materials [31–33].

1.2.4 Poly-Hydroxy Alkanoate-Based Natural Materials

Biodegradable polyester PHA is drawn from various hydroxy alkanoates via microbial fermentation as well as could be used in diverse applications, such as agricultural, medical as well as packaging industries. PHA begins with hydroxyalkanoate monomers, which are polyester family members. For example, these materials have low melting points, high crystallinity of thermoplastic elastomer molecules with thermoplastic polymers, excellent biocompatibility, as well as superior resistance to UV light. All of these properties of PHA have been governed by the monomer configuration. For food packaging with a short shelf life, PHA, as well as PHB, is the most likely candidate. Both P3HB-*co*-3HV polymers and P3HB homogeneous polymers are naturally occurring forms of polyhydroxybutyrate (PHB). Because they are polymeric granules, PHAs in bacteria serve as an energy storage medium, much like starch and fat do in plants as well as animals, respectively [34–36].

Compared to other non-polymers like polyethylene, the mechanical properties of PHB with 70% crystallinity were superior. The lamellar structure of PHB is the reason for its water permeability, barrier properties, as well as excellent aromatic behavior, which all contribute to its use in food packaging. As a result of these experimental studies, many researchers have attempted to use PHA/PHB in diverse applications despite its lower mechanical as well as barrier characteristics than PHA. When PHB was mixed with PLA as well as catechins have been added through the melt handling, the mechanical characteristics of the mixtures were evaluated. According to the findings, adding PHB–catechin combinations to plasticized PLA enhanced its mechanical characteristics, making it an excellent candidate for use in the containers of fatty foods [37].

1.2.5 Polyglycolide-Based Natural Materials

Ring-opening polymerization structure comprising a cyclic lactone as well as glycolide is used to make polyglycolide. The crystallinity index is around 50%, and as a result, it is insoluble in a variety of organic solvents. Its melting point is 222–226 °C, and its glass transition temperature is 37–42 °C. Polyglycolide is a strong material. Its biomedical implementations are limited, nevertheless, owing to its poor solubility as well as high acid-producing deterioration rate. The result is the development of caprolactone, trimethylene carbonate glycolide, or lactide, copolymers for healthcare devices [38, 39].

1.2.6 Polycarbonate-Based Natural Materials

It's a polymer with something like an elevated molecular weight that's easy to mold and bend. Two chemicals, glycolide and dioxanone, were combined to create copolymers. The copolymerization of propylene oxide as well as CO_2 results in polypropylene carbonate. For example, polycarbonate is easy to use and has a high degree of impact resistance. In the past, it has been combined with numerous different polymeric materials as a traditional method of use. The company sells a polyester carbonate called poly(oligo)(tetramethylene succinate)-*co*(tetramethylene carbonate). Carbonate addition to polyoligotetra-methylene succinate might well have induced crystal structure disorder, reducing its melting temperature, as well as attempting to make it more vulnerable to enzymatic as well as microbial attacks than polyolefins. This copolyester carbonate is more microbially degradable than either of its constituent elements [40, 41].

1.2.7 Soy-Based Bio-degradable Polymers

Polymer removal must have got to be the most pressing environmental issue for scientists. This has sparked a new wave of research that aims to use sustainable agricultural materials like starch or protein to create biopolymers. There are numerous advantages to using soybeans, including their low cost, wide range of applications, as well as appropriateness for the production of biodegradable plastics. PCL and polyethylene terephthalate (PBT) seem to be two other biopolymers commonly utilized within biodegradable materials for fiber reinforcement.

1.2.8 Polyurethanes

With so many applications in new methodologies like elastomers, adhesives, foams, fabrics, and coatings, that polyurethane has developed, it is no surprise that the material has become so widely used. There are numerous distinct physical and chemical properties to this particular polymer substance. The chemical composition of polyurethanes affects the biodegradation process. A suitable soft part can be used to halt or slow down the deterioration process. Polyether-based polyurethanes are completely biodegradable. If the polyol is polyester, polyurethanes are voluntarily biodegradable [42, 43].

1.2.9 Polyanhydrides

Polyanhydrides are being studied by a number of scientists, who discovered that the hydrolyzable locations in the recurring unit make them interesting biodegradable materials. There are few uses for aliphatic homo-polyanhydrides because of their elevated crystallinity index as well as rapid degradation. Polyanhydride degradation can be slowed by altering the polymer's hydrophobic and hydrophilic components [44]. The hydrophobicity of the polymer's diacid building blocks contributed to the polymer's slower degradation. They have been extensively studied in the field of biomaterials because of their hydrophobic aromatic comonomers. Polyanhydrides with diverse linkages, including ester, ether, along with urethane, are being made because of the huge assortment of diacid monomers available. For medical implementations, anhydride–amide copolymers were also established to improve the mechanical characteristics of polyanhydrides [45, 46].

1.3 Biodegradable Polymer Blends and Composites

1.3.1 Polylactic Acid and Polyethylene Blends

With compatibilizers, the impact opposition of the material has been even greater, while tensile properties including elongation, tensile strength, as well as Young's modulus had been relatively low in compatible as well as noncompatible substances, respectively, than in genuine PLA. PLA/PE blends have been made by Raghavan and Emekalam [47], and the degradation of the blends was studied in relation to the addition of starch. Filler materials were added to PLA/PE blends to increase Young's modulus as well as lower the stress as well as strain levels, according to a study of the mechanical characteristics of the blends. The mechanical but also thermal characteristics of PLA and polyethylene blends have been only tested by a small number of authors. The tensile properties of the blend lessened without a rise in thermal stability, as well as the blend's compatibility, was poor [48].

1.3.2 PLA and Acrylobutadiene Styrene (ABS) Blends

As a result of ABS's mechanical properties including tensile strength, impact strength, as well as tensile modulus, new blends with unique attributes have become more commonplace. Synthetic polymer blends with SANGMA and ETPB were also produced via the inclusion of ethyl triphene phosphinium bromide as a catalyst. SANGMA had been an essential responsive alignment for PLA/ABS blends with ETPB as a catalyst, as demonstrated by a rise in rubber particle distribution and improved resistance to impact loads as well as strain with a negligible deficit of tensile modulus as well as strength especially in comparison with pure PLA composites [49].

1.3.3 PCL and Polyethylene Blends

Many researchers used an internal mixer to organize the PCL/PE blend as well as appraise phase inversion during compounding. The mixture is inconsistent in the range of compatibilities that was tested. Maleic anhydride was used to prepare PCL as well as low-density polyethylene (LDPE) blends, which were then compared to PCL and block polyethylene glycol (PEG) blends. When compared to PCL/LDPE blends, the latter's mechanical characteristics have been superior, but the former was much more compatible [50].

1.3.4 PCL and Polyvinyl Chloride Blends

Thermal stabilizer dibasic lead phthalate (DLP) has been shown to affect PCL phase dispersion in PCL/PVC polymer blends. PCL, as well as PVC polymer blend solution rheology, was then happened. The H bond here between two chains of PCL and PVC resulted in complete compatibility in the blend. Few other experiments also affirmed their excellent suitability; a thermal property evaluation showed that perhaps the mixture had single Tg, which was in connection with the mechanical characteristics of the generated mixture, which also had shown that the combination break elongation rises as the PCL material is continued to increase [51].

1.3.5 TPS and Polypropylene Blends

Glycerol-containing TPS/PP polymer mixtures prepared and analyzed by a single screw extruder machine are few as well as far between. Shear-thinning behavior in conventional production machines indicated that the blends could be processed. In addition, the lubricating effect of glycerol on the material as well as the capillary rheometer dies decreased the mixture's viscosity as the glycerol content rose. Young's modulus significantly increased while strain reduced as TPS as well as glycerol content in the blend enhanced. This was revealed by the mixture's mechanical characteristics. Experiments on plasticizing biodiesel glycerol as well as glycerol used in the production of TPS as well as PP blends were few and far between; when the TPS content was enhanced, the tensile strength was reduced. The study found that the clay-modified TPS, as well as PP blends, had better mechanical characteristics than unmodified blends because they contained biodegradable components and had good mechanical attributes [52, 53].

1.3.6 TPS/PE Blends

Research on TPS as well as PE blends has been conducted by a small number of researchers. As a result of this method, two extruders were linked together, producing TPS and blends of TPS with glycerol as plasticizers. The physical, mechanical, as well as thermal characteristics of the formulated mixtures are being analyzed. When compared to other polyethylene blends, this one's thermal stability suffered because of the incompatibility of the two components. The strain of such mixture has been found to be comparable to that of PE; however, its modulus was found to be lower. Only a few studies looked at the mechanical as well as thermal characteristics of TPS/LDPE blends when ethylene, as well as vinyl acetate blend copolymers, were utilized as plasticizers; glycerol has also been used. Increased ethylene, as well as vinyl acetate substance in the mixture, resulted in improved mechanical properties as well as thermal stability [50, 54].

1.3.7 Poly(Butylene Succinate) Blends

PBS is an environment-friendly, biodegradable polyester with excellent thermal and chemical resistance. A polymer, they're a member of the group (alkenedicarboxylate). Plant-based fibers and fillers were added to PBS to improve its properties as well as lower its manufacturing costs. Rice straw fiber composites with amino acids as coupling agents have been studied by a small number of scientists. A binding agent comprising amino clusters resulted in composites with exceptional mechanical characteristics. The mechanical properties of PBS-reinforced coir fiber composites were examined by other researchers who used a 5% NaOH alkali treatment. An increment in fiber volume fraction resulted in a rise in tensile properties, while the strain at break was reduced. Research into the crystallization of PBS/cotton stalk bast fiber composites has revealed that cotton stalk bast fibers serve as both nucleating agents as well as defensive measures to chain segment transport all through crystallization [55, 56].

Flame retardant microencapsulated ammonium polyphosphate was studied and compared to magnesium hydroxide as well as aluminum hydroxide in terms of the thermal properties of composites. Only a few experimenters used melt mixing to produce PBS/bamboo fiber composites. In aspects of flame retardant properties, ammonium polyphosphate beat out magnesium hydroxide as well as aluminum hydroxide. Few studies have examined how sisal fiber content affects the rheological properties of PBS supplemented with sisal fiber composites. Shear-thinning was observed in the composites, with viscosity decreasing as the shear rate increased. Furthermore, a non-Newtonian composite index (n) reduces as the fiber content increases, suggesting that perhaps the composite viscosity is stable over a wide range of shear speeds [57].

1.4 Properties of Natural Materials for Food Packaging

Any food packaging material is expected to possess various characteristics such as barrier, thermal, mechanical, and biodegradable properties. Selection of natural materials based on all the above properties is the prevalent area of materials research. These materials offer improved gas barriers, antioxidants, antimicrobial, and light-blocking effects along with the inherent characteristics of the bio-based polymers.

1.4.1 Barrier Properties

Several of the requirements as well as a critical factor for biodegradable materials being used in food containers as well as other areas, which include the biomedical field, is that they have a higher moisture boundary property. However, TPS films have an increasing water vapor permeability (WVP). Even though starch is naturally hydrophilic, even before merging to glycerol, the bulging of the network could indeed retain a considerable amount of liquid. This bulging compromises the matrix's integrity of the structure, resulting in inadequate barrier properties [58]. The degradable films' excessive moisture permeability results in exterior trashiness. The sophisticated association between both the polymer matrix as well as protective characteristics are determined by a number of variables, including the matrix's structure, polarity, crystallinity, molecular weight, as well as the type of reinforcement. Moisture transmission among both food as well as the surrounding atmosphere results in spoilage; thus, food should be as resistant to WVP as feasible. The ASTM D570-81 standard way of determining a material's waterproofing requires curing prior to immersing weighed (Wi) samples in a specified volume of deionized water for 24 hours at ambient temperature. The specimens should then be eliminated as well as the moisture removed prior to weighing (Wf) [59, 60].

Relative humidity (RH) has noticeable impact on aquatic uptake of TPS layers. Aquatic acceptance and mechanical characteristics of resources at various RH are crucial for simulating the nature of initial packaging layers that are utilized to stock healthier food items (both veg and non-veg), etc. Aquatic acceptance depends on the group of plasticizers which are utilized while handling, which is added in a reasonable investigation on water uptake for TPS created from glycerol and bio-based isosorbide as plasticizers with corn starch as medium (TPSG and TPSI, respectively). At 75% of RH 52, TPSG was observed to hold aquatic acceptance of 25.7%, and TPSI which holds 22.8%. At 50% RH, TPSG possessed an aquatic acceptance of 10.4%, while TPSI possessed an 8.8% water uptake. When the RH was lowered to 25%, the moisture absorption values decreased further to 5.5% and 4.5%, respectively. The oxygen permeability of these materials varied insignificantly up to RH 75%, at which point it increased exponentially. Chitosan and chitin were found to have a significant impact on WVP values when added to TPS. The water vapor pressure of control TPS film was determined to be 1.3360 g/s m Pa. On the other hand, WVP values of 0.8760 and 0.5960 g/s m Pa indicate that the addition of chitin to TPS enhanced difficult characteristics more than the addition of chitosan, owing to the higher concentration of acetyl groups in the chitin structure. The majority of the published research on starch-based wrapping focuses on reducing WVP through the use of various fillers. Nonetheless, there is considerable potential for developing an intelligent packaging film that incorporates dynamic nanofillers [61–63].

1.4.2 Biodegradation Properties

A bio-based degradable polymer is defined as a polymer that degrades initially as a result of microorganism metabolic activity. Polymeric materials degrade primarily as a result of the bioactivity of microbes such as microorganisms, plankton, and germs. Amylases and glucosidases are enzymes that can attack and degrade starch. Nature provides a specific team of enzymes capable of attacking specific types of polymers. In general, three distinct classes of enzymes deteriorate a lignocellulosic polymer into glucose units: endo-cellulases, exo-cellulases, and cellobiohydrolases [64, 65]. These three classes of biocatalysts are collectively referred to as cellulases; even so, each class is capable of attacking a particular format of the polymer. No enzyme is capable of degrading the polymer effectively on its own. Bacteria produce a set of enzymatic necessary for polymer degradation by utilizing the organic matter in their surroundings. Biodegradable polymers fall into two categories: (i) proteolytic enzymes biodegradability polymeric materials (e.g. biopolymer, carbohydrate, glucans, etc.) and (ii) photo- or thermo-oxidizable polymers. Abiotic oxidative and biodegradative reactions occur at a higher rate in the presence of concentrated humidity than in the absence of saturated humidity. When bio-based polymers are released into the environment following their use, they are entirely degraded by microorganisms found in soil, saltwater, rivers, streams, and sewage. They have no negative impacts on the environment and contribute to the reduction of the greenhouse effect [66].

Few authors demonstrated complete dissolution of starch-based films containing glycerol, agar, and Sorbian mono-oleate in 30 days using an indoor soil composting method. The soil contained a diverse population of bacteria and fungi (*Staphylococcus* spp., *Salmonella* spp., *Streptococcus* spp., *Moraxella* sp., *Bacillus* sp., *Aspergillus* sp., and *Penicillin* sp.). Bacteria were counted at 30 3 106 to 43 3 106 CFU/g of samples, while fungi were counted at 18 3 103 to 23 3 103 CFU/g of soil. The number of bacteria associated with degradation was discovered to be 29.76 3 106, while the number of fungi was found to be 16.93 3 103. Microbes expanded in response to available growth and water resources. The glycerol content of the film affects microbial growth because it promotes swelling, which results in movement of chemical species of water and thus microbial growth. Microorganisms used starch as their sole carbon source, resulting in the destruction caused by the layers. Morphological observations indicate that the surface begins to erode after 10 days and completely after 20 days of biodegradation. The effect of variables on the biodegradation of starch films was investigated using a three-level Box–Behnken response design. The results indicated that the presence of water within the microstructure facilitates the entrance and improvement of microorganisms [67, 68].

1.4.3 Consequences of Storage Time

Molds, pultrusion, injection molding, and other processes are used to convert starches to thermoplastics. Liquid, polymeric, and other diluents are frequently added to aid in the decomposition of starch. Following processing, it has been demonstrated that TPS age and recrystallize into a variety of particle morphologies based on the making and storage circumstances. Aging is defined as the identified physical and/or chemical modifications in the characteristics of a PE as a feature of storage time when the polymer is kept at a constant temperature, under no stress, and unaffected by external parameters. Thus, aging of carbohydrates is a critical phenomenon that must be considered prior to application [69, 70]. A significant disadvantage of using starch is that TPS products deteriorate in reliability, appropriateness, and shelf life over time due to starch retrogradation. It is the process by which TPS's mechanical characteristics change as a result of recrystallization. The recrystallization process is triggered by macromolecules' proclivity for hydrogen bond formation during the evaporation of water and/or other cleaning agents. This process can be classified as amylose recrystallization or amylopectin irreversible crystallization. Retrogradation is also known as the lengthy recrystallization of amylopectin due to the slower rate of reversible recrystallization of amylose [71–73].

Aging has a complex influence on the development and consistency of TPS. Retrogradation occurs in TPS over time and is dependent on the type of plasticizer used (Figure 1.2). This degradation process affects the material's properties and applications. Crystallinity values vary according to storage duration, heat, humidity levels, and plasticizer composition. Additionally, methods for determining the retrogradation degree, such as X-ray diffraction analysis, are discussed. The effects of retrogradation on TPS properties such as tensile, elongation, and modulus are discussed. The rigidity of the product, as demonstrated by an increment in Young's modulus, was correlated with amylopectin reordering away from the starchy component, as demonstrated by an increase in B-type degree of structural order in a solid. These radical variations in the TPS suggest that as starch chains age, they become less mobile. According to the published literature, when various sets of

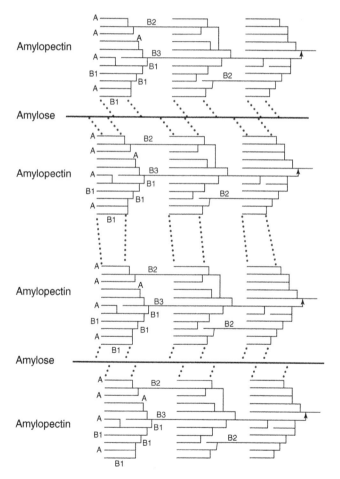

Figure 1.2 Retrogradation mechanism of starch solution. Source: Niranjana Prabhu and Prashantha [74]/with permission of John Wiley & Sons.

plasticizers and aging are used, starch-based materials exhibit big variations in material characteristics, and each plasticizer may be helpful for very particular purposes [74–78].

1.5 Environmental Impact of Food Packaging Materials

Implementation of inorganic food packaging films is a paramount factor that has to be considered to avoid the environmental hazards which it causes. Though the fabricated composites satisfy the regulatory limits for a food packaging material and have been proven to be an effective packaging material, their end environmental effect at the time of disposal is also to be taken into account [79]. Very few studies were carried out during the earlier stages on the detrimental environmental hazards caused by inorganic packaging materials over the environment. On contrary, during recent times, LCA has been used as a unified and systematic tool to determine the environmental effects of using an inorganic food packaging material through numerous experiments by considering various product lifecycle stages such as the raw materials used for the production of composite films, the process of film manufacturing, time of usage and method of disposal [80, 81].

From various studies, it could be stated that the environmental effects of inorganic polymer–metal packaging materials depend on factors such as degree of filler incorporation, method of manufacturing and synthesis of metal fillers, and the initial effect of food storage upon the environment. As migration capability and ionic or particulate mobility of the food packaging material plays a significant role in determining the environmental and toxic impacts, regulations regarding the migration evaluation of the food packaging material have to be implemented strictly to avoid adverse environmental effects [82, 83]. All these parameters are deemed to be important if the disposability of the food packaging material is below the par level. All these effects could readily be mitigated if natural materials are employed for food packaging and a few disadvantages like human toxicity, environmental burden, and disposal problems can be overcome by using those natural materials. It is also important to check the activity of the natural material employed for food packaging since it involves food consumption by human beings [84].

1.6 Conclusion

Evolution of natural, biodegradable materials in place of synthetic materials for food packaging applications has become the order of research today. Since plastic-based packaging materials accumulate to a greater extent through landfills and cause serious environmental pollution, manufacturing of biodegradable food packaging materials from natural and renewable materials is deemed to be necessary. All these materials have their importance since they were derived from natural and renewable sources. Research has to be oriented toward the commercial aspects of food packaging materials in such a way that they provide real-time information that helps for the commercialization of natural food packaging materials.

TPS, PLA, PHB, chitosan, and cellulose-based materials are used in various food packaging industries which in turn enhances the farmers economy who produces the raw materials. If the farmers themselves use all these natural-based materials in applications like compostable materials for agricultural and horticultural fields, bags, and mulching films, they will be beneficial to the farmers also. Usage of these materials poses various advantages like reduction in toxic gas release, decreased use of nonbiodegradable materials, and mitigation of environmental pollution. Cost is another important factor to be considered while designing an application based on natural materials for food packaging. Various researches have to be focused on reducing the cost of utilization of these natural material-based packaging materials since the cost is higher than the petroleum-based plastic materials during the current scenario. Such researches not only strive to reduce the cost of these natural materials but also enhance the applicability of natural materials for food packaging. As the development and application of natural materials need a multidisciplinary approach, efforts by various specialists from fields like microbiology, environmental science, chemistry, chemical engineering, polymer engineering, and other industrial giants who enforce the regulation on the use of natural materials have to meet at a concurrent point for the implementation and commercialization of natural biodegradable and eco-friendly food packaging materials.

References

1 Espitia, P.J.P., Du, W.X., de Jesús Avena-Bustillos, R. et al. (2014). Edible films from pectin: physical-mechanical and antimicrobial properties – a review. *Food Hydrocolloids* 35: 287–296.
2 Ramesh, M., Deepa, C., Kumar, L.R. et al. (2020). Life-cycle and environmental impact assessments on processing of plant fibres and its bio-composites: a critical review. *Journal of Industrial Textiles* https://doi.org/10.1177/1528083720924730.
3 Mahalik, N.P. and Nambiar, A.N. (2010). Trends in food packaging and manufacturing systems and technology. *Trends in Food Science & Technology* 21 (3): 117–128.
4 Muizniece-Brasava, S., Dukalska, L., and Kantike, I. (2011). Consumer's knowledge and attitude to traditional and environmentally friendly food packaging materials in market of Latvia. *The 6th Baltic Conference on Food Science and Technology "FoodBalt-2011"*, Jelgava, Latvia (5–6 May 2011).
5 Tavassoli-Kafrani, E., Shekarchizadeh, H., and Masoudpour-Behabadi, M. (2016). Development of edible films and coatings from alginates and carrageenans. *Carbohydrate Polymers* 137: 360–374.
6 Sedayu, B.B., Cran, M.J., and Bigger, S.W. (2019). A review of property enhancement techniques for carrageenan-based films and coatings. *Carbohydrate Polymers* 216: 287–302.

7 Balaji, D., Ramesh, M., Kannan, T. et al. (2021). Experimental investigation on mechanical properties of banana/snake grass fiber reinforced hybrid composites. *Materials Today: Proceedings* 42: 350–355. https://doi.org/10.1016/j.matpr.2020.09.548.

8 Sanyang, M.L., Ilyas, R.A., Sapuan, S.M., and Jumaidin, R. (2018). Sugar palm starch-based composites for packaging applications. In: *Bionanocomposites for Packaging Applications* (ed. M. Jawaid and S.K. Swain), 125–147. Cham: Springer.

9 Youssef, A.M., Assem, F., Essam, M. et al. (2019). Development of a novel bionanocomposite material and its use in packaging of Ras cheese. *Food Chemistry* 270: 467–475.

10 Youssef, A.M., El-Sayed, S.M., El-Sayed, H.S. et al. (2018). Novel bionanocomposite materials used for packaging skimmed milk acid coagulated cheese (Karish). *International Journal of Biological Macromolecules* 115: 1002–1011.

11 Ramesh, M., Deepa, C., Tamil Selvan, M. et al. (2020). Mechanical and water absorption properties of *Calotropis gigantea* plant fibers reinforced polymer composites. *Materials Today: Proceedings* 46: 3367–3372. https://doi.org/10.1016/j.matpr.2020.11.480.

12 Youssef, A.M. and El-Sayed, S.M. (2018). Bionanocomposites materials for food packaging applications: concepts and future outlook. *Carbohydrate Polymers* 193: 19–27.

13 Bumbudsanpharoke, N. and Ko, S. (2015). Nano-food packaging: an overview of market, migration research, and safety regulations. *Journal of Food Science* 80: R910.

14 Youssef, A.M. (2013). Polymer nanocomposites as a new trend for packaging applications. *Polymer – Plastics Technology and Engineering* 52: 635–660.

15 Yam, K.L., Takhistov, P.T., and Miltz, J. (2005). Intelligent packaging: concepts and applications. *Journal of Food Science* 70: R1–R10.

16 Cazón, P., Velazquez, G., Ramírez, J.A., and Vázquez, M. (2017). Polysaccharide-based films and coatings for food packaging: a review. *Food Hydrocolloids* 68: 136–148.

17 Bhuvaneswari, V., Priyadharshini, M., Deepa, C. et al. (2021). Deep learning for material synthesis and manufacturing systems: a review. *Materials Today: Proceedings* 46 (9): 3263–3269. https://doi.org/10.1016/j.matpr.2020.11.351.

18 Siracusa, V., Rocculi, P., Romani, S., and Dalla Rosa, M. (2008). Biodegradable polymers for food packaging: a review. *Trends in Food Science & Technology* 19 (12): 634–643.

19 Hoque, M.Z., Akter, F., Hossain, K.M. et al. (2010). Isolation, identification and analysis of probiotic properties of *Lactobacillus* spp. from selective regional yoghurts. *World Journal of Dairy & Food Sciences* 5 (1): 39–46.

20 Mostafavi, F.S. and Zaeim, D. (2020). Agar-based edible films for food packaging applications – a review. *International Journal of Biological Macromolecules* 159: 1165–1176.

21 Wani, A.A., Singh, P., Shah, M.A. et al. (2013). Physico-chemical, thermal and rheological properties of starches isolated from newly released rice cultivars grown in Indian temperate climates. *LWT – Food Science and Technology* 53 (1): 176–183.

22 Lin, S., Chen, L., Huang, L. et al. (2015). Novel antimicrobial chitosan–cellulose composite films bioconjugated with silver nanoparticles. *Industrial Crops and Products* 70: 395–403.

23 Ramesh, M., Rajeshkumar, L., Balaji, D., and Bhuvaneswari, V. (2021). Green composite using agricultural waste reinforcement. In: *Green Composites. Materials Horizons: From Nature to Nanomaterials* (ed. S. Thomas and P. Balakrishnan), 21–34. Singapore: Springer https://doi.org/10.1007/978-981-15-9643-8_2.

24 Philip, S., Keshavarz, T., and Roy, I. (2007). Polyhydroxyalkanoates: biodegradable polymers with a range of applications. *Journal of Chemical Technology and Biotechnology* 247: 233–247.

25 Mohanty, A.K., Misra, M., and Hinrichsen, G. (2000). Biofibres, biodegradable polymers and biocomposites: an overview. *Macromolecular Materials and Engineering* 276–277: 1–24.

26 Ncama, K., Magwaza, L., Mditshwa, A., and Zeray Tesfay, S. (2018). Plant-based edible coatings for managing postharvest quality of fresh horticultural produce: a review. *Food Packaging and Shelf Life* 16: 157–167.

27 Murmu, S.B. and Mishra, H.N. (2018). The effect of edible coating based on arabic gum, sodium caseinate and essential oil of cinnamon and lemon grass on guava. *Food Chemistry* 245: 820–828.

28 Shit, S.C. and Shah, P. (2014). Edible polymers: challenges and opportunities. *Journal of Polymers* 2014: 1–13.

29 Ramesh, M. and Rajeshkumar, L. (2021). Technological advances in analyzing of soil chemistry. In: *Applied Soil Chemistry* (ed. Inamuddin, M.I. Ahamed, R. Boddula, and T. Altalhi), 61–78. Beverly, MA: Wiley – Scrivener Publishing LLC.

30 Chan, C.M., Vandi, L.-J., Pratt, S. et al. (2018). Composites of wood and biodegradable thermoplastics: a review. *Polymer Reviews* 58 (3): 444–494.

31 Ramesh, M., Rajesh Kumar, L., Khan, A., and Asiri, A.M. (2020). Self-healing polymer composites and its chemistry. In: *Self-Healing Composite Materials*, 415–427. Woodhead Publishing.

32 Mazuki, N.F., Nagao, Y., Kufian, M.Z., and Samsudin, A.S. (2020). The influences of PLA into PMMA on crystallinity and thermal properties enhancement-based hybrid polymer in gel properties. *Materials Today: Proceedings* https://doi.org/10.1016/j.matpr.2020.11.037.

33 Lee, L.-T., He, S.-P., and Huang, C.-F. (2020). Enhancement of crystallization behaviors in quaternary composites containing biodegradable polymer by supramolecular inclusion complex. *Crystals* 10 (12): 1137.

34 Thirmizir, M.Z.A., Mohd Ishak, Z.A., and Salim, M.S. (2020). Compatibilization and crosslinking in biodegradable thermoplastic polyester blends. In: *Reactive and Functional Polymers Volume Two* (ed. T.J. Gutiérrez), 23–89. Cham: Springer.

35 Abioye, A.A. and Obuekwe, C.C. (2020). Investigation of the biodegradation of low-density polyethylene-starch bi-polymer blends. *Results in Engineering* 5: 100090.

36 Ramesh, M., Maniraj, J., and Rajesh Kumar, L. (2021). Biocomposites for energy storage. *Biobased Composites: Processing, Characterization, Properties, and Applications* 123–142. https://doi.org/10.1002/9781119641803.ch9.

37 Zarrintaj, P., Saeb, M.R., Jafari, S.H., and Mozafari, M. (2020). Application of compatibilized polymer blends in biomedical fields. In: *Compatibilization of Polymer Blends* (ed. A.R. Ajitha and S. Thomas), 511–537. Elsevier.

38 Moustafa, H., Youssef, A.M., Darwish, N.A., and Abou-Kandil, A.I. (2019). Eco-friendly polymer composites for green packaging: future vision and challenges. *Composites Part B: Engineering* 172: 16–25.

39 Ramesh, M. and Rajeshkumar, L. (2018). Wood flour filled thermoset composites. In: *Thermoset Composites: Preparation, Properties and Applications*, vol. 38, 33–65. Materials Research Foundations https://doi.org/10.21741/9781945291876-2.

40 Priyadharshini, M., Balaji, D., Bhuvaneswari, V. et al. (2022). Fiber reinforced composite manufacturing with the aid of artificial intelligence – a state-of-the-art review. *Archives of Computational Methods in Engineering* https://doi.org/10.1007/s11831-022-09775-y.

41 Pranamuda, H., Chollakup, R., and Tokiwa, Y. (1999). Degradation of polycarbonate by a polyester degrading strain, *Amycolatopsis* sp. strain HT-6. *Applied and Environmental Microbiology* 65: 4220–4222.

42 Kim, B.K., Seo, J.W., and Jeong, H.M. (2003). Morphology and properties of waterborne polyurethane/clay nanocomposites. *European Polymer Journal* 39: 85–91.

43 Bandyopadhyay, J. and Ray, S.S. (2019). Are nanoclay-containing polymer composites safe for food packaging applications?—an overview. *Journal of Applied Polymer Science* 136 (12): 47214.

44 Leong, K.W., Brott, B.C., and Langer, R. (1985). Biodegradable polyanhydrides as drug carrier matrices: characterization, degradation and release characteristics. *Journal of Biomedical Materials Research* 19: 941–955.

45 Ramesh, M. and Kumar, L.R. (2020). Bioadhesives. In: *Green Adhesives* (ed. R. Inamuddin, M.I. Boddula, and A.A.M. Ahamed), 45–167. https://doi.org/10.1002/9781119655053.

46 Haghighi, H., Licciardello, F., Fava, P. et al. (2020). Recent advances on chitosan-based films for sustainable food packaging applications. *Food Packaging and Shelf Life* 26: 100551.

47 Raghavan, D. and Emekalam, A. (2001). Characterization of starch/polyethylene and starch/polyethylene/poly(lactic acid) composites. *Polymer Degradation and Stability* 72: 509–517.

48 Balakrishnan, H., Hassan, A., Wahit, M., and Mechanical, J. (2010). Thermal, and morphological properties of polylactic acid/linear low density polyethylene blends. *Journal of Elastomers and Plastics* 42: 223–239.

49 Li, Y. and Shimizu, H. (2009). Improvement in toughness of poly(l-lactide) (PLLA) through reactive blending with acrylonitrile–butadiene–styrene copolymer (ABS): morphology and properties. *European Polymer Journal* 45 (3): 738–746.

50 Hamad, K., Kaseem, M., Ko, Y.G., and Deri, F. (2014). Biodegradable polymer blends and composites: an overview. *Polymer Science, Series A* 56 (6): 812–829.

51 Chiu, F.-C. and Min, K. (2000). Miscibility, morphology and tensile properties of vinyl chloride polymer and poly(ε-caprolactone) blends. *Polymer International* 49 (2): 223–234.

52 Kaseem, M., Hamad, K., and Deri, F. (2012). Rheological and mechanical properties of polypropylene/thermoplastic starch blend. *Polymer Bulletin* 68 (4): 1079–1091.

53 Ramesh, M., Rajeshkumar, L., and Balaji, D. (2021). Aerogels for insulation applications. In: *Aerogels II: Preparation, Properties and Applications*, vol. 98 (ed. Inamuddin), 57–76. Millersville, PA: Materials Research Foundations https://doi.org/10.21741/9781644901298-4.

54 Ramesh, M., Rajeshkumar, L., Deepa, C. et al. (2021). Impact of silane treatment on characterization of *Ipomoea staphylina* plant fiber reinforced epoxy composites. *Journal of Natural Fibers* https://doi.org/10.1080/15440478.2021.1902896.

55 Nam, T.H., Ogihara, S., and Kobayashi, S. (2012). Interfacial, mechanical and thermal properties of coir fiber-reinforced poly(lactic acid) biodegradable composites. *Advanced Composite Materials* 21 (1): 103–122.

56 Ramesh, M., Rajeshkumar, L., and Balaji, D. (2021). Mechanical and dynamic properties of ramie fiber reinforced composites. In: *Mechanical and Dynamic Properties of Biocomposites* (ed. R. Nagarajan, S.M.K. Thiagamani, S. Krishnasamy, and S. Siengchin), 275–322. Beverly, MA: Wiley.

57 Bin, T., Jin-ping, Q., Liu, L.-m. et al. (2011). Non-isothermal crystallization kinetics and dynamic mechanical thermal properties of poly(butylene succinate) composites reinforced with cotton stalk bast fibers. *Thermochimica Acta* 525 (1–2): 141–149.

58 Rychter, P., Kot, M., Bajer, K. et al. (2016). Utilization of starch films plasticized with urea as fertilizer for improvement of plant growth. *Carbohydrate Polymers* 137: 127–138.

59 Ramesh, M., Rajeshkumar, L., Balaji, D. et al. (2021). Self-healable conductive materials. In: *Self-Healing Smart Materials* (ed. Inamuddin, M.I. Ahamed, R. Boddula, and T.A. Altalhi), 297–320. Beverly, MA: Wiley https://doi.org/10.1002/9781119710219.ch11.

60 Ghosh, S. (2016). Biodegradation study of polyethylene-based biocomposites and bionanocomposites. *Polyethylene-Based Biocomposites and Bionanocomposites* 345–364.

61 Ramesh, M., Rajeshkumar, L., and Saravanakumar, R. (2021). Mechanically-induced self-healable materials. In: *Self-Healing Smart Materials* (ed. Inamuddin, M.I. Ahamed, R. Boddula, and T.A. Altalhi), 379–404. Beverly, MA: Wiley https://doi.org/10.1002/9781119710219.ch15.

62 Battegazzore, D., Bocchini, S., Nicola, G. et al. (2015). Isosorbide, a green plasticizer for thermoplastic starch that does not retrograde. *Carbohydrate Polymers* 119: 78–84.

63 Ramesh, M., Rajeshkumar, L., and Balaji, D. (2021). Influence of process parameters on the properties of additively manufactured fiber-reinforced polymer composite materials: a review. *Journal of Materials Engineering and Performance* 30 (7): 4792–4807. https://doi.org/10.1007/s11665-021-05832-y.

64 Zhang, S., He, Y., Yin, Y., and Jiang, G. (2019). Fabrication of innovative thermoplastic starch bio-elastomer to achieve high toughness poly(butylene succinate) composites. *Carbohydrate Polymers* 206: 827–836.

65 Ramesh, M., Deepa, C., Niranjana, K. et al. (2021). Influence of Haritaki (*Terminalia chebula*) nano-powder on thermo-mechanical, water absorption and morphological properties of Tindora (*Coccinia grandis*) tendrils fiber reinforced epoxy composites. *Journal of Natural Fibers* https://doi.org/10.1080/15440478.2021.1921660.

66 Prakash Maran, J., Sivakumar, V., Thirugnanasambandham, K., and Sridhar, R. (2014). Degradation behavior of biocomposites based on cassava starch buried under indoor soil conditions. *Carbohydrate Polymers* 101: 20–28.

67 Prachayawarakorn, J. and Pomdage, W. (2014). Effect of carrageenan on properties of biodegradable thermoplastic cassava starch/low-density polyethylene composites reinforced by cotton fibers. *Materials & Design* 61: 264–269.

68 Mohankumar, D., Amarnath, V., Bhuvaneswari, V. et al. (2021). Extraction of plant based natural fibers – a mini review. *IOP Conference Series: Materials Science and Engineering* 1145: 012023. https://doi.org/10.1088/1757-899X/1145/1/012023.

69 Mościcki, L., Mitrus, M., Wójtowicz, A. et al. (2012). Application of extrusion-cooking for processing of thermoplastic starch (TPS). *Food Research International* 47 (2): 291–299.

70 Ramesh, M., Deepa, C., Rajeshkumar, L. et al. (2021). Influence of fiber surface treatment on the tribological properties of *Calotropis gigantea* plant fiber reinforced polymer composites. *Polymer Composites* https://doi.org/10.1002/pc.26149.

71 Nazrin, A., Sapuan, S.M., Zuhri, M.Y.M. et al. (2020). Nanocellulose reinforced thermoplastic starch (TPS), polylactic acid (PLA), and polybutylene succinate (PBS) for food packaging applications. *Frontiers in Chemistry* 8: 213.

72 Frost, K., Barthes, J., Kaminski, D. et al. (2011). Thermoplastic starch–silica–polyvinyl alcohol composites by reactive extrusion. *Carbohydrate Polymers* 84 (1): 343–350.

73 Bhuvaneswari, V., Rajeshkumar, L., and Nimel Sworna Ross, K. (2021). Influence of bioceramic reinforcement on tribological behaviour of aluminium alloy metal matrix composites: experimental study and analysis. *Journal of Materials Research and Technology* https://doi.org/10.1016/j.jmrt.2021.09.090.

74 Niranjana Prabhu, T. and Prashantha, K. (2018). A review on present status and future challenges of starch based polymer films and their composites in food packaging applications. *Polymer Composites* 39 (7): 2499–2522.

75 Ramesh, M., Rajeshkumar, L., and Bhoopathi, R. (2021). Carbon substrates: a review on fabrication, properties and applications. *Carbon Letters* 31: 557–580. https://doi.org/10.1007/s42823-021-00264-z.

76 Devarajan, B., Saravanakumar, R., Sivalingam, S. et al. (2021). Catalyst derived from wastes for biofuel production: a critical review and patent landscape analysis. *Applied Nanoscience* https://doi.org/10.1007/s13204-021-01948-8.

77 Ambigaipalan, P., Hoover, R., Donner, E., and Liu, Q. (2013). Retrogradation characteristics of pulse starches. *Food Research International* 54 (1): 203–212.

78 Ramesh, M., Balaji, D., Rajeshkumar, L. et al. (2021). Tribological behavior of glass/sisal fiber reinforced polyester composites. In: *Vegetable Fiber Composites and their Technological Applications*, Composites Science and Technology (ed. M. Jawaid and A. Khan), 445–459. Singapore: Springer https://doi.org/10.1007/978-981-16-1854-3_20.

79 Pourzahedi, L., Vance, M., and Eckelman, M.J. (2017). Life cycle assessment and release studies for 15 nanosilver-enabled consumer products: investigating hotspots and patterns of contribution. *Environmental Science & Technology* 51: 7148–7158. https://doi.org/10.1021/acs.est.6b05923.

80 Felix Sahayaraj, A., Muthukrishnan, M., Ramesh, M., and Rajeshkumar, L. (2021). Effect of hybridization on properties of tamarind (*Tamarindus indica* L.) seed nano-powder incorporated jute-hemp fibers reinforced epoxy composites. *Polymer Composites* https://doi.org/10.1002/pc.26326.

81 Videira-Quintela, D., Martin, O., and Montalvo, G. (2021). Recent advances in polymer-metallic composites for food packaging applications. *Trends in Food Science & Technology* 109: 230–244. https://doi.org/10.1016/j.tifs.2021.01.020.

82 Ramesh, M. and Rajeshkumar, L. (2021). Case-studies on green corrosion inhibitors. In: *Sustainable Corrosion Inhibitors*, vol. 107, 204–221. Materials Research Foundations https://doi.org/10.21741/9781644901496-9.

83 Senturk Parreidt, T., Müller, K., and Schmid, M. (2018). Alginate-based edible films and coatings for food packaging applications. *Foods* 7 (10): 170.

84 Sánchez-Safont, E.L., Aldureid, A., Lagarón, J.M. et al. (2018). Biocomposites of different lignocellulosic wastes for sustainable food packaging applications. *Composites Part B: Engineering* 145: 215–225.

2

Plant Extracts-Based Food Packaging Films

Aris E. Giannakas

University of Patra, Department of Food Science & Technology, G. Seferi st. 2, Agrinio 30100, Greece

2.1 Introduction

In the last few years, the role of packaging in the food sector becomes more and more vital as it not only protects the food but also can increase food shelf life. Circular economy [1], sustainability [2–5], and material nanotechnology [6, 7] trends drive researchers in the use of renewable feedstock, to decrease negative environmental effects. An efficient carbon-neutral alternative is the use of renewable biomass wastes, food by-products, and food side streams for food packaging applications with efficient recyclability (mechanically, chemically, or through microbial degradation). In this direction in the food packaging sector, it is suggested: (i) the replacement of petroleum-based packaging films with bio-based, biodegradable polymers and biopolymers and (ii) the use of natural additives in food packaging, such as plant extracts that can be derived from food industry by-products and side streams. Plant extracts have received greater attention [8, 9] as they contain high concentrations of phenolic components that possess strong antioxidant activities and thus are safer than synthetic antioxidants such as butyl-hydroxytoluene (BHT), butyl-hydroxyanisole (BHA), and *tert*-butyl hydroxyquinone (TBHQ) and their effect on consumer health are unclear [10]. In advance, some plant extracts have pH indicator properties and thus can be used not only as antioxidant and antimicrobial agents in active packaging films but also as food spoilage sensory agents for smart packaging applications. This contributes to the current demand, reported by Food and Agriculture Organization (FAO) and the food industry, for fresh food products that are natural and easy to consume, have specific advantages for health, are microbiologically safe and lack synthetic additives [11]. In this regard, in most cases different types of natural biopolymers, including polysaccharides, proteins, and lipids, have been used in the development of packaging films and edible

---

Natural Materials for Food Packaging Application, First Edition.
Edited by Jyotishkumar Parameswaranpillai, Aswathy Jayakumar,
E. K. Radhakrishnan, Suchart Siengchin, and Sabarish Radoor.
© 2023 WILEY-VCH GmbH. Published 2023 by WILEY-VCH GmbH.

coatings. Also, mainly biodegradable polymers and less synthetic polymers have been used as substrates for such bioactive extract carriers. Such packaging films and edible coatings not only supply moisture, mechanical, and gas barriers for food but also act as carriers for natural bioactive substances with antimicrobial and antioxidant activities. The incorporation of such natural bioactive substances in packaging films can effectively retard oxidation and delay microbial spoilage and thus extend the food shelf life.

This chapter revises the literature in the last five years on the development of novel active packaging films that incorporate natural additives from plant extracts. Specific goals of this chapter are as follows: (i) the effect of plant extracts on the structural and techno-functional properties of obtained active packaging films such as mechanical, thermomechanical, and barrier properties; (ii) the effects of plant extracts on bioactive properties (antioxidant, antimicrobial, and antibrowning) of packaging films; and (iii) the effects of active edible films with natural additives on the bioactive properties and quality of fresh food products.

2.2 Extraction Methods for Plant Extracts

An extract is a substance made by extracting a part of raw material, often by using a solvent such as ethanol, oil, or water. Extracts may be sold as tinctures, absolutes, or in powder form. Commonly, a plant component to be extracted is cut into small pieces or ground into powder. The powder is dissolved in a solvent (water, alcohol, acid, or mixture of solvents) to obtain a solution with a completely dissolved solute. The method of using water is called hydrous extraction and the use of alcohol or acid is called solvent extraction. In many cases, plant extract needs to be extracted with suitable solvents to separate unwanted materials and preserving bioactive compounds. The most common method used to extract polyphenolic compounds from plant extracts is the traditional maceration extraction and Soxhlet extraction method. In this respect, hydrous ethanol is commonly used as the extraction solvent. Meanwhile, other extraction solvents, including water, polyethylene glycol, olive oil, methanol, and ethyl acetate as well as their mixtures, are sometimes considered [12, 13]. The extraction yield and composition of polyphenolic compounds from plant extracts are greatly affected by extraction conditions (e.g. extraction solvent, solid–liquid ratio, temperature, pH value, and time) [13]. Several modern assisted extraction methods, including high hydrostatic pressure extraction, microwave-assisted extraction, and ultrasound extraction, are synergically used with maceration extraction, providing higher extraction yield with a shorter operation time.

Microwave-assisted extraction method has obtained, over the last years, increasing attention from analytical chemists and it has been successfully utilized for the extraction of various contaminants from different foods [14]. It uses microwave energy to facilitate the partition of analytes from the sample matrix into the solvent [14, 15]. Microwave radiation interacts with dipoles of polar and polarizable materials such as solvent and a plant sample resulting in heating near the surface of the materials and heat is transferred by conduction. Dipole rotation of the molecules induced by microwave electromagnetic disrupts hydrogen bonding enhancing the migration of dissolved ions and promoting solvent penetration into the matrix [14].

Ultrasound-assisted extraction or sonication extraction method makes use of ultrasound (from 20 to 2000 kHz) [16] and thus increases the surface contact between solvents and samples, alters and disrupts the physical and chemical properties of the sample, and facilitates the release of compounds and enhancing mass transport of the solvents into the plant cells.

Accelerated solvent extraction is a more efficient solvent extraction method compared to maceration extraction method and makes use of minimum amount of solvent compared to maceration extraction method [17].

Supercritical carbon dioxide extraction (CO_2) extraction is an alternative method that is suitable for large-scale production without toxic solvent residues. CO_2 is an ideal solvent for the extraction of polyphenolic compounds from plant extracts since it is nontoxic and easy to remove from the final products [18]. The low operation temperature of supercritical CO_2 extraction is beneficial to preserve the chemical structure and bioactivity of polyphenolic compounds in the extract [19].

In some plants, some phytochemicals in their matrices are dispersed in the cell cytoplasm, and secondary metabolites are retained in the polysaccharide–lignin network by hydrogen or hydrophobic bonding and are not accessible with a solvent extraction process. In these cases, enzyme-assisted extraction is applied. Enzymatic pretreatment has been considered as an effective way to release bounded compounds and increase overall yield [20]. Specific enzymes such as cellulase, α-amylase, and pectinase added during extraction enhance recovery by breaking the cell wall and hydrolyzing the structural polysaccharides and lipid bodies [21, 22].

2.3 Research Investigation of Bibliographic Data

Our bibliographic research investigation for the topic of plant extract-based food packaging films based on the scientific databases of Elsevier, American Chemical Society (ACS), Springer, Wiley and Royal Society of Chemistry, and the open access scientific databases MDPI and Frontiers. We use the keywords "plant extracts" and "packaging films" while the search extended to the last five years (2017–2021). Our investigation for the year 2021 ends on 8th of August 2021. Representative results of our search findings are shown in column bar diagram of Figure 2.1.

In Figure 2.1 it is depicted clearly that the number of available online scientific manuscripts on the topic of plant extract-based food packaging films is increased rapidly in the last five years. This fact shows the increased interest of researchers in the field. It must be underlined that the published papers for the year 2021 are almost the same with those of the year 2020 and it is expected to be higher because the investigation results for the year 2021 ended on the 8th of August 2021. Moreover, as it is illustrated in Figure 2.2 among available online manuscripts in the topic chitosan (21.4%), starch (17.9%), and cellulose (16.4%) correspondingly are the most

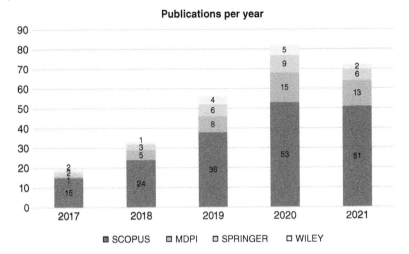

Figure 2.1 Column bar graphical presentation of the total number of publications in the field of plant extract-based food packaging films in the last five years.

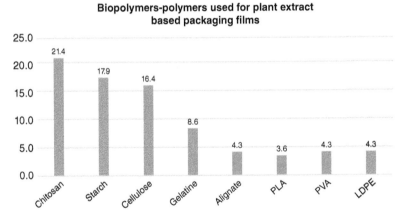

Figure 2.2 Percent number of available online studies according to the biopolymer or polymer used as matrix for the natural extract-based packaging films.

investigated biopolymers as matrices for plant extract additives. Gelatin and alginate-based packaging films are also promising biopolymers as resulted in Figure 2.2. There are also some studies with poly-lactide acid (PLA)-based bio-based films, polyvinyl alcohol (PVA)-based biodegradable films and low-density polyethylene (LDPE)-based polymer-based films. Thus, it is clearly shown that biopolymers are the most investigated and promising matrices as carriers of bioactive plant extract additives.

In the next pages, they have revised the most important studies with plant-extract-based food packaging films according to the categorization made in the column bar of Figure 2.2.

2.4 Chitosan Plant Extract-Based Food Packaging Films

Chitin is the second most abundant biopolymer after cellulose [23, 24] and acts as a structural material. It is present in the exoskeletons of crustaceans, insects, and cell walls of fungi. Chitin is composed mainly of (1–4)-linked 2-acetamido-2-deoxy-β-D-glucose monomers. Chitosan is the deacetylated derivative of chitin and is majorly composed of properties such as the degree of acetylation (DA), nitrogen composition, N/C ratio, molecular size, and polydispersity, which are dependent on the source of chitin. In general, chitin has a DA above 90%, nitrogen content of 7%, and N/C ratio of 0.146. Chitosan has a DA of less than 40% and nitrogen content greater than 7% [25]. Solubility, biocapability, film formation, antimicrobial properties, and good barrier properties make chitosan one of the most promising future candidates for food packaging applications [26–29]. Table 2.1 listed the studies referred to chitosan plant extract-based active packaging films of the last five years. Representative results of the studies are shown here below.

Table 2.1 Publications with chitosan plant extract-based food packaging films in last five years.

a/a	Year	Title	References
1	2021	Characterization of chitosan film incorporated with curcumin extract	[30]
2	2021	Chitosan-based antioxidant films incorporated with pine needles (*Cedrus deodara*) extract for active food packaging applications	[31]
3	2021	Development of red apple pomace extract/chitosan-based films reinforced by TiO_2 nanoparticles as a multifunctional packaging material	[32]
4	2021	Efficacy of pomegranate phenolic extract and chitosan as an edible coating for shelf life extension of Indian white shrimp during refrigerated storage	[33]
5	2020	Development of nanochitosan-based active packaging films containing free and nanoliposome caraway (*Carum carvi.* L) seed extract	[34]
6	2020	Chitosan-based biodegradable active food packaging film containing Chinese chive (*Allium tuberosum*) root extract for food application	[35]
7	2020	Multifunctional bionanocomposite films based on konjac glucomannan/chitosan with nano-ZnO and mulberry anthocyanin extract for active food packaging	[36]
8	2020	Reduction in spoilage microbiota and cyclopiazonic acid mycotoxin with chestnut extract enriched chitosan packaging: stability of inoculated gouda cheese	[37]

(Continued)

Table 2.1 (Continued)

a/a	Year	Title	References
9	2020	Ionic liquid-mediated recovery of carotenoids from the *Bactris gasipaes* fruit waste and their application in food-packaging chitosan films	[38]
10	2020	Active chitosan–chestnut extract films used for packaging and storage of fresh pasta	[39]
11	2019	Incorporation of pink pepper residue extract into chitosan film combined with a modified atmosphere packaging: effects on the shelf life of salmon fillets	[40]
12	2019	Development of multifunctional food packaging films based on chitosan, TiO_2 nanoparticles, and anthocyanin-rich black plum peel extract	[41]
13	2019	Mango leaf extract incorporated chitosan antioxidant film for active food packaging	[42]
14	2019	Preparation and characterization of Piper Betle Linn. leaf extract incorporated chitosan films as potential active food packaging materials	[43]
15	2017	Development and characterization of bioactive edible films from spider crab (*Maja crispata*) chitosan incorporated with Spirulina extract	[44]
16	2017	Physical properties of chitosan films incorporated with natural antioxidants	[45]
17	2017	Mathematical modeling of gallic acid release from chitosan films with grape seed extract and carvacrol	[46]
18	2017	Antimicrobial and physical properties of chitosan films incorporated with turmeric extract	[47]

Curcumin is a phenolic compound derived from turmeric roots (*Curcuma longa* L.). Rachtanapun et al. [30] incorporated curcumin extract into chitosan films and studied their properties. The results suggested that the addition of curcumin extracts into chitosan films increased yellowness and light barriers of obtained films. FTIR experiments showed a good interplay between the phenolic compounds of the curcumin extract molecules and the chitosan. This interplay improved the mechanical properties and reduced the moisture content, water solubility, and water barrier of the obtained films. The antioxidant activity of the films was increased by increasing curcumin extract content.

Pine needles (*Cedrus deodara*) comprise curcuminoid pigments, which are natural phenolic compounds. Kadam et al. [31] incorporated 5%, 10%, and 20% (v/w) pine needle extract into chitosan films and studied the barrier properties, color, and microstructures of the obtained films. XRD analysis showed an excellent compatibility between chitosan and pine needle extract molecules. The films containing pine needle extract at the concentration of 10% and 20% showed high antioxidant activity. The water and oxygen barrier properties of obtained films increased as the pine needle extract content increased from 5% to 20%.

Sabu et al. [33] developed chitosan films coated with pomegranate phenolic extract. The content of pomegranate phenolic extract in chitosan films varied from 0.5% to 1%. They used the obtained chitosan–pomegranate phenolic extract to coat over the fresh shrimp samples to study the extension in shelf life. Their results suggested that the edible chitosan–pomegranate phenolic extract coating increased the shelf life of refrigerated shrimps up to 10 days.

Homayounpour et al. [34] prepared biocompatible active films based on nano-chitosan incorporated with free and nanoliposome caraway seed extract. FTIR spectroscopy revealed new interactions between nano-chitosan and nanoliposomes. Scanning electron microscopy (SEM) exhibited a homogenous structure and nearly smooth surface morphology with a good dispersion for the obtained nano-chitosan–nanoliposome films. Yellowness (b value) of obtained films decreased while whiteness (L value) index decreased. The incorporation of nanoliposomes into nano-chitosan films also improved the mechanical flexibility and the eater barrier.

Riaz et al. [35] developed novel chitosan based-films incorporated with Chinese chive root extract via the solution casting method. Chinese chive (*Allium tuberosum*) is an alliaceous perennial green vegetable of the Liliaceae family. Chinese chive root extract was added into chitosan at different contents varying from 1% to 5% in w/w. SEM analysis revealed that higher extract concentration led to the formation of agglomerates within the films. The addition of Chinese chive root extract resulted in a decrease in the tensile properties of the films. Besides the water solubility, swelling degree, and water vapor permeability were significantly decreased in obtained films. The antioxidant activity of obtained chitosan films was increased significantly as the concentration of the added Chinese chive root extract increased.

Kõrge et al. [37] developed active chitosan-based films, by blending chitosan with fibrous chestnut extract (*Castanea sativa* Mill.), which is an extract rich in tannins. The obtained active films were used as packaging films for Gouda cheese. The active films with chestnut extract were effective as they did not break down around the cheese and showed protective properties against bacteria.

In another work, Kõrge et al. [40] prepared chitosan films incorporated with a tannins-rich chestnut extract. The obtained film showed enhanced mechanical, antioxidant, and antibacterial properties. The obtained active film applied in packing and storage of filled fresh pasta. Active components of tannins-rich chestnut extract films prevented microbial growth on the food surface during the entire 60 days.

Fresh salmon safety and quality is a major concern of consumers. Merlo et al. [40] prepared chitosan films incorporated with pink pepper residue extract films. This film was tested as packaging film in the modified atmosphere of skinless salmon fillets by monitoring the quality properties during refrigerated storage at 2 °C for 28 days. The results showed that chitosan film and chitosan film incorporated with pink pepper extract significantly reduced lipid oxidation relative to the control and thus extended the shelf life of salmon.

Rambabu et al. [42] developed novel chitosan films modified with mango leaf extract. The obtained active mango leaf extract–chitosan films were thicker, denser, and hydrophobic than pure chitosan film. Mango leaf extract-modified chitosan films showed enhanced tensile strength and antioxidant activity and were suggested as a novel active film for cashew nuts preservation.

2.5 Starch/Extract-Based Food Packaging Films

The demand for more sustainable and bio-sourced food packaging materials [48] increased the interest in the development of novel starch-based food packaging films. Starch is one of the most promising biopolymer candidates to replace the conventional plastics in food packaging due to its biodegradability, relative abundance, chemical inertness, and resistance to degradation, as well as its excellent film-forming capabilities. Another advantage of starch against chitosan and other polysaccharides is its ability to become a thermoplastic material upon the addition of food-safe plasticizers such as polyols (glycerol, sorbitol, xylitol, etc.), sugars (glucose, fructose, etc.), and water [49, 50]. The most common plasticizer glycerol, is a waste byproduct and an economic burden of an ever-growing biodiesel market. Plasticized starch can be directly used for developing packaging films through extrusion process which is the most applied process in food packaging films industry. Table 2.2 listed the studies referred to starch plant extract-based active packaging films of the last five years. Representative results of the references in Table 2.2 are shown here below.

Guo et al. [51] evaluated the effects of super-chilled storage ($-1.3\,°C$) combined with starch film packaging containing different contents of sea buckthorn pomace extract (SSF, 0%, 1%, 2%, and 3% w/w) on the quality of chilled beef. The release kinetics, microstructure, and mechanical properties of the film were also measured to investigate its suitability for super-chilled storage. The results concluded that the developed starch–sea buckthorn pomace extract film is an effective packaging material for beef at super-chilling temperatures.

Microencapsulation is a novel technique that can turn polyphenols of black tea residue into a valuable functional ingredient for different food applications. Rajapaskha et al. [52] used free or microencapsulated black tea extract with pectin–sodium caseinate spent and incorporated it in a cassava starch film. Their results showed that the added spent black tea extract forms hydrogen bonds with starch and results in a less elastic film with improved mechanical properties and with significant antioxidant activity. The release of antioxidant polyphenols was increased into both aqueous and fatty food simulants. Besides, the active film exhibited better barrier properties against UV light and water vapor than the pure starch film.

The ethyl acetate fraction of the crude bark extract of *Euphorbia umbellata* presented high amounts of phenolic compounds with concentration-dependent antioxidant potential and antimicrobial activities against *Staphylococcus aureus* and *Escherichia coli*. Rosas et al. [53] added this fraction into biodegradable starch-based films. Developed films showed maintaining significant antioxidant and antimicrobial activities.

Ceballos et al. [54] developed native and hydrolyzed cassava starch/yerba mate extract films by extrusion molding method. They showed that hydrolyzed starch needed less specific energy to gelatinize in the extrusion process while film with native starch – 10 wt% of extract was more hydrophobic and tensile resistant. Hydrolyzed

Table 2.2 Publications with starch plant extract-based food packaging films in last five years.

a/a	Year	Title	References
1	2021	Changes in chilled beef packaged in starch film containing sea buckthorn pomace extract and quality changes in the film during super-chilled storage	[51]
2	2021	Development and characterization of functional starch-based films incorporating free or microencapsulated spent black tea extract	[52]
3	2021	Starch-based biodegradable active packaging with *Euphorbia umbellata* (PAX) Bruyns bioactive extract	[53]
4	2020	Effect of yerba mate extract on the performance of starch films obtained by extrusion and compression molding as active and smart packaging	[54]
5	2020	Novel biodegradable starch film for food packaging with antimicrobial chicory root extract and phytic acid as a cross-linking agent	[55]
6	2020	Impact of olive extract addition on corn starch-based active edible films properties for food packaging applications	[56]
7	2019	Development of teff starch films containing camu–camu (*Myrciaria dubia* Mc. Vaugh) extract as an antioxidant packaging material	[57]
8	2019	Cowpea starch films containing maqui berry extract and their application in salmon packaging	[58]
9	2019	Antioxidant properties of rye starch films containing rosehip extract and their application in packaging of chicken breast	[59]
10	2018	Application of adzuki bean starch in antioxidant films containing cocoa nibs extract	[60]
12	2017	Active and smart biodegradable packaging based on starch and natural extracts	[61]

starch with 20 wt% of extract film showed the highest tensile toughness. The incorporation of yerba mate extract led to active and smart biodegradable materials.

Baek et al. [58] prepared a new antioxidant film with cowpea starch and maqui berry extract and achieved delayed lipid oxidation in the salmon samples during storage at 4 °C.

Go end Song [59] developed a new biodegradable film by adding different concentrations (0.4, 0.7, and 1.0 g per 100 ml) of rosehip extract into rye starch. The novel film with 1.0% rosehip extract was used to wrap chicken breast fillets. Chicken samples wrapped with this active film had lower peroxide and thiobarbituric acid reactive substances (TBARS) values than chicken samples wrapped with film not containing rosehip extract as well as the non-packaged control, suggesting that lipid oxidation in the chicken breast is inhibited by the inclusion of rosehip extract. In advance, this study showed that as in most cases the added rosehip extract increased the mechanical properties and optical properties of the films.

2.6 Cellulose and Cellulosic Derivatives-Based Food Packaging Films Modified with Plant Extract

Cellulose is considered to be one of the most abundant biopolymers and therefore shows considerable promise as a raw material for packaging applications [62]. Carboxymethylcellulose (CMC) is a cellulose derivative with carboxymethyl substituents ($-CH_2COOH$) bound to some of the hydroxyl groups of the cellulose. CMC can be used as an effective additive to improve product quality and processing properties in various fields, including the food, cosmetic, paper, and textile industries [63]. Methylcellulose is a polysaccharide with the ability to form cohesive thin films with excellent mechanical strength and moderate oxygen permeability. It has a strong potential for the immobilization of natural substances to produce smart packaging for foods [64]. Hydroxypropyl methylcellulose (HPMC) is a biodegradable, highly viscoelastic cellulose ether with film-forming properties that have the potential to use as a food packaging film [65]. Cellulose of bacterial origin is a material with extraordinary and differentiated properties from other polysaccharide-based polymers that are gaining special interest for applications in the food industry [66]. Table 2.3 listed the studies referred to cellulose and cellulosic derivatives modified

Table 2.3 Publications with cellulose and cellulosic derivatives plant extract-based food packaging films in last five years.

a/a	Year	Title	References
1	2020	Biodegradable and active-intelligent films based on methylcellulose and jambolão (*Syzygium cumini*) skins extract for food packaging	[64]
2	2021	Preparation of pH-sensitive food packaging film based on konjac glucomannan and hydroxypropyl methyl cellulose incorporated with mulberry extract	[65]
3	2021	Development of intelligent/active food packaging film based on TEMPO-oxidized bacterial cellulose containing thymol and anthocyanin-rich purple potato extract for shelf life extension of shrimp	[67]
4	2021	Biobased films of nanocellulose and mango leaf extract for active food packaging: supercritical impregnation versus solvent casting	[68]
5	2021	Cellulose nanofiber-based nanocomposite films reinforced with zinc oxide nanorods and grapefruit seed extract	[69]
6	2018	Carboxymethyl cellulose from renewable rice stubble incorporated with Thai rice grass extract as a bioactive packaging film for green tea	[70]
7	2017	Preparation and properties of cellulose nanocomposite films with in situ generated copper nanoparticles using *Terminalia catappa* leaf extract	[71]
8	2017	Development of bacterial cellulose-based slow-release active films by incorporation of *Scrophularia striata* Boiss. extract	[72]
9	2017	Development of biopolymer composite films using a microfluidization technique for carboxymethylcellulose and apple skin particles	[63]

2.6 Cellulose and Cellulosic Derivatives-Based Food Packaging Films Modified with Plant Extract

plant extract-based active packaging films of the last five years. Representative results of these references in Table 2.4 are highlighted here below.

Jambolão (*Syzygium cumini*) is an edible fruit from a tropical tree, native of India, that is widespread nowadays in different regions of Brazil. Jambolão fruits are berries with length around 1.5 and 3.5 cm and dark purple color [64].

Table 2.4 Publications with chitosan composite plant extract-based food packaging films in last five years.

a/a	Year	Title	References
1	2020	Effects of peanut shell and skin extracts on the antioxidant ability, physical and structure properties of starch–chitosan active packaging films	[73]
2	2020	Effect of grapefruit seed extract ratios on functional properties of corn starch–chitosan bionanocomposite films for active packaging	[74]
3	2020	Biodegradable zein active film containing chitosan nanoparticle encapsulated with pomegranate peel extract for food packaging	[75]
4	2018	Biodegradable rice starch/carboxymethyl chitosan films with added propolis extract for potential use as active food packaging	[76]
5	2017	Antimicrobial, optical and mechanical properties of chitosan–starch films with natural extracts	[77]
6	2017	Antioxidant edible films based on chitosan and starch-containing polyphenols from thyme extracts	[78]
7	2017	The properties of chitosan and gelatin films incorporated with ethanolic red grape seed extract and *Ziziphora clinopodioides* essential oil as biodegradable materials for active food packaging	[79]
8	2019	Development of active packaging based on chitosan–gelatin blend films functionalized with Chinese hawthorn (*Crataegus pinnatifida*) fruit extract	[80]
9	2018	Influence of *Codium tomentosum* extract in the properties of alginate and chitosan edible films	[81]
10	2017	Development and characterization of chitosan or alginate-coated low-density polyethylene films containing *Satureja hortensis* extract	[82]
11	2021	Characterization of chitosan–hydroxypropyl methylcellulose blend films enriched with nettle or sage leaf extract for active food packaging applications	[83]
12	2021	Eco-friendly natural extract-loaded antioxidative chitosan/polyvinyl alcohol-based active films for food packaging	[84]
13	2019	Development of grapefruit seed extract-loaded poly(ε-caprolactone)/chitosan films for antimicrobial food packaging	[85]
14	2021	The effect of electrospun polycaprolactone nonwovens containing chitosan and propolis extracts on fresh pork packaged in linear low-density polyethylene films	[86]
15	2019	Novel LDPE/chitosan rosemary and melissa extract nanostructured active packaging films	[87]

Da Silva Filipini et al. [64] produced active and pH-sensitive films by the incorporation of jambolão skins extract (0%, 10%, 30%, and 50%) in methylcellulose films. The obtained films offer the combined benefits to increase food shelf life and indicate product freshness. Overall, the incorporation of jambolão skins extracts in methylcellulose films improved mechanical and barrier performance by increasing the cohesiveness of the matrix structures. Color changes occurred in films by altering pH, because of the pH-sensitivity of anthocyanins structure. The active-intelligent films produced were biodegraded in sea water in 2 days and soil in 15 days.

Glucomannan is a water-soluble polysaccharide that is considered as a dietary fiber. It is a hemicellulose component in the cell walls of some plant species. Glucomannan is a food additive used as an emulsifier and thickener. Zhou et al. [65] prepared a pH-sensitive food packaging film based on konjac glucomannan and HPMC incorporated with mulberry extracts. FT-IR and XRD analysis revealed that there is good molecular interplay among the three components resulted in an improvement of the mechanical properties and the UV resistance. The best antioxidant and antibacterial properties were obtained when the addition of mulberry extract in the composite film was 20%. In addition, the composite film allows real-time visual monitoring of fish freshness.

Bastande et al. [68] prepared nanofibrillated cellulose (NFC) films loaded with mango leaf extract and showed that the supercritical solvent impregnation method is beneficial against solvent casting film-processing method for UV-light protection, antioxidant properties, and antimicrobial activity against gram-positive and gram-negative bacteria.

Rodsamran and Sothornvit [70] used a commercial CMC and a CMC from rice stubble to form CMC-film incorporated with rice grass extract microencapsulated powder. The bioactive ability of this film was tested on the quality of green tea (GTE) during storage compared with commercial high-density polyethylene film. The CMC with rice grass extract microencapsulated film effectively prevented dried GTE from lipid oxidation. Moreover, the total phenolic content of GTE packed in CMC with rice grass extract microencapsulated film, which was equivalent to that packed in high-density polyethylene, having the highest acceptability score from the sensory evaluation.

2.7 Gelatin and Alginate/Plant Extract-Based Food Packaging Films

Among other biopolymers with film forming properties, gelatin has attracted the attention for the preparation of biodegradable food packaging materials due to its low cost, renewability, high transparency, biodegradability, biocompatibility, and excellent film-forming properties [88, 89]. Gelatin is derived with chemical modification of the animal protein collagen, mainly obtained from waste produced during animal processing. By products of fish, pork, and bovine processing are the most usually used sources for gelatin production.

Alginates are natural hydrophilic polysaccharide biopolymers mainly extracted from marine brown algae. In the form of films or coatings, they exhibit: good film-forming properties, low permeability to O_2 and vapors, flexibility, water solubility, and gloss while being tasteless and odorless [90]. In the last few years, alginates are combined with additives such as organic acids, essential oils, plant extracts, bacteriocins, and nanomaterials, they contribute to the retention of moisture, reduction in shrinkage, retardation of oxidation, inhibition of color and texture degradation, reduction in microbial load, enhancement of sensory acceptability, and minimization of cooking losses. Thus, it could be stated that alginates are promising candidates for future active, intelligent, and green packaging technologies.

Recently, Ucak et al. [91] studied the microbiological, chemical quality, and sensorial changes of rainbow trout fillets coated with gelatin films supplemented with different contents (2%, 8%, 16%), of propolis extract (PE) as a source of polyphenols during 15 days of refrigerated storage ($4 \pm 1\,°C$). According to peroxide value (PV) and TBARS assays, lipid oxidation was delayed in the fillets coated with gelatin films incorporated with propolis extracts compared with the control and pure gelatin-coated fillets. The results clearly showed that propolis extract can be recommended to be used as a natural antioxidant and antimicrobial additive with gelatin films to maintain rainbow trout fillet quality.

Espino-Manzano et al. [92] produced a water-in-oil (W/O) nanoemulsion by applying ultrasound to xoconostle extract and orange oil. Then they incorporated this bioactive nanoemulsion into gelatine films in different proportions 1:0 (control), 1:0.10, 1:0.25, 1:0.50, 1:0.75, and 1:1 (gelatine:nanoemulsion) and gave active films with significant antioxidant activity. The color of the films was influenced by the incorporation of nanoemulsions, showing that it was significantly different from the control. Mechanical properties were affected by the incorporation of nanoemulsified bioactive compounds into gelatine films. The obtained films presented changes in strength and flexibility.

Luo et al. [93] developed a novel green and bioactive films with sodium alginate incorporated with guava leaf extracts. Seven formulations were performed with different sodium alginate: Results showed that guava leaf extracts create intermolecular bonding with sodium alginate and thus could greatly enhance the antioxidant activity, antibacterial activity, tensile strength, and water solubility of the sodium alginate film as well as the water barrier property while inducing a decrease in the moisture content and elongation at the break.

2.8 Composites/Plant Extract-Based Food Packaging Films

To overcome some disadvantages of most biopolymers such as weak mechanical properties and wettability, the researchers use technologies to develop a new class of materials called biopolymers composites blends [94]. In these novel composite biopolymers matrices, the researchers investigate the incorporation and the performance of plant extracts. In the following pages, we reviewed the literature for composite plant extract-based packaging films and we categorized the founded studies in chitosan composites, starch composites, cellulose composites, and other composites plant extract-based packaging films.

2.8.1 Chitosan Composites/Plant Extract-Based Food Packaging Films

In the last five years, a significant number of studies is found in the literature that referred to chitosan composites/plant extract-based active packaging films. These studies are listed in the next Table 2.4. Representative results of these studies are highlighted here below.

Meng et al. [73] studied, the antioxidant ability of peanut shell and peanut skin extracts and their effects on the physical and structural properties of starch–chitosan films. The results showed that the DPPH radical scavenging ability of peanut skin extracts was significantly higher than the peanut shell extracts. This is proposed to be due to the rich rutin and 4-O-caffeoulquinic acid existed in the peanut skin extracts. When added the peanut skin and shell extracts into the starch–chitosan film, the apparent viscosity of film forming solution at $100\,s^{-1}$ decreased. Moreover, water vapor permeability and swelling of film decreased with the addition of peanut skin and shell extracts. Two peanut extracts also increased the color L^* and opacity of film. The tensile strength of film increased with the addition of peanut skin extracts and decreased with peanut shell extracts. The addition of two extracts also increased the endothermic temperature of the starch–chitosan film.

Jha [74] investigated the effect of different ratios of grapefruit seed extract on functional properties of bionanocomposite films prepared with corn starch incorporated with chitosan–nanoclay. Experimental results exhibited that the addition of grapefruit seed extract properly dispersed with starch combined with chitosan bionanocomposite films. The presence of grapefruit seed extract from 0% to 1.5% v/v increased crystallinity and tensile strength while decreased in elongation at break and water barrier. Bread samples packed, synthetic plastic exhibited the proliferation of fungal growth in six days, whereas bread packed with the starch/chitosan/1.5% v/v grapefruit seed extract bionanocomposite film showed strong antifungal effect for 20 days.

Lozano-Navarro et al. [77] incorporated extracts from cranberry, blueberry, beetroot, pomegranate, oregano, pitaya, and resveratrol (from grapes) in chitosan–starch composite films. The films with beetroot, cranberry, and blueberry extracts demonstrated the best antimicrobial activity against various bacteria and fungi in comparison with unmodified chitosan–starch film. They concluded that the addition of antioxidants improved the antimicrobial performance of these films. These polymers combined with the extracts effectively inhibit or reduce microorganism growth from human and environmental contact; therefore, previous sterilization could be unnecessary in comparison with traditional plastics. The presence of extracts decreased transmittance percentages at 280 and 400 nm, as well as the transparency values while increasing their opacity values, providing better UV–vis light barrier properties. Despite diminished glass transition temperatures (T_g), the values obtained are still adequate for food packaging applications.

Kan et al. [80] extracted polyphenols from the fruits of Chinese hawthorn and further added them into chitosan–gelatin blend films and investigated the microstructure, physical, mechanical, barrier, and antioxidant properties of the films. Results showed epicatechin, chlorogenic acid, and procyanidin B2 were the main polyphenols in the extract of hawthorn fruits. The inner microstructure of chitosan–gelatin blend films became more compact when the extract was incorporated. The intermolecular interactions between the film matrix and the extract were through hydrogen bonding and electrostatic interactions. The incorporation of the extract remarkably increased the thickness, tensile strength, and elongation at break of chitosan–gelatin blend films. However, the moisture content, water vapor permeability, and light transmittance of chitosan–gelatin blend films were significantly reduced by the addition of the extract. Moreover, chitosan–gelatin blend films containing the extract exhibited potent free radical scavenging ability.

Augusto et al. [81] investigated the effect of *Codium tomentosum* seaweed extract (SE) in alginate and chitosan edible films. In the obtained novel active alginate films, the incorporation of extract decreased film solubility, water barrier, and mechanical elasticity, and showed no effect on thermal properties.

Bigi et al. [83] fabricated chitosan (CS)/HPMC blend enriched with nettle and sage leaf extracts. The successful incorporation of leaf extract was confirmed by SEM and FTIR analysis while the obtained active films showed improvement in UV–visible light barrier property, water barrier, and antioxidant activity.

Rahmani et al. [82] developed novel bilayer active films with improved antioxidant activity based on alginate or chitosan-coated LDPE containing different concentrations of summer savory extract (SSE). LDPE's surface energy was increased via the cold atmospheric plasma method. Results showed that plasma treatment increased oxygen-containing polar group, surface roughness, decreased water contact angle of LDPE surface, and resulted in enhanced adhesion between polysaccharide coating and LDPE. Tensile strength of both alginate and chitosan coated-LDPE was increased, while chitosan-coated LDPE films had lower water solubility than alginate-coated LDPE films.

Vargas Romero et al. [86] have developed active linear low-density polyethylene (LLDPE) films coated with a polycaprolactone/chitosan nonwoven and LLDPE films coated with a polycaprolactone/chitosan nonwoven fortified with Colombian propolis extract. The active LLDPE films were evaluated for the preservation of fresh pork loin (longissimus dorsi) chops during refrigerated storage at 4 °C for up to 20 days. The incorporation of the propolis-containing nonwoven layer provided antioxidant and antimicrobial properties to LLDPE film, as evidenced by improved color stability, no differences in lipid oxidation, and a delay of four days for the onset of bacteria growth of pork chops during the refrigerated storage period.

Giannakas et al. [87] combined LDPE with chitosan (CS) to develop a novel food packaging film with an increased oxygen permeability barrier and higher antimicrobial activity. Two essential oil extracts rosemary (RO) and melissa (MO) were added to this packaging matrix to improve its antioxidant properties and vanish food odor problems. The experimental results have indicated that the final material exhibited advanced properties. More specifically, chitosan addition was observed to lead to an enhanced oxygen and water vapor permeability barrier while the extracted essential oil addition led to enhanced tensile strength and antioxidant properties.

2.8.2 Starch Composites/Extract-Based Food Packaging Films

Table 2.5 listed the last five-year publications available in the literature which are referred to as starch composite active films with other polymers or biopolymers incorporated with plant extracts. PVA is a semicrystalline, water-soluble synthetic polymer, with good biodegradability, excellent chemical resistance, good mechanical properties linked to the presence of OH groups, and their hydrogen bond formation ability [50, 102]. PVA has been widely utilized for the preparation of blends and composites with several natural, renewable polymers. PVA blended successfully with starch [49] and can give an excellent composite matrix for plant extract incorporation [95–97].

Recently, Kumar et al. [95] fabricated a PVA–starch blend incorporated with pineapple peel extract. The novel pineapple peel extract-based films displayed good antioxidant properties with enhanced thermal properties.

Mustafa et al. [96] developed active and intelligent food packaging films based on biodegradable polymers such as PVA and starch, incorporated with natural

Table 2.5 Publications with starch composites plant extract-based food packaging films in last five years.

a/a	Year	Title	References
1	2021	Pineapple peel extract incorporated poly(vinyl alcohol)-corn starch film for active food packaging: preparation, characterization and antioxidant activity	[95]
2	2020	PVA/starch/propolis/anthocyanins rosemary extract composite films as active and intelligent food packaging materials	[96]
3	2021	Development and characterization of PVA–starch incorporated with coconut shell extract and sepiolite clay as an antioxidant film for active food packaging applications	[97]
4	2020	Influence of pinhão starch and natural extracts on the performance of thermoplastic cassava starch/PBAT extruded blown films as a technological approach for bio-based packaging material	[98]
5	2020	Antioxidant and antibacterial activities of cassava starch and whey protein blend films containing rambutan peel extract and cinnamon oil for active packaging	[99]
6	2020	Active meat packaging from thermoplastic cassava starch containing sappan and cinnamon herbal extracts via LLDPE-blown film extrusion	[100]
7	2017	Intelligent pH indicator film composed of agar/potato starch and anthocyanin extracts from purple sweet potato	[101]

additives, that is, propolis extract (PE) and anthocyanin. Boric acid was used as a cross-linker. The results showed good compatibility of film mixture and an improvement of the mechanical strength for films containing 20% PE. Moreover, films showed a significant inhibition, against *E. coli* and methicillin-resistant *S. aureus* and a great color response against different pH ranging from 2 to 14. Finally, a food spoilage test was performed by using pasteurized milk and the films responded visibly by changing color and protected milk from spoilage.

Tanwar et al. [97] fabricated novel active PVA and corn starch (ST) films incorporated with 3%, 5%, 10%, and 20% (v/v) coconut shell extract and sepiolite clay. The addition of sepiolite clay to PVA improved mechanical and thermal properties, while coconut shell extract addition improves water barrier, antioxidant activity, and the color of the active films. Finally, this PVA–corn starch-based active film successfully improved the oxidative stability of packaged soybean oil.

Müller et al. [98] produced films by using the blown extrusion method from blends made with cassava and pinhão thermoplastic starch, compostable polyester (poly(butylene adipate-*co*-terephthalate, PBAT), and natural extracts (rosemary and GTE). The effect of the incorporation of the extracts and the type of starch added in the film properties were investigated following the mixture design (2^3) approach. Regression models and response surface curves were generated to predict the film properties. The incorporation of the extracts decreased the lightness parameter and the films produced with GTE extract were more opaque than those made with rosemary. Starch/rosemary blends were more flexible, while the extract type did not have a significant effect on tensile strength. Film elongation (ELO) ranged from 520% to 719% comparable to some synthetic polymers. The water vapor permeability was improved in approximately 14% with addition of the extracts.

Thermoplastic cassava starch containing sappan and cinnamon herbal extracts via LLDPE-blown film extrusion were developed by Khumkomgool et al. [100]. Their study concluded that active films with sappan and cinnamon gave antimicrobial and preserved redness of beef.

2.8.3 Other Composites Plant Extract-Based Food Packaging Films

In the recent literature are founded some studies with cellulose-based composite active films with other polymers [103] and biopolymers [104] incorporated with plant extracts. There are also some publication; plant extract-based active composite films with PLA [105, 106], gelatin [107, 108] and alginate [109].

Kanatt [103] developed active/smart CMC–PVA composite films containing rose petal extract for fish packaging. Rose petal extract having excellent bioactivity, was used as the functional component in CMC–PVA packaging film. Due to its high anthocyanin content, rose petal extract could be used as a pH indicator, as its color changed from bright red (pH 2) to yellow (pH 10). The developed film was used for packing Indian Mackerel and assessing its quality during chilled storage. Mackerel packed in neat film spoiled in 3 days while that in rose petal extract containing films had shelf life of 12 days. Total bacterial count, pH, total volatile basic nitrogen and thiobarbituric acid reactive substance of fish in control were higher as compared to in rose petal extract films. On spoilage rose petal extract containing films showed visible color response from red to green.

Multifunctional CMC/agar-based smart films were fabricated by combining cellulose nanocrystals (CNCs) separated from onion peel and shikonin isolated from the roots of *Lithospermum erythrorhizon* [104]. The CNC had a nanosized crystal structure with a length and width of 225.6 ± 42.4 and 10.2 ± 2.2 nm, respectively, having an aspect ratio of 22 : 1. The integration of CNCs significantly improved mechanical, water resistance, and optical properties. The addition of shikonin provided functional properties such as pH sensitivity, antioxidant, and antimicrobial properties without significant change in the water barrier, and thermal stability of the film. The films also exhibited high light transmittance with increased UV light barrier properties.

A new active coating was developed by Arrieta et al. [110] by using *Cucumis metuliferus* fruit extract as antioxidant additive with the aim of obtaining an easy way to functionalize LDPE films for food packaging applications. Cellulose acetate and cellulose acetate incorporated with *C. metuliferus* fruit extract film-forming solutions were successfully coated onto the surface of LDPE, showing good adhesion in the final bilayer structure. The obtained results suggest that the cellulose acetate incorporated with *C. metuliferus* fruit extract-based coating can be used to easily introduce active functionality to typically used LDPE at industrial level and enhance its oxygen barrier, without affecting the high transparency, revealing their potential application in the active food packaging sector to extend the shelf life of packaged food by prevention of lipid oxidation of fatty food or by prevention fruit browning.

Wrona et al. [105] developed a novel active film material based on HPMC-containing poly(lactic acid) nanoparticles loaded with antioxidant GTE extract. Emulsification–solvent evaporation technique was used to fabricate and control the size of poly(lactic acid) nanoparticles and thus enable a controlled release of the GTE extract from the HPMC-based film. A obtained composite films characterized on their physicochemical properties and the release of the GTE extract was confirmed by migration studies in 50% v/v ethanol/water food simulant. The study suggests that the HPMC-poly(lactic acid) nanoparticles-based active film could potentially be used for extending the shelf life of food products with high-fat content.

Alirezalu et al. [109] studied the effects of combining a polyamide–alginate casing incorporated with nisin and ε-polylysine nanoparticles and a mixed plant extract as ingredient to improve the shelf life and safety of frankfurter-type sausage. The mixed plant extract composed of olive leaves (OLE), GTE and stinging nettle extracts (SNE) in equal rates were studied. The film characteristics and microbiological properties of sausage samples were evaluated. Sausage samples were packaged in polyethylene bags (vacuum condition) and analyzed during 45 days of storage at 4 °C. Control sausages were also treated with 120 ppm sodium nitrite. Polyamide–alginate films containing 100 ppm nisin and 500 ppm ε-polylysine nanoparticles had the highest ultimate tensile strength compared to other films. However, 100 ppm nisin and 500 ε-polylysine nanoparticles decreased water vapor permeability of films. The results also revealed that nisin nanoparticles had significantly ($p < 0.05$) low inhibitory effects against *E. coli*, *S. aureus*, molds and yeasts, and total viable counts compared to control and ε-polylysine nanoparticles. Furthermore, 1000 ppm ε-polylysine nanoparticles displayed the highest antimicrobial activity. Based on the obtained results, the films containing ε-polylysine nanoparticle could be considered as a promising packaging for frankfurter-type sausages.

2.9 Conclusion

From all the above revised online available studies for plant extract-based food packaging films we conclude in: (i) most available studies are referred to biopolymers based plant extract based packaging films, (ii) chitosan, starch, and cellulosic derivatives are the most investigated biopolymers as polymer carriers for plant extracts, (iii) solution blending method is the most used method for casting plant extract-based packaging films, (iv) there are limited studies referred to extrusion preparation method mainly for thermoplastic starch based plant extract based packaging films, (v) limited studies also are referred to bilayer films with biopolymers extract based coating into polyethylene films, (vi) physicochemical characterization of films mainly with XRD and or FTIR spectroscopy shows a good performance of most biopolymer matrices with plant extract molecules which is affected by the content of plant extract used, (vii) interplay between biopolymers and plant extract molecules reflected in the enhancement of the mechanical thermomechanical and barrier properties of obtained films, (viii) in most cases color changes in the obtained films it is resulted from the incorporation of plant extract molecules into biopolymers, (ix) UV blocking properties and pH-sensitive properties are obtained in some studies in accordance with the kind of plant extract used, (x) in all available studies antioxidant properties are studied while antimicrobial properties studied in some of them, (xi) in some studied the shelf-life extension of various foods such as chicken breast, pork meat, salmon fillets, and shrimps by monitoring the deceleration of lipid oxidation.

The field of extract-based food packaging films is promising and much more efforts must be devoted in investigating biopolymers such as alginates, gelatin, PLA, polyhydroxybutyrate (PHB), and polyhydroxyalkanoate (PHA) as potential novel and sustainable carriers of various plant extracts. In addition, researchers must give emphasis in developing active films with methods applied in packaging industry such as extrusion process, bilayer films, and coatings or suggesting novel methods for applying the use of active biopolymers plant extract films in food industry. The performance of such active films in the extension of shelf life of various foods and fruits is another field which is needed to be more investigated more in the next few years.

Acknowledgment

This chapter has been financed by the funding program "MEDICUS," of the University of Patras.

References

1. Hamam, M., Chinnici, G., Di Vita, G. et al. (2021). Circular economy models in agro-food systems: a review. *Sustainability* 13 (6): 3453.
2. Holden, N.M., White, E.P., Lange, M.C., and Oldfield, T.L. (2018). Review of the sustainability of food systems and transition using the internet of food. *npj Science of Food* 2 (1): 18.
3. Otto, S., Strenger, M., Maier-Nöth, A., and Schmid, M. (2021). Food packaging and sustainability – consumer perception vs. correlated scientific facts: a review. *Journal of Cleaner Production* 298: 126733.
4. Nemat, B., Razzaghi, M., Bolton, K., and Rousta, K. (2019). The role of food packaging design in consumer recycling behavior—a literature review. *Sustainability* 11 (16): 4350.
5. Guillard, V., Gaucel, S., Fornaciari, C. et al. (2018). The next generation of sustainable food packaging to preserve our environment in a circular economy context. *Frontiers in Nutrition* 5: 121.
6. Duncan, T.V. (2011). Applications of nanotechnology in food packaging and food safety: barrier materials, antimicrobials and sensors. *Journal of Colloid and Interface Science* 363 (1): 1–24.
7. Singh, T., Shukla, S., Kumar, P. et al. (2017). Application of nanotechnology in food science: perception and overview. *Frontiers in Microbiology* 8: 1501.
8. Ncama, K., Magwaza, L.S., Mditshwa, A., and Tesfay, S.Z. (2018). Plant-based edible coatings for managing postharvest quality of fresh horticultural produce: a review. *Food Packaging and Shelf Life* 16: 157–167.
9. Mir, S.A., Dar, B.N., Wani, A.A., and Shah, M.A. (2018). Effect of plant extracts on the techno-functional properties of biodegradable packaging films. *Trends in Food Science & Technology* 80: 141–154.
10. Salmas, C.E., Giannakas, A.E., Baikousi, M. et al. (2021). Development of poly (l-lactic acid)/chitosan/basil oil active packaging films via a melt-extrusion process using novel chitosan/basil oil blends. *Processes* 9 (1): 88.
11. Silva-Weiss, A., Ihl, M., Sobral, P.J.A. et al. (2013). Natural additives in bioactive edible films and coatings: functionality and applications in foods. *Food Engineering Reviews* 5 (4): 200–216.
12. Yong, H. and Liu, J. (2021). Active packaging films and edible coatings based on polyphenol-rich propolis extract: a review. *Comprehensive Reviews in Food Science and Food Safety* 20 (2): 2106–2145.
13. Alara, O.R., Abdurahman, N.H., and Ukaegbu, C.I. (2021). Extraction of phenolic compounds: a review. *Current Research in Food Science* 4: 200–214.
14. Moret, S., Conchione, C., Srbinovska, A., and Lucci, P. (2019). Microwave-based technique for fast and reliable extraction of organic contaminants from food, with a special focus on hydrocarbon contaminants. *Foods* 8 (10): 503.
15. Nour, A.H., Oluwaseun, A.R., Nour, A.H. et al. (2021). *Microwave-Assisted Extraction of Bioactive Compounds (Review)*. IntechOpen.

16 Chemat, F., Rombaut, N., Sicaire, A.-G. et al. (2017). Ultrasound assisted extraction of food and natural products. Mechanisms, techniques, combinations, protocols and applications. A review. *Ultrasonics Sonochemistry* 34: 540–560.

17 Richter, B.E., Jones, B.A., Ezzell, J.L. et al. (1996). Accelerated solvent extraction: a technique for sample preparation. *Analytical Chemistry* 68 (6): 1033–1039.

18 Díaz-Reinoso, B., Moure, A., Domínguez, H., and Parajó, J.C. (2006). Supercritical CO_2 extraction and purification of compounds with antioxidant activity. *Journal of Agricultural and Food Chemistry* 54 (7): 2441–2469.

19 de Andrade Lima, M., Andreou, R., Charalampopoulos, D., and Chatzifragkou, A. (2021). Supercritical carbon dioxide extraction of phenolic compounds from potato (*Solanum tuberosum*) peels. *Applied Sciences* 11 (8): 3410.

20 Gligor, O., Mocan, A., Moldovan, C. et al. (2019). Enzyme-assisted extractions of polyphenols – a comprehensive review. *Trends in Food Science & Technology* 88: 302–315.

21 Cheng, X., Bi, L., Zhao, Z., and Chen, Y. (2015). Advances in enzyme assisted extraction of natural products. In: *3rd International Conference on Material, Mechanical and Manufacturing Engineering (IC3ME 2015)*, 371–375. Atlantis Press.

22 Wang, L., Liu, F., Li, T. et al. (2018). Enzyme assisted extraction, purification and structure analysis of the polysaccharides from naked pumpkin seeds. *Applied Sciences* 8 (10): 1866.

23 Muzzarelli, R.A.A., Boudrant, J., Meyer, D. et al. (2012). Current views on fungal chitin/chitosan, human chitinases, food preservation, glucans, pectins and inulin: a tribute to Henri Braconnot, precursor of the carbohydrate polymers science, on the chitin bicentennial. *Carbohydrate Polymers* 87 (2): 995–1012.

24 Ngo, D. and Kim, S. (2014). Antioxidant effects of chitin, chitosan, and their derivatives. In: *Marine Carbohydrates: Fundamentals and Applications, Part B*, vol. 73, 15–31. Elsevier Inc.

25 Ravi Kumar, M.N.V. (2000). A review of chitin and chitosan applications. *Reactive and Functional Polymers* 46 (1): 1–27.

26 Aider, M. (2010). Chitosan application for active bio-based films production and potential in the food industry: review. *LWT – Food Science and Technology* 43 (6): 837–842.

27 Cazón, P. and Vázquez, M. (2019). Applications of chitosan as food packaging materials. In: *Sustainable Agriculture Reviews 36: Chitin and Chitosan: Applications in Food, Agriculture, Pharmacy, Medicine and Wastewater Treatment* (ed. G. Crini and E. Lichtfouse), 81–123. Cham: Springer International Publishing.

28 Elsabee, M.Z. and Abdou, E.S. (2013). Chitosan based edible films and coatings: a review. *Materials Science and Engineering C* 33 (4): 1819–1841.

29 Priyadarshi, R. and Rhim, J.-W. (2020). Chitosan-based biodegradable functional films for food packaging applications. *Innovative Food Science & Emerging Technologies* 62: 102346.

30 Rachtanapun, P., Klunklin, W., Jantrawut, P. et al. (2021). Characterization of chitosan film incorporated with curcumin extract. *Polymers* 13 (6): 963.

31 Kadam, A.A., Singh, S., and Gaikwad, K.K. (2021). Chitosan based antioxidant films incorporated with pine needles (*Cedrus deodara*) extract for active food packaging applications. *Food Control* 124: 107877.

32 Lan, W., Wang, S., Zhang, Z. et al. (2021). Development of red apple pomace extract/chitosan-based films reinforced by TiO_2 nanoparticles as a multifunctional packaging material. *International Journal of Biological Macromolecules* 168: 105–115.

33 Sabu, S., Xavier, K.A.M., and Sasidharan, A. (2021). Efficacy of pomegranate phenolic extract and chitosan as an edible coating for shelf life extension of Indian white shrimp during refrigerated storage. *Journal of Packaging Technology and Research* 5 (2): 59–67.

34 Homayounpour, P., Shariatifar, N., and Alizadeh-Sani, M. (2021). Development of nanochitosan-based active packaging films containing free and nanoliposome caraway (*Carum carvi.* L) seed extract. *Food Science & Nutrition* 9 (1): 553–563.

35 Riaz, A., Lagnika, C., Luo, H. et al. (2020). Chitosan-based biodegradable active food packaging film containing Chinese chive (*Allium tuberosum*) root extract for food application. *International Journal of Biological Macromolecules* 150: 595–604.

36 Sun, J., Jiang, H., Wu, H. et al. (2020). Multifunctional bionanocomposite films based on konjac glucomannan/chitosan with nano-ZnO and mulberry anthocyanin extract for active food packaging. *Food Hydrocolloids* 107: 105942.

37 Kõrge, K., Šeme, H., Bajić, M. et al. (2020). Reduction in spoilage microbiota and cyclopiazonic acid mycotoxin with chestnut extract enriched chitosan packaging: stability of inoculated gouda cheese. *Foods* 9 (11): 1645.

38 de Souza Mesquita, L.M., Martins, M., Maricato, É. et al. (2020). Ionic liquid-mediated recovery of carotenoids from the *Bactris gasipaes* fruit waste and their application in food-packaging chitosan films. *ACS Sustainable Chemistry & Engineering* 8 (10): 4085–4095.

39 Kõrge, K., Bajić, M., Likozar, B., and Novak, U. (2020). Active chitosan–chestnut extract films used for packaging and storage of fresh pasta. *International Journal of Food Science & Technology* 55 (8): 3043–3052.

40 Merlo, T.C., Contreras-Castillo, C.J., Saldaña, E. et al. (2019). Incorporation of pink pepper residue extract into chitosan film combined with a modified atmosphere packaging: effects on the shelf life of salmon fillets. *Food Research International* 125: 108633.

41 Zhang, X., Liu, Y., Yong, H. et al. (2019). Development of multifunctional food packaging films based on chitosan, TiO_2 nanoparticles and anthocyanin-rich black plum peel extract. *Food Hydrocolloids* 94: 80–92.

42 Rambabu, K., Bharath, G., Banat, F. et al. (2019). Mango leaf extract incorporated chitosan antioxidant film for active food packaging. *International Journal of Biological Macromolecules* 126: 1234–1243.

43 Thuong, N.T., Ngoc Bich, H.T., Thuc, C.N.H. et al. (2019). Preparation and characterization of piper Betle Linn. leaf extract incorporated chitosan films as potential active food packaging materials. *ChemistrySelect* 4 (27): 8150–8157.

44 Balti, R., Mansour, M.B., Sayari, N. et al. (2017). Development and characterization of bioactive edible films from spider crab (*Maja crispata*) chitosan incorporated with Spirulina extract. *International Journal of Biological Macromolecules* 105: 1464–1472.

45 Souza, V.G.L., Fernando, A.L., Pires, J.R.A. et al. (2017). Physical properties of chitosan films incorporated with natural antioxidants. *Industrial Crops and Products* 107: 565–572.

46 Rubilar, J.F., Cruz, R.M.S., Zuñiga, R.N. et al. (2017). Mathematical modeling of gallic acid release from chitosan films with grape seed extract and carvacrol. *International Journal of Biological Macromolecules* 104: 197–203.

47 Kalaycıoğlu, Z., Torlak, E., Akın-Evingür, G. et al. (2017). Antimicrobial and physical properties of chitosan films incorporated with turmeric extract. *International Journal of Biological Macromolecules* 101: 882–888.

48 Lauer, M.K. and Smith, R.C. (2020). Recent advances in starch-based films toward food packaging applications: physicochemical, mechanical, and functional properties. *Comprehensive Reviews in Food Science and Food Safety* 19 (6): 3031–3083.

49 Katerinopoulou, K., Giannakas, A., Grigoriadi, K. et al. (2014). Preparation and characterization of acetylated corn starch-(PVOH)/clay nanocomposite films. *Carbohydrate Polymers* 102 (1): 216–222.

50 Katerinopoulou, K., Giannakas, A., Barkoula, N.-M., and Ladavos, A. (2019). Preparation, characterization, and biodegradability assessment of maize starch-(PVOH)/clay nanocomposite films. *Starch* 71 (1–2): 1800076. Wiley Online Library.

51 Guo, Z., Ge, X., Gou, Q. et al. (2021). Changes in chilled beef packaged in starch film containing sea buckthorn pomace extract and quality changes in the film during super-chilled storage. *Meat Science* 182: 108620.

52 Rajapaksha, S.W. and Shimizu, N. (2021). Development and characterization of functional starch-based films incorporating free or microencapsulated spent black tea extract. *Molecules* 26 (13): 3898.

53 Rosas, M.R., Lemes, B.M., Minozzo, B.R. et al. (2021). Starch-based biodegradable active packaging with *Euphorbia umbellata* (PAX) Bruyns bioactive extract. *Journal of Packaging Technology and Research* 5 (2): 97–106.

54 Ceballos, R.L., Ochoa-Yepes, O., Goyanes, S. et al. (2020). Effect of yerba mate extract on the performance of starch films obtained by extrusion and compression molding as active and smart packaging. *Carbohydrate Polymers* 244: 116495.

55 Jaśkiewicz, A., Budryn, G., Nowak, A., and Efenberger-Szmechtyk, M. (2020). Novel biodegradable starch film for food packaging with antimicrobial chicory root extract and phytic acid as a cross-linking agent. *Foods* 9 (11): 1696.

56 García, A.V., Álvarez-Pérez, O.B., Rojas, R. et al. (2020). Impact of olive extract addition on corn starch-based active edible films properties for food packaging applications. *Foods* 9 (9): 1339.

57 Ju, A. and Song, K.B. (2019). Development of teff starch films containing camu-camu (*Myrciaria dubia* Mc. Vaugh) extract as an antioxidant packaging material. *Industrial Crops and Products* 141: 111737.

58 Baek, S.-K., Kim, S., and Song, K.B. (2019). Cowpea starch films containing maqui berry extract and their application in salmon packaging. *Food Packaging and Shelf Life* 22: 100394.

59 Go, E.-J. and Song, K.B. (2019). Antioxidant properties of rye starch films containing rosehip extract and their application in packaging of chicken breast. *Starch – Stärke* 71 (11–12): 1900116.

60 Kim, S., Baek, S.-K., Go, E., and Song, K.B. (2018). Application of adzuki bean starch in antioxidant films containing cocoa nibs extract. *Polymers* 10 (11): 1210.

61 Medina-Jaramillo, C., Ochoa-Yepes, O., Bernal, C., and Famá, L. (2017). Active and smart biodegradable packaging based on starch and natural extracts. *Carbohydrate Polymers* 176: 187–194.

62 Stark, N.M. (2016). Opportunities for cellulose nanomaterials in packaging films: a review and future trends. *Journal of Renewable Materials* 4 (5): 313–326.

63 Choi, I., Chang, Y., Shin, S.-H. et al. (2017). Development of biopolymer composite films using a microfluidization technique for carboxymethylcellulose and apple skin particles. *International Journal of Molecular Sciences* 18 (6): 1278.

64 da Silva Filipini, G., Romani, V.P., and Guimarães Martins, V. (2020). Biodegradable and active-intelligent films based on methylcellulose and jambolão (*Syzygium cumini*) skins extract for food packaging. *Food Hydrocolloids* 109: 106139.

65 Zhou, N., Wang, L., You, P. et al. (2021). Preparation of pH-sensitive food packaging film based on konjac glucomannan and hydroxypropyl methyl cellulose incorporated with mulberry extract. *International Journal of Biological Macromolecules* 172: 515–523.

66 Cazón, P. and Vázquez, M. (2021). Bacterial cellulose as a biodegradable food packaging material: a review. *Food Hydrocolloids* 113: 106530.

67 Wen, Y., Liu, J., Jiang, L. et al. (2021). Development of intelligent/active food packaging film based on TEMPO-oxidized bacterial cellulose containing thymol and anthocyanin-rich purple potato extract for shelf life extension of shrimp. *Food Packaging and Shelf Life* 29: 100709.

68 Bastante, C.C., Silva, N.H.C.S., Cardoso, L.C. et al. (2021). Biobased films of nanocellulose and mango leaf extract for active food packaging: supercritical impregnation versus solvent casting. *Food Hydrocolloids* 117: 106709.

69 Roy, S., Kim, H.C., Panicker, P.S. et al. (2021). Cellulose nanofiber-based nanocomposite films reinforced with zinc oxide nanorods and grapefruit seed extract. *Nanomaterials* 11 (4): 877.

70 Rodsamran, P. and Sothornvit, R. (2018). Carboxymethyl cellulose from renewable rice stubble incorporated with Thai rice grass extract as a bioactive packaging film for green tea. *Journal of Food Processing and Preservation* 42 (9): e13762.

71 Muthulakshmi, L., Rajini, N., Nellaiah, H. et al. (2017). Preparation and properties of cellulose nanocomposite films with in situ generated copper nanoparticles using *Terminalia catappa* leaf extract. *International Journal of Biological Macromolecules* 95: 1064–1071.

72 Sukhtezari, S., Almasi, H., Pirsa, S. et al. (2017). Development of bacterial cellulose based slow-release active films by incorporation of *Scrophularia striata* Boiss. extract. *Carbohydrate Polymers* 156: 340–350.

73 Meng, W., Shi, J., Zhang, X. et al. (2020). Effects of peanut shell and skin extracts on the antioxidant ability, physical and structure properties of starch-chitosan active packaging films. *International Journal of Biological Macromolecules* 152: 137–146.

74 Jha, P. (2020). Effect of grapefruit seed extract ratios on functional properties of corn starch-chitosan bionanocomposite films for active packaging. *International Journal of Biological Macromolecules* 163: 1546–1556.

75 Cui, H., Surendhiran, D., Li, C., and Lin, L. (2020). Biodegradable zein active film containing chitosan nanoparticle encapsulated with pomegranate peel extract for food packaging. *Food Packaging and Shelf Life* 24: 100511.

76 Suriyatem, R., Auras, R.A., Rachtanapun, C., and Rachtanapun, P. (2018). Biodegradable rice starch/carboxymethyl chitosan films with added propolis extract for potential use as active food packaging. *Polymers* 10 (9): 954.

77 Lozano-Navarro, J.I., Díaz-Zavala, N.P., Velasco-Santos, C. et al. (2017). Antimicrobial, optical and mechanical properties of chitosan–starch films with natural extracts. *International Journal of Molecular Sciences* 18 (5): 997.

78 Talón, E., Trifkovic, K.T., Nedovic, V.A. et al. (2017). Antioxidant edible films based on chitosan and starch containing polyphenols from thyme extracts. *Carbohydrate Polymers* 157: 1153–1161.

79 Shahbazi, Y. (2017). The properties of chitosan and gelatin films incorporated with ethanolic red grape seed extract and *Ziziphora clinopodioides* essential oil as biodegradable materials for active food packaging. *International Journal of Biological Macromolecules* 99: 746–753.

80 Kan, J., Liu, J., Yong, H. et al. (2019). Development of active packaging based on chitosan-gelatin blend films functionalized with Chinese hawthorn (*Crataegus pinnatifida*) fruit extract. *International Journal of Biological Macromolecules* 140: 384–392.

81 Augusto, A., Dias, J.R., Campos, M.J. et al. (2018). Influence of *Codium tomentosum* extract in the properties of alginate and chitosan edible films. *Foods* 7 (4): 53.

82 Rahmani, B., Hosseini, H., Khani, M. et al. (2017). Development and characterisation of chitosan or alginate-coated low density polyethylene films containing *Satureja hortensis* extract. *International Journal of Biological Macromolecules* 105: 121–130.

83 Bigi, F., Haghighi, H., Siesler, H.W. et al. (2021). Characterization of chitosan-hydroxypropyl methylcellulose blend films enriched with nettle or sage leaf extract for active food packaging applications. *Food Hydrocolloids* 120: 106979.

84 Annu, Ali, A., and Ahmed, S. (2021). Eco-friendly natural extract loaded antioxidative chitosan/polyvinyl alcohol based active films for food packaging. *Heliyon* 7 (3): e06550.

85 Wang, K., Lim, P.N., Tong, S.Y., and Thian, E.S. (2019). Development of grapefruit seed extract-loaded poly(ε-caprolactone)/chitosan films for antimicrobial food packaging. *Food Packaging and Shelf Life* 22: 100396.

86 Vargas Romero, E., Lim, L.-T., Suárez Mahecha, H., and Bohrer, B.M. (2021). The effect of electrospun polycaprolactone nonwovens containing chitosan and propolis extracts on fresh pork packaged in linear low-density polyethylene films. *Foods* 10 (5): 1110.

87 Giannakas, A., Salmas, C., Leontiou, A. et al. (2019). Novel LDPE/chitosan rosemary and melissa extract nanostructured active packaging films. *Nanomaterials* 9 (8): 1105.

88 Said, N.S., Howell, N.K., and Sarbon, N.M. (2021). A review on potential use of gelatin-based film as active and smart biodegradable films for food packaging application. *Food Reviews International* https://doi.org/10.1080/87559129.2021.1929298.

89 Riahi, Z., Priyadarshi, R., Rhim, J.-W., and Bagheri, R. (2021). Gelatin-based functional films integrated with grapefruit seed extract and TiO_2 for active food packaging applications. *Food Hydrocolloids* 112: 106314.

90 Kontominas, M.G. (2020). Use of alginates as food packaging materials. *Foods* 9 (10): 1440.

91 Ucak, I., Khalily, R., Carrillo, C. et al. (2020). Potential of propolis extract as a natural antioxidant and antimicrobial in gelatin films applied to rainbow trout (*Oncorhynchus mykiss*) fillets. *Foods* 9 (11): 1584.

92 Espino-Manzano, S.O., León-López, A., Aguirre-Álvarez, G. et al. (2020). Application of nanoemulsions (W/O) of extract of *Opuntia oligacantha* C.F. Först and orange oil in gelatine films. *Molecules* 25 (15): 3487.

93 Luo, Y., Liu, H., Yang, S. et al. (2019). Sodium alginate-based green packaging films functionalized by guava leaf extracts and their bioactivities. *Materials* 12 (18): 2923.

94 Rajeswari, A., Stobel Christy, E.J., and Pius, A. (2021). Chapter 5 – Biopolymer blends and composites: processing technologies and their properties for industrial applications. In: *Biopolymers and Their Industrial Applications* (ed. S. Thomas, S. Gopi and A. Amalraj), 105–147. Elsevier.

95 Kumar, P., Tanwar, R., Gupta, V. et al. (2021). Pineapple peel extract incorporated poly(vinyl alcohol)-corn starch film for active food packaging: preparation, characterization and antioxidant activity. *International Journal of Biological Macromolecules* 187: 223–231.

96 Mustafa, P., Niazi, M.B.K., Jahan, Z. et al. (2020). PVA/starch/propolis/anthocyanins rosemary extract composite films as active and intelligent food packaging materials. *Journal of Food Safety* 40 (1): e12725.

97 Tanwar, R., Gupta, V., Kumar, P. et al. (2021). Development and characterization of PVA-starch incorporated with coconut shell extract and sepiolite clay as an antioxidant film for active food packaging applications. *International Journal of Biological Macromolecules* 185: 451–461.

98 Müller, P.S., Carpiné, D., Yamashita, F., and Waszczynskyj, N. (2020). Influence of pinhão starch and natural extracts on the performance of thermoplastic cassava starch/PBAT extruded blown films as a technological approach for bio-based packaging material. *Journal of Food Science* 85 (9): 2832–2842.

99 Chollakup, R., Pongburoos, S., Boonsong, W. et al. (2020). Antioxidant and antibacterial activities of cassava starch and whey protein blend films containing rambutan peel extract and cinnamon oil for active packaging. *LWT – Food Science and Technology* 130: 109573.

100 Khumkomgool, A., Saneluksana, T., and Harnkarnsujarit, N. (2020). Active meat packaging from thermoplastic cassava starch containing sappan and cinnamon herbal extracts via LLDPE blown-film extrusion. *Food Packaging and Shelf Life* 26: 100557.

101 Choi, I., Lee, J.Y., Lacroix, M., and Han, J. (2017). Intelligent pH indicator film composed of agar/potato starch and anthocyanin extracts from purple sweet potato. *Food Chemistry* 218: 122–128.

102 Giannakas, A., Vlacha, M., Salmas, C. et al. (2016). Preparation, characterization, mechanical, barrier and antimicrobial properties of chitosan/PVOH/clay nanocomposites. *Carbohydrate Polymers* 140: 408–415.

103 Kanatt, S.R. Active/smart carboxymethyl cellulose-polyvinyl alcohol composite films containing rose petal extract for fish packaging. *International Journal of Food Science & Technology* 56 (11): 5753–5761.

104 Roy, S., Kim, H.-J., and Rhim, J.-W. (2021). Synthesis of carboxymethyl cellulose and agar-based multifunctional films reinforced with cellulose nanocrystals and shikonin. *ACS Applied Polymer Materials* 3 (2): 1060–1069.

105 Wrona, M., Cran, M.J., Nerín, C., and Bigger, S.W. (2017). Development and characterisation of HPMC films containing PLA nanoparticles loaded with green tea extract for food packaging applications. *Carbohydrate Polymers* 156: 108–117.

106 Shavisi, N., Khanjari, A., Basti, A.A. et al. (2017). Effect of PLA films containing propolis ethanolic extract, cellulose nanoparticle and *Ziziphora clinopodioides* essential oil on chemical, microbial and sensory properties of minced beef. *Meat Science* 124: 95–104.

107 Roy, S. and Rhim, J.-W. (2021). Preparation of gelatin/carrageenan-based color-indicator film integrated with shikonin and propolis for smart food packaging applications. *ACS Applied Bio Materials* 4 (1): 770–779.

108 Figueroa-Lopez, K.J., Castro-Mayorga, J.L., Andrade-Mahecha, M.M. et al. (2018). Antibacterial and barrier properties of gelatin coated by electrospun polycaprolactone ultrathin fibers containing black pepper oleoresin of interest in active food biopackaging applications. *Nanomaterials* 8 (4): 199.

109 Alirezalu, K., Yaghoubi, M., Poorsharif, L. et al. (2021). Antimicrobial polyamide-alginate casing incorporated with nisin and ε-polylysine nanoparticles combined with plant extract for inactivation of selected bacteria in nitrite-free frankfurter-type sausage. *Foods* 10 (5): 1003.

110 Arrieta, M.P., Garrido, L., Faba, S. et al. (2020). *Cucumis metuliferus* fruit extract loaded acetate cellulose coatings for antioxidant active packaging. *Polymers* 12 (6): 1248.

3

Essential Oils in Food Packaging Applications

Madhushree Hegde[1], Akshatha Chandrashekar[1], Mouna Nataraja[1], Niranjana Prabhu[1], Jineesh A. Gopi[2], and Jyotishkumar Parameswaranpillai[2]

[1] *M.S. Ramaiah University of Applied Sciences, Department of Chemistry, Faculty of Mathematical and Physical Sciences, New BEL Road, MSR Nagar, Bengaluru, Karnataka, 560054, India*
[2] *Alliance University, Division of Chemistry, Department of Science, Anekal - Chandapura Road, Bengaluru, Karnataka, 562106, India*

3.1 Introduction

Packaging has a vital role in food industries. Packaging should be done following the regular changes or updates in the food packaging regulations and compliances, which is essential for domestic and export markets. Nowadays, food packaging is also based on demography, habits, trends, and innovation. The improper selection and type of packaging resulted in several issues that negatively affect the acceptance of food products in global markets [1]. Due to the rise in the consumer's knowledge and preferences toward food quality and safety, research on food packaging is attaining great momentum. The introduction of new and novel packaging technologies is intensified in the last decade [2]. The objective of food packaging is to keep food fresh in an effective way that ensures both industry necessities and the desires of a consumer, by maintaining food safety with less impact on the environment [3]. The nature and type of packaging material have vital roles in the protection of the food product. A range of materials such as glass, metals, wood, cardboard, and polymers are used for food packaging [4]. Polymers are the face of modern food packaging and are widely used for the packaging of almost all food products. The major advantages of polymers as packaging materials are their versatility, lightweight, low energy consumption, cost-effectiveness on large-scale manufacturing, and longer duration of food protection and preservation. They are used in the form of rigid containers, flexible films, closures, and external coatings to paperboard cartons, glass, and metal jars to avoid tampering due to environmental conditions. Emphasis

---

Natural Materials for Food Packaging Application, First Edition.
Edited by Jyotishkumar Parameswaranpillai, Aswathy Jayakumar,
E. K. Radhakrishnan, Suchart Siengchin, and Sabarish Radoor.
© 2023 WILEY-VCH GmbH. Published 2023 by WILEY-VCH GmbH.

on discovering alternatives to existing petroleum-based polymers and laying importance on reduced environmental impact led to the development of biopolymers for food packaging. Based on functionality, food packaging is categorized into ergonomic packaging, informative packaging, active packaging, and intelligent packaging. The quality of food and its packaging and the interactions between them determines the overall quality of the packaged food. Manufacturers must select suitable packaging materials to fulfill the requirements for efficient food packaging applications. For flexible packaging materials such as plastic materials, migration of O_2 and moisture are a major concern. In glass and metal packaging materials, this is not a big concern due to their inertness and low permeability.

As per the Framework Regulation on Food Contact Materials (1935/2004), active materials are used for enhancing the shelf life and quality of food products and intelligent materials are for monitoring the surroundings of the food products or the characteristics of the food products [5]. In active packaging, active compounds are incorporated in the packaging material, which can interact with the perishable food product. The functional groups on the active compounds delays or stop the deterioration of the food by acting as antioxidants, antimicrobial, scavengers for O_2 and CO_2, etc. The active compound can be incorporated in the packaging material, or on a label, or placed in the packet. Active compounds may release substances such as preservatives, antioxidants, and flavoring agents into the food [6]. Active packaging systems function as oxygen scavengers, antimicrobial agents, CO_2 absorbing and releasing agents, moisture absorbers, and ethylene scavengers. Active packaging is categorized into chemo-active and bioactive, based on the nature of additives added. There is a trend toward the use of natural products over synthetic products. Therefore, natural extracts such as essential oils and polyphenols are being widely used in food packaging materials [7]. This chapter is focused to discuss the use of essential oil as an additive in food packaging materials. The influence of essential oil on chemical, physical, antimicrobial, and antioxidant properties are discussed in this book chapter. Also, the challenges associated with the use of essential oil as an additive in food packaging materials and the prospects have been emphasized.

3.2 Chemistry and Classification of Essential Oils

Essential oils are isolated from different parts of plants and the name of the essential oil depend on the plants from which the oil is extracted [8, 9]. Methods such as solvent extraction, mechanical processes, steam distillation, and dry distillation are used for the extraction of essential oils [10]. Generally essential oils possess higher density compared to oils, low solubility in water, and higher solubility in ethers and alcohols. They are usually colorless in nature and possess unique odors as well. The odor of the essential oil is mainly due to the presence of a small quantity of aliphatic compounds with oxygenated functional groups [11]. Essential oils are different in terms of their functional groups, and they exist as aldehydes, amines, ketones, epoxides, alcohols, and sulfides [12]. Essential oils can be classified into terpenes and non-terpenoid hydrocarbons based on their chemical components. The majority of essential oils are based on monoterpenes. The chemistry of various essential oils was discussed by Blowman et al. [13] and Zuzarte et al. [14]. Types of essential oils used in packaging materials, their source, and a few methods of extraction of these essential oils are listed in Table 3.1.

Blowman et al. [13] compiled chemical structures of several essential oil constituents, which are shown in Figure 3.1.

Table 3.1 Essential oils used in the preparation of food packaging films, their sources, and methods of extraction.

Sl no.	Essential oil	Source	Methods of extraction	References
1	Neem oil	*Azadirachta indica*	Microwave-assisted extraction, solvent, and supercritical fluid extraction	[15–18]
2	Lavender oil	*Lavandula angustifolia*, *Lavandula X intermedia* var.	Microwave-assisted headspace solid-phase extraction steam and hydro-distillation, solvent-free microwave extraction, subcritical water-hydro-distillation, and supercritical-fluid extraction	[19–23]
3	Clove oil	*Syzygium aromaticum*	Steam distillation, ohmic heating-assisted hydro-distillation, CO_2-assisted extraction, microwave, and ultrasound-assisted extractions, and supercritical fluid extraction	[24–26]
4	Tea tree	*Melaleuca alternifolia*, *Melaleuca bracteata*	Solvent extraction, steam distillation, hydro-distillation, and supercritical fluid extraction	[27–29]
5	Cinnamon oil	*Cinnamomum cassia*, *Cinnamomum zeylanicum*, *Cinnamomum tamala*, *Cinnamomum burmannii*, *Cinnamomum pauciflorum*	Hydro-distillation, ultrasound-enhanced subcritical water extraction, microwave-assisted extraction, and sono-hydro distillation	[30–33]
6	Rosemary oil	*Rosmarinus officinalis* L.	Hydro-distillation and microwave-assisted hydro-distillation, supercritical CO_2-assisted extraction, steam distillation, organic solvent extraction, gravity methods, and controlled instantaneous decompression	[34–39]

(Continued)

Table 3.1 (Continued)

Sl no.	Essential oil	Source	Methods of extraction	References
7	Oregano oil	*Origanum vulgare* L.	Solvent-free microwave extraction, hydro-distillation, ohmic-assisted hydro-distillation, and subcritical water extraction	[40–43]
8	Thyme oil	*Thymus vulgaris* L., *Thymbra spicata* L.	Ohmic-assisted hydro-distillation, and subcritical water extraction	[44, 45]
9	Eucalyptus oil	*Eucalyptus globulus*	Subcritical-water extraction, supercritical fluid extraction, and hydro-distillation	[46, 47]
10	*Pistacia lentiscus* L. essential oil	*Pistacia lentiscus* L.	Supercritical CO_2 extraction	[48]

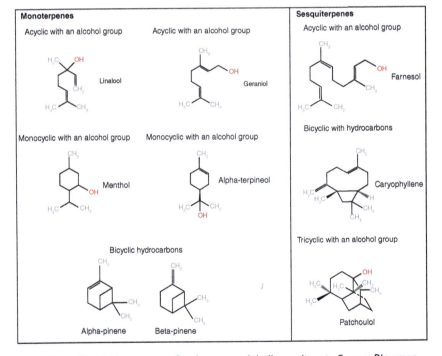

Figure 3.1 Chemical structures of various essential oil constituents. Source: Blowman et al. [13] / Hindawi / Licensed under CC BY 4.0.

3.3 Essential Oils in Food Packaging Applications

Essential oils possess antimicrobial and antioxidant properties that enable them to be useful in improving the shelf life of food products. Essential oils can reduce or eradicate the presence of microorganisms or retard the oxidation of lipids in food products. To enhance the stability of essential oils, they are encapsulated with liposomes, polymers, and lipid nanoparticles [49]. Essential oils can be used in food packaging, which is film or coating. Films are usually thin sheets that are used as covers, and wrappers in food packaging. Coatings are applied on the outer surface of the food products [50]. Polymer matrices are generally used as carriers for essential oil in food packaging films. Among the polymers, biopolymers such as proteins and polysaccharides are widely studied as packaging films that can carry essential oils. Varghese et al. [51] discussed various techniques such as casting, extrusion, and compression molding for the fabrication of food packaging films that contains essential oils.

3.3.1 Effect of Essential Oil on the Mechanical, Barrier, and Other Physical Properties of Food Packaging Materials

3.3.1.1 Tensile Properties

Tensile strength and elongation are the two important mechanical properties of a food packaging material. The incorporation of essential oil into polymer creates a heterogenous film structure, which leads to discontinuity in the film structure. This results in the reduction of tensile properties of the polymer films [52]. Song et al. [53] explored the effect of lemon essential oil on the tensile properties of corn and wheat starch film and found that the addition of lemon essential oil decreased the tensile strength of the film but enhanced its elongation at break. The rise in the elongation at break of the film is due to the presence of liquid oil droplets, which can be easily deformed. Peng et al. [54] explored the effect of lemon essential oil on the tensile properties of chitosan films. Results showed that both tensile strength and elongation at break of chitosan films were reduced with the incorporation of lemon essential oil. Similar observations are obtained with corn starch films containing orange essential oil [55]. Contrary to these studies, Ojagh et al. [56] found that the tensile strength of chitosan films was enhanced by the addition of cinnamon essential oil (CEO) into the films. A similar observation is found in CEO-incorporated chitosan-based films [57]. The difference in these observations related to tensile properties is due to the difference in film-forming materials, and the interactions between essential oil and the polymer components. Bastos et al. [58] prepared cellulose ester-based films incorporated with lemongrass, rosemary pepper, and basil essential oils with a loading of 10% and 20% (v/w). This study revealed that essential oils acted as plasticizers and increased elongation at break. On the other hand, the tensile strength and Young's modulus of the cellulose ester were reduced with the incorporation of essential oils. Few studies showed that the incorporation of essential oil does not affect the tensile characteristics of the polymer films. For example, the addition of CEO into alginate biocomposite films exhibited no effect on the tensile strength of alginate films [59].

3.3.1.2 Barrier Properties

Improved gas and water barrier properties in polymers for food packaging aids in preserving food characteristics and enhancing the shelf life. The water vapor permeability (WVP) of films is an important characteristic that determines their suitability for food packaging. One of the objectives of the food packaging film is to reduce the moisture transfer between the food and its surroundings. Song et al. [53] found that the addition of lemon essential oil reduced the WVP of corn and wheat starch films. This may be due to the presence of lipid globules that enhanced the tortuosity factor of vapor-diffusion path in the film along with less interaction with the water. Similar observations have been made by various researchers where the WVP of films was increased by the presence of essential oil [60–62]. Some studies are contrary to this and showed an increase in the WVP upon the addition of essential oil [63]. The decreased film continuity and the creation of small cavities by essential oil are considered the reasons for this phenomenon.

Studies revealed that the addition of essential oil enhances the oxygen permeability of polymer films. For example, the presence of CEO in sago starch film enhanced the oxygen permeability of polymer films [64]. The incorporation of 3% (v/v) *Zataria multiflora* Boiss (ZEO) or *Mentha pulegium* (MEO) essential oils increased the oxygen permeability of corn starch polymer films [60]. Incorporation of carvacrol, citral, and α-terpineol essential oils into poly(butylene adipate terephthalate) and poly(lactic acid) decreased crystallinity and increased amorphous phase and this led to enhanced oxygen permeability of polymer films [65]. Similar observations are also listed by several researchers [66, 67]. Bonilla et al. [68] analyzed the effect of essential oil on oxygen permeability and its effect on the quality of stored pork meat products. Contrary to these observations, cinnamon, clove, and oregano essential oils (OEOs) decreased the oxygen permeability of starch and gelatin films [69]. The mixture of zinc oxide nanoparticles and fennel essential oil (FEO) decreased the water vapor and oxygen permeability [70]. There are also few studies where the addition of essential oils does not affect the barrier characteristics of polymer films [71, 72].

3.3.1.3 Other Physical Properties

Color of the polymer films is usually altered by the presence of essential oil. The nature of the essential oil determines the color. For example, the addition of *Z. multiflora* Boiss essential oil did not change the color of chitosan films [73]. A similar observation is made when garlic essential oil was incorporated into alginate films [74]. On the other hand, the addition of thyme oil in Hake protein films imparts transparent yellow color [75]. The incorporation of tarragon and coriander essential induces yellow color to hake protein films [76]. Similarly, the incorporation of CEO induces slight yellow color to chitosan films [56]. Essential oils reduce the transparency of the polymer films. This is mainly because of the presence of essential oil droplets in the films, which causes light scattering [57, 76]. The addition of essential oil reduces the gloss of the polymer films due to the increase in heterogeneity and surface roughness in the polymer film. The incorporation of essential oils changes the water solubility of polymer films. *Zataria multiflora* Boiss essential oil and *Mentha pulegium* essential oil reduce the solubility of soybean polysaccharide (SSPS) films due to the interactions between polysaccharide chains and essential oils [77]. Similarly, the addition of OEO reduced the water solubility of sweet potato starch films [78]. The incorporation of nano-emulsioned ginger essential oil (GEO) into gelatin-based films decreased the moisture content and water solubility [79]. The incorporation of Rosemary essential oil (1.5% v/v) reduced the water solubility of chitosan films by 25% due to the interaction between hydrophilic groups of chitosan and rosemary essential oil [80]. Lemon grass oil can also decrease the water solubility of chitosan films [81]. Similarly, Lavender essential oil (OEL) decreases the starch, furcellaran, and gelatin films solubility in water [82]. The presence of OEO in blends of rice starch/fish protein reduced the film solubility to 8% [83].

The microstructure of polymer films will be altered by the addition of essential oil [84]. Flórez et al. [85] analyzed the influence of sandalwood essential oil (SEO) at different loadings on the microstructure of chitosan films using scanning electron microscopy and the results are shown in Figure 3.2. SEM images show that the

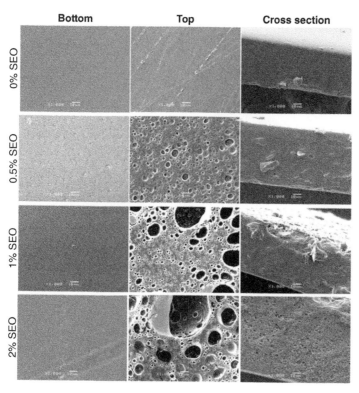

Figure 3.2 SEM images of the SEO-chitosan films using malic acid as solvent. Source: Flórez et al. [85] / ELSEVIER / CC BY 4.0.

homogeneity of the chitosan films was disturbed by the incorporation of SEO and this is due to the oil droplets trapped in the polysaccharide network. The heterogeneity of the films also depends on the amount of essential oil. In another study, a small amount of essential oil maintains the smoothness of the films and higher loading leads to bubble formation on the film surface and eventually leads to a rough surface [86].

3.3.2 Antioxidant Properties of Essential Oil Incorporated Food Packaging Materials

Antioxidants possess antioxidant properties due to their ability to stop the oxidation of products [87]. Deterioration of food products is mainly due to lipid oxidation and this oxidation process decreases the shelf life and induces rancidity of the food products. These processes negatively affect the safety and nutritional values of food products. Essential oils are ideal candidates as antioxidants over synthetic antioxidants. Essential oils possess the capability to retard the oxidation processes through different modes of action which are listed by Rodriguez-Garcia et al. [88] and Rehman et al. [89]. Those modes of action are shown in Figure 3.3.

Steps in the mechanism against oxidation by essential oils are discussed by Rehman et al. [89] and shown in Figure 3.4.

The antioxidant property of the films was analyzed using various methods such as ABTS (2,2'-azinobis (3-ethylbenzothiazoline-6-sulfonic acid) diammonium salt) method, DPPH (2,2-diphenyl-1-picrylhydrazyl) free radical capture method, ferric-reducing antioxidant power (FRAP), Ferrous ion chelating activity (FIC) assay

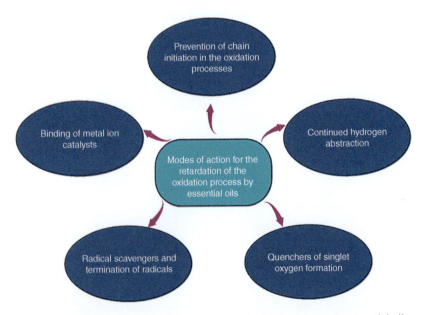

Figure 3.3 Mode of actions for the retardation of oxidation process by essential oils.

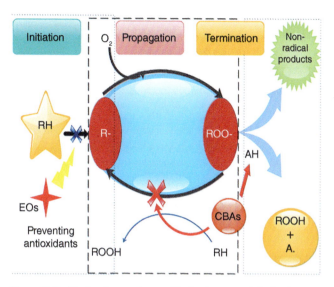

Figure 3.4 Mechanism against oxidation by essential oils three steps, RH = substrate, R˙ = alkyl radical, ROO– = peroxyl radical, ROOH = hydroperoxide, CBAs = chain breaking antioxidants. Source: Rehman et al. [89] / with permission of Elsevier.

method, Trolox Equivalent Antioxidant Capacity (TEAC) method, Folin-Ciocalteu method, etc. [90]. A few studies on the antioxidant effects of essential oils on polymer films are listed in Table 3.2.

Table 3.2 Studies on the antioxidant properties of essential oil-incorporated polymer films and membranes.

Sl no.	Essential oil	Polymer films	Antioxidant properties	References
1	Cinnamon bark essential oil	Chitosan films	Cinnamon bark essential oil enhanced the antioxidant activity of chitosan films by 6.0–14.5-fold	[91]
2	Licorice essential oil	Carboxymethyl xylan film	Licorice essential oil provided scavenging of free radicals and inhibited lipid peroxidation in food at a storage temperature of 25 °C for more than 20 days	[92]
3	Rosemary and *Aloe vera* oil	Cellulose acetate membrane	ABTS and DPPH methods proved that an increase in the essential oil content improved the free radical scavenger activity of the membrane	[93]
4	*Origanum onites* L. essential oil (OOEO)	Gelatin films	8% OOEO incorporated gelatin film showed 41.77% inhibition activity in the DPPH method and showed 83.13% radical inhibition activity in the ABTS method	[94]

(Continued)

Table 3.2 (Continued)

Sl no.	Essential oil	Polymer films	Antioxidant properties	References
5	*Salvia officinalis* essential oil	Potato starch films with 1.5% Zedo gum	Starch–Zedo gum films and essential oil incorporated Starch–Zedo gum films showed antioxidant activity by 62.64% and 68.35% respectively	[95]
6	Oregano essential oil	Hydroxypropyl methylcellulose	DPPH and ABTS methods were used to evaluate the antioxidant property of the composite films, and the results showed that 5% (v/v) oregano essential oil enhanced the antioxidant activity by 46% (DPPH method) and 69% (ABTS method) higher compared to control films	[96]
7	Caraway essential oil	Chitosan biofilm	The antioxidant activity of chitosan film was improved from 28.96% to around 80% by the addition of caraway oil to the chitosan films (after 24 h)	[97]
8	Eugenol oil	Gelatin/chitosan films	Trolox-equivalent-antioxidant-capacity test showed that the inclusion of eugenol essential oil improved the antioxidant nature of gelatine/chitosan films	[98]
9	Tea tree essential oil (TTEO)	Chitosan	Chitosan film with TTEO exhibited excellent antioxidant activity determined by both ABTS and DPPH methods	[99]
10	Oregano essential oil	Soy protein concentrate films	Soy protein concentrated films' antioxidant activity was increased by the addition of both free and microencapsulated oregano essential oil	[100]
11	*Litsea cubeba* oil	Chitosan–starch films	The presence of essential oil increased the DPPH radical scavenging ability of films from 20.67% to 52.34%	[101]
12	Oregano essential oil	Polylactic acid (PLA)	Incorporation of higher amount of essential oil (5% and 10%) showed higher antioxidant activity compared to neat PLA films and 2% essential oil incorporated PLA films	[102]

3.3.3 Antibacterial Properties of Essential Oil Incorporated Food Packaging Materials

Degradation is mainly due to the growth of microorganisms and food-borne pathogens. These microorganisms enhance the oxidation of lipids and other constituents in the food. Also, they can change the organoleptic properties of food. These food pathogens and the toxins generated during the degradation of food cause illness to consumers [103, 104]. Antimicrobial packaging can be the solution to these problems and can enhance the food shelf time. Essential oils which contain various bioactive compounds act as antimicrobial agents [105]. Essential oils act as antimicrobial agents through various mechanisms [106] which are shown in Figure 3.5.

Various in vitro methods such as disc diffusion methods, agar wells methods, agar dilution methods, and broth dilution methods are used to measure the antibacterial activity of essential oils [107]. The diffusion of antimicrobial compounds to the food product is limited because essential oil is included in the food packaging film. This helps in reducing the change in the flavor of the product due to the existence of excess amount of essential oil in the food product [105]. Table 3.3 lists studies related to the use of essential oils as antibacterial agents in food packaging applications.

Wang et al. [115] studied the antibacterial characteristics of green Sichuan pepper essential oil incorporated starch-chitosan-konjac glucomannan blend films (SCK). Incorporation of 2.5% Sichuan pepper essential oil into films (SCKZ) exhibited excellent antibacterial characteristics against *Staphylococcus aureus* and *Escherichia coli* with inhibition zones of up to 21.72 ± 1.25 and 18.73 ± 0.74 cm respectively and the results are shown in Figure 3.6.

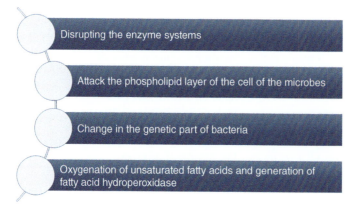

Figure 3.5 Various antimicrobial mechanisms of essential oils.

Table 3.3 Studies on antibacterial properties of essential oil-incorporated polymer films.

Sl no.	Essential oil	Polymer films	Characteristics	References
1	*Thymus piperella*	Chitosan films	A study using the agar disc diffusion method showed that *Thymus piperella* essential oil induces antibacterial property to chitosan film against *Serratia marcescens* and *Listeria innocua* bacteria	[105]
2	Rosemary and Aloe Vera oil	Cellulose acetate membranes	Growth turbidity test showed that an increase in rosemary and aloe vera oil content increases the antibacterial property of cellulose acetate membrane against *Escherichia coli* (*E. coli*) and *Bacillus subtilis* (*B. subtilis*) and the effect is more prominent against *B. subtilis* in comparison to *E. coli*	[93]
3	Cinnamon essential oil	Carboxymethyl cellulose (CMC)–polyvinyl alcohol (PVA) based films	Polymer films with 1.5% and 3% of cinnamon essential oil showed complete inhibition to *Penicillium digitatum* in *in vitro* and *in vivo* tests	[108]
4	Lemon and sage oil	Quinoa as polymer in edible films	Coating material prepared by incorporating sage oil in edible quinoa film showed excellent antibacterial properties against various bacteria and can be effectively used to improve the shelf life of rainbow trout fillets	[109]
5	Oregano essential oil	Hydroxypropyl methylcellulose	Disc diffusion test showed that hydroxypropyl methylcellulose (HPMC) incorporated with oregano essential oil nano-emulsion (ORNE) have good antibacterial property against *Salmonella typhimurium* with a large inhibition zone of 47.5 mm	[96]
6	Apricot (*Prunus armeniaca*) kernel essential oil (AKEO)	Chitosan films	The presence of AKEO in chitosan films showed excellent antibacterial property against *E. coli* and *B. subtilis* bacteria. This is due to *N*-methyl-2-pyrrolidone (NMP) in the AKEO which interacts with the cell membrane of the bacteria and disintegrates the bacteria followed by the leakage of intracellular fluids	[110]

Sl no.	Essential oil	Polymer films	Characteristics	References
7	*Litsea cubeba* oil (LEO)	Chitosan films containing hardleaf oatchestnut starch	Agar diffusion test showed that LEO-induced inhibition in the growth of *S. aureus* and *E. coli* in chitosan films. The effect is more prominent against *S. aureus* bacteria than *E. coli* bacteria due to the difference in their cell wall structures	[101]
8	Oregano essential oil (OEO)	PLA films	Evaluation of the antibacterial property of OEO-containing PLA films was done using the dilution method and the results showed that the film containing 10% OEO possess excellent antibacterial property against *Salmonella enterica*	[102]
9	Cinnamon essential oil (CEO)	Films from chitosan and carboxymethyl cellulose	CEO-incorporated films showed antimicrobial properties against *Listeria monocytogenes* and *Pseudomonas aeruginosa*. The hydrophobic nature of CEO allows it to penetrate the phospholipid bilayer of bacteria which leads to the destruction of the cell wall of the bacteria	[111]
10	Cinnamon essential oil	Chitosan/gelatin-films	The viable cell colony count method was used to study the antibacterial property of the films. The study showed that the incorporation of CEO enhanced the antibacterial activity against *E. coli* and *L. monocytogenes*	[112]
11	Rosemary and ginger essential oil	chitosan films	The antimicrobial property of films was tested using the viable cell colony count method and the study showed that rosemary oil enhanced the antimicrobial property of chitosan film against *E. coli* and *E. faecalis*. Also ginger essential oil enhanced the antimicrobial property of chitosan film against *E. faecalis* and *S. aureus*	[113]
12	Oregano essential oil	Poly(butylene adipate-co-terephthalate) (PBAT)	Oregano essential oil incorporated PBAT film showed a reduction in the total coliforms, *Staphylococcus aureus*, and psychrotrophic microorganisms in the microbiological assays-based experiments. These films have the potential to be used in the active packaging of fish fillets	[114]

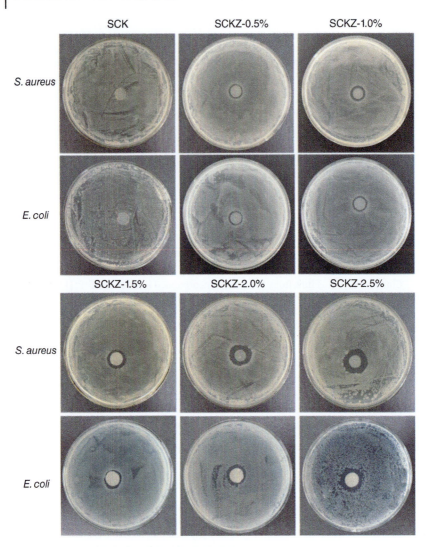

Figure 3.6 Antibacterial activity of composite films against *S. aureus* and *E. coli*. Source: Wang et al. [115]/ with permission from Elsevier.

The addition of certain essential oil increases the antifungal properties of the food packaging films. For example, a 4% incorporation of CEO-inhibited *Aspergillus flavus* growth in the polyethylene terephthalate (PET) film [116]. Basil oil is an antifungal agent and it enhances the antifungal properties of chitosan films against food-pathogenic fungi such as *Aspergillus niger*, *A. flavus*, and *Penicillium* sp. [117]. Chein et al. [118] studied the antifungal properties of chitosan films with various essential oils against *A. flavus* and *Penicillium citrinum*. In this study, they found that CEO-incorporated chitosan films are suitable for controlling the fungal growth in peanuts.

3.4 Challenges and Future Trends Associated with the Use of Essential Oil in Food Packaging Applications and Future Trends

Higher cost of essential oils, the lower thermal stability of essential oils during extrusion and molding processes of the films, difficulty in controlling the release of essential oil into the food products, etc., are constraints in the use of essential oil in food packaging applications [51]. There are some toxicity issues related to the use of essential oil due to the presence of aldehyde and phenol groups in the essential oils [119]. Therefore, attention to be given to the use of essential oil in active food packaging applications.

Nano-emulsification techniques were used to prepare essential oil-filled polymer matrix to reduce the volatilization losses and this method is evolving as good tool in food packaging industry. Encapsulation techniques are getting great attention in the field of incorporation of essential oil in the polymer which will solve a few problems related to the use of essential oils. For example, Silva et al. [120] prepared cyclodextrin nanosponges that encapsulated coriander essential oil and this helps in the controlled oil release. Recent studies showed that essential oils can be encapsulated into nanometric surfactant micelles which will allow the controlled release of antioxidants [121–123]. Intelligent packaging, in which the essential oil can be released with the help of an external stimulus is a probable choice for the problems associated with the uncontrolled migration of essential oils into food products. In this approach, the release of essential oil can be done at a later stage which is just before the start of spoilage of food products. The external stimulus can be environmental conditions such as pH, temperature, magnetic field, etc. [124]. For example, Marturano et al. [125] developed a polyethylene/PLA film containing thyme essential oil. They encapsulated thyme essential oil in crosslinked polyamide shells which are photo-responsive. The release of thyme oil can be controlled with UV light of 360 nm wavelength. In another work, Marturano et al. [125] developed light-triggered release of active ingredients of essential oil from polymeric nanosized capsules for active packaging applications.

3.5 Conclusions

Spoilage of food and the loss of quality of food are serious concerns and there are several methods to avoid these and active food packaging is one of them. Foods are prone to microbial attack and oxidation. The incorporation of agents that can reduce oxidation and prevent the growth of microbes is an effective approach in active food packaging. Essential oils are natural materials with antioxidant and antimicrobial properties. Casting, extrusion, and different molding techniques are used to prepare essential oil-incorporated polymeric films for food packaging applications. The incorporation of essential oil alters the physical, chemical, barrier, antioxidant, and antimicrobial properties of the films. Due to the reported toxicity of essential oil, it is important to consider the balance between the efficacy and toxicity of essential oils. Novel preparation methods such as encapsulation, nanoemulsification, electrospinning, and controlled release of essential oil using external stimulus, etc., can solve the constraints related to the use of essential oil in food packaging.

References

1 Gordon, A. and Williams, R. (2020). The role and importance of packaging and labeling in assuring food safety, quality and regulatory compliance of export products II: packaging & labeling considerations. In: *Food Safety and Quality Systems in Developing Countries: Volume III: Technical and Market Considerations* (ed. A. Gordon), 285–341. Academic Press.
2 Kalpana, S., Priyadarshini, S.R., Maria Leena, M. et al. (2019). Intelligent packaging: trends and applications in food systems. *Trends in Food Science and Technology* 93: 145–157.
3 Deshwal, G.K. and Panjagari, N.R. (2019, 2019). Review on metal packaging: materials, forms, food applications, safety and recyclability. *Journal of Food Science and Technology* 57 (7): 2377–2392.
4 Piergiovanni, L. and Limbo, S. (2016). Introduction to food packaging materials. In: *Food Packaging Materials* (ed. L. Piergiovanni and S. Limbo), 1–3. Springer.
5 Lee, S.J. and Rahman, A.T.M.M. (2014). Intelligent packaging for food products. In: *Innovations in Food Packaging*, 2e (ed. J.H. Han), 171–209. Academic Press.
6 Carrizo, D., Taborda, G., Nerín, C., and Bosetti, O. (2016). Extension of shelf life of two fatty foods using a new antioxidant multilayer packaging containing green tea extract. *Innovative Food Science and Emerging Technologies* 33: 534–541.
7 Vinceković, M., Viskić, M., Jurić, S. et al. (2017). Innovative technologies for encapsulation of Mediterranean plants extracts. *Trends in Food Science and Technology* 69: 1–12.
8 Ríos, J.L. (2016). Essential oils: what they are and how the terms are used and defined. In: *Essential Oils in Food Preservation, Flavor and Safety* (ed. V.R. Preedy), 3–10. Academic Press.
9 El Asbahani, A., Miladi, K., Badri, W. et al. (2015). Essential oils: from extraction to encapsulation. *International Journal of Pharmaceutics* 483 (1–2): 220–243.
10 Aziz, Z.A.A., Ahmad, A., Setapar, S.H.M. et al. (2018). Essential oils: extraction techniques, pharmaceutical and therapeutic potential – a review. *Current Drug Metabolism* 19 (13): 1100–1110.
11 Bhavaniramya, S., Vishnupriya, S., Al-Aboody, M.S. et al. (2019). Role of essential oils in food safety: antimicrobial and antioxidant applications. *Grain & Oil Science and Technology* 2 (2): 49–55.
12 Sharma, S., Barkauskaite, S., Jaiswal, A.K., and Jaiswal, S. (2021). Essential oils as additives in active food packaging. *Food Chemistry* 343: 128403.

13 Blowman, K., Magalhães, M., Lemos, M.F.L. et al. (2018). Anticancer properties of essential oils and other natural products. *Evidence-Based Complementary and Alternative Medicine* 2018: 3149362.

14 Zuzarte, M. and Salgueiro, L. (2015). Essential oils chemistry. In: *Bioactive Essential Oils and Cancer* (ed. D.P. de Sousa), 19–61. Springer Cham.

15 Chaudhary, M.F., Ashraf, A., Waseem, M. et al. (2021). Neem oil. In: *Green Sustainable Process for Chemical and Environmental Engineering and Science Plant-Derived Green Solvents: Properties and Applications* (ed. Inamudiin, R. Boddula and A.M. Asiri), 57–73. Elsevier.

16 Nde, D.B., Boldor, D., and Astete, C. (2015). Optimization of microwave assisted extraction parameters of neem (*Azadirachta indica* A. Juss) oil using the Doehlert's experimental design. *Industrial Crops and Products* 65: 233–240.

17 Blum, F.C., Singh, J., and Merrell, D.S. (2019). in vitro activity of neem (*Azadirachta indica*) oil extract against *Helicobacter pylori*. *Journal of Ethnopharmacology* 232: 236–243.

18 Mendonça, F.M.R., Polloni, A.E., Junges, A. et al. (2019). Encapsulation of neem (*Azadirachta indica*) seed oil in poly(3-hydroxybutyrate-*co*-3-hydroxyvalerate) by SFEE technique. *Journal of Supercritical Fluids* 152: 104556.

19 Liu, X., Fu, J., Wang, L., and Wang, C. (2021). Polydimethylsiloxane/ZIF-8@GO sponge headspace solid-phase extraction followed by GC-MS for the analysis of lavender essential oil. *Analytical Biochemistry* 622: 114167.

20 Filly, A., Fabiano-Tixier, A.S., Louis, C. et al. (2016). Water as a green solvent combined with different techniques for extraction of essential oil from lavender flowers. *Comptes Rendus Chimie* 19 (6): 707–717.

21 Radwan, M.N., Morad, M.M., Ali, M.M., and Wasfy, K.I. (2020). A solar steam distillation system for extracting lavender volatile oil. *Energy Reports* 6: 3080–3087.

22 Adaşoğlu, N., Dinçer, S., and Bolat, E. (1994). Supercritical-fluid extraction of essential oil from Turkish lavender flowers. *Journal of Supercritical Fluids* 7 (2): 93–99.

23 Çelebi Uzkuç, N.M., Uzkuç, H., Berber, M.M. et al. (2021). Stabilisation of lavender essential oil extracted by microwave-assisted hydrodistillation: characteristics of starch and soy protein-based microemulsions. *Industrial Crops and Products* 172: 114034.

24 Moyler, D.A. (1993). Extraction of essential oils with carbon dioxide. *Flavour and Fragrance Journal* 8 (5): 235–247.

25 Gonzalez-Rivera, J., Duce, C., Campanella, B. et al. (2021). In situ microwave assisted extraction of clove buds to isolate essential oil, polyphenols, and lignocellulosic compounds. *Industrial Crops and Products* 161: 113203.

26 Jadhav, J.J., Jadeja, G.C., and Desai, M.A. (2022). Effect of extraction techniques on the yield, composition, and quality of clove (*Syzygium aromaticum*) essential oil. In: *Clove (Syzygium aromaticum)* (ed. M.F. Ramadan), 485–500. Academic Press.

27 Baker, G.R., Lowe, R.F., and Southwell, I.A. (2000). Comparison of oil recovered from tea tree leaf by ethanol extraction and steam distillation. *Journal of Agricultural and Food Chemistry* 48 (9): 4041–4043.

28 Yasin, M., Younis, A., Ramzan, F. et al. (2021). Extraction of essential oil from river tea tree (*Melaleuca bracteata* F. Muell.): antioxidant and antimicrobial properties. *Sustainability* 13 (9): 4827.

29 Wong, V., Wyllie, S.G., Cornwell, C.P., and Tronson, D. (2001). Supercritical fluid extraction (SFE) of monoterpenes from the leaves of *Melaleuca alternifolia* (tea tree). *Molecules* 6 (2): 92–103.

30 Wang, R., Wang, R., and Yang, B. (2009). Extraction of essential oils from five cinnamon leaves and identification of their volatile compound compositions. *Innovative Food Science and Emerging Technologies* 10 (2): 289–292.

31 Guo, J., Yang, R., Gong, Y.S. et al. (2021). Optimization and evaluation of the ultrasound-enhanced subcritical water extraction of cinnamon bark oil. *LWT – Food Science and Technology* 147: 111673.

32 Kallel, I., Hadrich, B., Gargouri, B. et al. (2019). Optimization of cinnamon (*Cinnamomum zeylanicum* Blume) essential oil extraction: evaluation of antioxidant and antiproliferative effects. *Evidence-Based Complementary and Alternative Medicine* 2019: 6498347.

33 Modi, P.I., Parikh, J.K., and Desai, M.A. (2019). Sonohydrodistillation: innovative approach for isolation of essential oil from the bark of cinnamon. *Industrial Crops and Products* 142: 111838.

34 Boutekedjiret, C., Bentahar, F., Belabbes, R., and Bessiere, J.M. (2003). Extraction of rosemary essential oil by steam distillation and hydrodistillation. *Flavour and Fragrance Journal* 18 (6): 481–484.

35 Presti, M.L., Ragusa, S., Trozzi, A. et al. (2005). A comparison between different techniques for the isolation of rosemary essential oil. *Journal of Separation Science* 28 (3): 273–280.

36 Chen, H., Gu, Z., Yang, L. et al. (2021). Optimization extraction of rosemary essential oils using hydrodistillation with extraction kinetics analysis. *Food Science & Nutrition* 9 (11): 6069–6077.

37 Bousbia, N., Abert Vian, M., Ferhat, M.A. et al. (2009). Comparison of two isolation methods for essential oil from rosemary leaves: hydrodistillation and microwave hydrodiffusion and gravity. *Food Chemistry* 114 (1): 355–362.

38 Zermane, A., Larkeche, O., Meniai, A.H. et al. (2016). Optimization of Algerian rosemary essential oil extraction yield by supercritical CO_2 using response surface methodology. *Comptes Rendus Chimie* 19 (4): 538–543.

39 Teshale, F., Narendiran, K., Beyan, S.M., and Srinivasan, N.R. (2022). Extraction of essential oil from rosemary leaves: optimization by response surface methodology and mathematical modeling. *Applied Food Research* 2 (2): 100133.

40 Bayramoglu, B., Sahin, S., and Sumnu, G. (2008). Solvent-free microwave extraction of essential oil from oregano. *Journal of Food Engineering* 88 (4): 535–540.

41 Moncada, J., Tamayo, J.A., and Cardona, C.A. (2016). Techno-economic and environmental assessment of essential oil extraction from Oregano (*Origanum vulgare*) and Rosemary (*Rosmarinus officinalis*) in Colombia. *Journal of Cleaner Production* 112: 172–181.

42 Hashemi, S.M.B., Nikmaram, N., Esteghlal, S. et al. (2017). Efficiency of Ohmic assisted hydrodistillation for the extraction of essential oil from oregano (*Origanum* *vulgare* subsp. *viride*) spices. *Innovative Food Science and Emerging Technologies* 41: 172–178.

43 Soto Ayala, R. and Luque de Castro, M.D. (2001). Continuous subcritical water extraction as a useful tool for isolation of edible essential oils. *Food Chemistry* 75 (1): 109–113.

44 Gavahian, M., Farahnaky, A., Javidnia, K., and Majzoobi, M. (2012). Comparison of ohmic-assisted hydrodistillation with traditional hydrodistillation for the extraction of essential oils from *Thymus vulgaris* L. *Innovative Food Science and Emerging Technologies* 14: 85–91.

45 Ozel, M.Z., Gogus, F., and Lewis, A.C. (2003). Subcritical water extraction of essential oils from *Thymbra spicata*. *Food Chemistry* 82 (3): 381–386.

46 Jiménez-Carmona, M.M. and Luque de Castro, M.D. (1999). Isolation of eucalyptus essential oil for GC-MS analysis by extraction with subcritical water. *Chromatographia* 50 (9): 578–582.

47 Bey-Ould Si Said, Z., Haddadi-Guemghar, H., Boulekbache-Makhlouf, L. et al. (2016). Essential oils composition, antibacterial and antioxidant activities of hydrodistillated extract of *Eucalyptus globulus* fruits. *Industrial Crops and Products* 89: 167–175.

48 Congiu, R., Falconieri, D., Marongiu, B. et al. (2002). Extraction and isolation of *Pistacia lentiscus* L. essential oil by supercritical CO_2. *Flavour and Fragrance Journal* 17 (4): 239–244.

49 Fernández-López, J. and Viuda-Martos, M. (2018). Introduction to the special issue: application of essential oils in food systems. *Foods* 7 (4): 56.

50 Ribeiro-Santos, R., de Melo, N.R., Andrade, M., and Sanches-Silva, A. (2017). Potential of migration of active compounds from protein-based films with essential oils to a food and a food simulant. *Packaging Technology and Science* 30 (12): 791–798.

51 Varghese, S.A., Siengchin, S., and Parameswaranpillai, J. (2020). Essential oils as antimicrobial agents in biopolymer-based food packaging – a comprehensive review. *Food Bioscience* 38: 100785.

52 Shojaee-Aliabadi, S., Hosseini, H., Mohammadifar, M.A. et al. (2013). Characterization of antioxidant-antimicrobial κ-carrageenan films containing *Satureja hortensis* essential oil. *International Journal of Biological Macromolecules* 52 (1): 116–124.

53 Song, X., Zuo, G., and Chen, F. (2018). Effect of essential oil and surfactant on the physical and antimicrobial properties of corn and wheat starch films. *International Journal of Biological Macromolecules* 107 (Part A): 1302–1309.

54 Peng, Y., Yin, L., and Li, Y. (2013). Combined effects of lemon essential oil and surfactants on physical and structural properties of chitosan films. *International Journal of Food Science and Technology* 48 (1): 44–50.

55 do Evangelho, J.A., da Silva Dannenberg, G., Biduski, B. et al. (2019). Antibacterial activity, optical, mechanical, and barrier properties of corn starch films containing orange essential oil. *Carbohydrate Polymers* 222: 114981.

56 Ojagh, S.M., Rezaei, M., Razavi, S.H., and Hosseini, S.M.H. (2010). Development and evaluation of a novel biodegradable film made from chitosan and cinnamon essential oil with low affinity toward water. *Food Chemistry* 122 (1): 161–166.

57 Hosseini, M.H., Razavi, S.H., and Mousavi, M.A. (2009). Antimicrobial, physical and mechanical properties of chitosan-based films incorporated with thyme, clove and cinnamon essential oils. *Journal of Food Processing & Preservation* 33 (6): 727–743.

58 Do Socorro Rocha Bastos, M., Da Silva Laurentino, L., Canuto, K.M. et al. (2016). Physical and mechanical testing of essential oil-embedded cellulose ester films. *Polymer Testing* 49: 156–161.

59 Frank, K., Garcia, C.V., Shin, G.H., and Kim, J.T. (2018). Alginate biocomposite films incorporated with cinnamon essential oil nanoemulsions: physical, mechanical, and antibacterial properties. *International Journal of Polymer Science* 2018: 1519407.

60 Ghasemlou, M., Aliheidari, N., Fahmi, R. et al. (2013). Physical, mechanical and barrier properties of corn starch films incorporated with plant essential oils. *Carbohydrate Polymers* 98 (1): 1117–1126.

61 Shojaee-Aliabadi, S., Hosseini, H., Mohammadifar, M.A. et al. (2014). Characterization of κ-carrageenan films incorporated plant essential oils with improved antimicrobial activity. *Carbohydrate Polymers* 101 (1): 582–591.

62 Pérez-Gago, M.B. and Krochta, J.M. (2001). Lipid particle size effect on water vapor permeability and mechanical properties of whey protein/beeswax emulsion films. *Journal of Agricultural and Food Chemistry* 49 (2): 996–1002.

63 Klangmuang, P. and Sothornvit, R. (2016). Barrier properties, mechanical properties and antimicrobial activity of hydroxypropyl methylcellulose-based nanocomposite films incorporated with Thai essential oils. *Food Hydrocolloids* 61: 609–616.

64 Arezoo, E., Mohammadreza, E., Maryam, M., and Abdorreza, M.N. (2020). The synergistic effects of cinnamon essential oil and nano TiO_2 on antimicrobial and functional properties of sago starch films. *International Journal of Biological Macromolecules* 157: 743–751.

65 Laorenza, Y. and Harnkarnsujarit, N. (2021). Carvacrol, citral and α-terpineol essential oil incorporated biodegradable films for functional active packaging of Pacific white shrimp. *Food Chemistry* 363: 130252.

66 Souza, V.G.L., Pires, J.R.A., Vieira, É.T. et al. (2019). Activity of chitosan-montmorillonite bionanocomposites incorporated with rosemary essential oil: from in vitro assays to application in fresh poultry meat. *Food Hydrocolloids* 89: 241–252.

67 Han, Y., Yu, M., and Wang, L. (2018). Physical and antimicrobial properties of sodium alginate/carboxymethyl cellulose films incorporated with cinnamon essential oil. *Food Packaging and Shelf Life* 15: 35–42.

68 Bonilla, J., Vargas, M., Atarés, L., and Chiralt, A. (2014). Effect of chitosan essential oil films on the storage-keeping quality of pork meat products. *Food and Bioprocess Technology* 7 (8): 2443–2450.

69 Acosta, S., Chiralt, A., Santamarina, P. et al. (2016). Antifungal films based on starch-gelatin blend, containing essential oils. *Food Hydrocolloids* 61: 233–240.

70 Babapour, H., Jalali, H., and Mohammadi Nafchi, A. (2021). The synergistic effects of zinc oxide nanoparticles and fennel essential oil on physicochemical, mechanical, and antibacterial properties of potato starch films. *Food Science & Nutrition* 9 (7): 3893–3905.

71 Rojas-Graü, M.A., Avena-Bustillos, R.J., Olsen, C. et al. (2007). Effects of plant essential oils and oil compounds on mechanical, barrier and antimicrobial properties of alginate–apple puree edible films. *Journal of Food Engineering* 81 (3): 634–641.

72 Jouki, M., Yazdi, F.T., Mortazavi, S.A., and Koocheki, A. (2014). Quince seed mucilage films incorporated with oregano essential oil: physical, thermal, barrier, antioxidant and antibacterial properties. *Food Hydrocolloids* 36: 9–19.

73 Moradi, M., Tajik, H., Razavi Rohani, S.M. et al. (2012). Characterization of antioxidant chitosan film incorporated with *Zataria multiflora* Boiss essential oil and grape seed extract. *LWT – Food Science and Technology* 46 (2): 477–484.

74 Pranoto, Y., Salokhe, V.M., and Rakshit, S.K. (2005). Physical and antibacterial properties of alginate-based edible film incorporated with garlic oil. *Food Research International* 38 (3): 267–272.

75 Pires, C., Ramos, C., Teixeira, G. et al. (2011). Characterization of biodegradable films prepared with hake proteins and thyme oil. *Journal of Food Engineering* 105 (3): 422–428.

76 Pires, C., Ramos, C., Teixeira, B. et al. (2013). Hake proteins edible films incorporated with essential oils: physical, mechanical, antioxidant and antibacterial properties. *Food Hydrocolloids* 30 (1): 224–231.

77 Sánchez-González, L., Chiralt, A., González-Martínez, C., and Cháfer, M. (2011). Effect of essential oils on properties of film forming emulsions and films based on hydroxypropylmethylcellulose and chitosan. *Journal of Food Engineering* 105 (2): 246–253.

78 Li, J., Ye, F., Lei, L., and Zhao, G. (2018). Combined effects of octenylsuccination and oregano essential oil on sweet potato starch films with an emphasis on water resistance. *International Journal of Biological Macromolecules* 115: 547–553.

79 Alexandre, E.M.C., Lourenço, R.V., Bittante, A.M.Q.B. et al. (2016). Gelatin-based films reinforced with montmorillonite and activated with nanoemulsion of ginger essential oil for food packaging applications. *Food Packaging and Shelf Life* 10: 87–96.

80 Abdollahi, M., Rezaei, M., and Farzi, G. (2012). Improvement of active chitosan film properties with rosemary essential oil for food packaging. *International Journal of Food Science and Technology* 47 (4): 847–853.

81 Han Lyn, F. and Nur Hanani, Z.A. (2020). Effect of lemongrass (*Cymbopogon citratus*) essential oil on the properties of chitosan films for active packaging. *Journal of Packaging Technology and Research* 4 (1): 33–44.

82 Jamróz, E., Juszczak, L., and Kucharek, M. (2018). Investigation of the physical properties, antioxidant and antimicrobial activity of ternary potato starch-furcellaran-gelatin films incorporated with lavender essential oil. *International Journal of Biological Macromolecules* 114: 1094–1101.

83 Romani, V.P., Prentice-Hernández, C., and Martins, V.G. (2017). Active and sustainable materials from rice starch, fish protein and oregano essential oil for food packaging. *Industrial Crops and Products* 97: 268–274.

84 Jamróz, E., Cabaj, A., Tkaczewska, J. et al. (2022). Incorporation of curcumin extract with lemongrass essential oil into the middle layer of triple-layered films based on furcellaran/chitosan/gelatin hydrolysates – in vitro and in vivo studies on active and intelligent properties. *Food Chemistry* 402: 134476.

85 Flórez, M., Cazón, P., and Vázquez, M. (2022). Active packaging film of chitosan and *Santalum album* essential oil: characterization and application as butter sachet to retard lipid oxidation. *Food Packaging and Shelf Life* 34: 100938.

86 Zhao, R., Guan, W., Zhou, X. et al. (2022). The physiochemical and preservation properties of anthocyanidin/chitosan nanocomposite-based edible films containing cinnamon-perilla essential oil pickering nanoemulsions. *LWT – Food Science and Technology* 153: 112506.

87 Amorati, R. and Valgimigli, L. (2018). Methods to measure the antioxidant activity of phytochemicals and plant extracts. *Journal of Agricultural and Food Chemistry* 66 (13): 3324–3329.

88 Rodriguez-Garcia, I., Silva-Espinoza, B.A., Ortega-Ramirez, L.A. et al. (2016). Oregano essential oil as an antimicrobial and antioxidant additive in food products. *Critical Reviews in Food Science and Nutrition* 56 (10): 1717–1727.

89 Rehman, A., Qunyi, T., Sharif, H.R. et al. (2021). Biopolymer based nanoemulsion delivery system: an effective approach to boost the antioxidant potential of essential oil in food products. *Carbohydrate Polymer Technologies and Applications* 2: 100082.

90 Atarés, L. and Chiralt, A. (2016). Essential oils as additives in biodegradable films and coatings for active food packaging. *Trends in Food Science and Technology* 48: 51–62.

91 López-Mata, M.A., Ruiz-Cruz, S., Silva-Beltrán, N.P. et al. (2015). Physicochemical and antioxidant properties of chitosan films incorporated with cinnamon oil. *International Journal of Polymer Science* 2015: 974506.

92 Luís, Â., Pereira, L., Domingues, F., and Ramos, A. (2019). Development of a carboxymethyl xylan film containing licorice essential oil with antioxidant properties to inhibit the growth of foodborne pathogens. *LWT – Food Science and Technology* 111: 218–225.

93 El Fawal, G.F., Omer, A.M., and Tamer, T.M. (2019). Evaluation of antimicrobial and antioxidant activities for cellulose acetate films incorporated with Rosemary and Aloe Vera essential oils. *Journal of Food Science and Technology* 56 (3): 1510–1518.

94 Kilinc, D., Ocak, B., and Özdestan-Ocak, Ö. (2020). Preparation, characterization and antioxidant properties of gelatin films incorporated with *Origanum onites* L. essential oil. *Journal of Food Measurement and Characterization* 15 (1): 795–806.

95 Pirouzifard, M., Yorghanlu, R.A., and Pirsa, S. (2019). Production of active film based on potato starch containing Zedo gum and essential oil of *Salvia officinalis* and study of physical, mechanical, and antioxidant properties. *Journal of Thermoplastic Composite Materials* 33 (7): 915–937.

96 Lee, J.Y., Garcia, C.V., Shin, G.H., and Kim, J.T. (2019). Antibacterial and antioxidant properties of hydroxypropyl methylcellulose-based active composite films incorporating oregano essential oil nanoemulsions. *LWT – Food Science and Technology* 106: 164–171.

97 Hromiš, N.M., Lazić, V.L., Markov, S.L. et al. (2015). Optimization of chitosan biofilm properties by addition of caraway essential oil and beeswax. *Journal of Food Engineering* 158: 86–93.

98 Bonilla, J., Poloni, T., Lourenço, R.V., and Sobral, P.J.A. (2018). Antioxidant potential of eugenol and ginger essential oils with gelatin/chitosan films. *Food Bioscience* 23: 107–114.

99 Cazón, P., Antoniewska, A., Rutkowska, J., and Vázquez, M. (2021). Evaluation of easy-removing antioxidant films of chitosan with *Melaleuca alternifolia* essential oil. *International Journal of Biological Macromolecules* 186: 365–376.

100 dos Santos Paglione, I., Galindo, M.V., de Medeiros, J.A.S. et al. (2019). Comparative study of the properties of soy protein concentrate films containing free and encapsulated oregano essential oil. *Food Packaging and Shelf Life* 22: 100419.

101 Zheng, K., Li, W., Fu, B. et al. (2018). Physical, antibacterial and antioxidant properties of chitosan films containing hardleaf oatchestnut starch and *Litsea cubeba* oil. *International Journal of Biological Macromolecules* 118: 707–715.

102 Llana-Ruiz-Cabello, M., Pichardo, S., Bermúdez, J.M. et al. (2016). Development of PLA films containing oregano essential oil (*Origanum vulgare* L. virens) intended for use in food packaging. *Food Additives & Contaminants: Part A* 33 (8): 1374–1386.

103 Viuda-Martos, M., Mohamady, M.A., Fernández-López, J. et al. (2011). in vitro antioxidant and antibacterial activities of essentials oils obtained from Egyptian aromatic plants. *Food Control* 22 (11): 1715–1722.

104 Saggiorato, A.G., Gaio, I., Treichel, H. et al. (2009). Antifungal activity of basil essential oil (*Ocimum basilicum* L.): evaluation in vitro and on an Italian-type sausage surface. *Food and Bioprocess Technology* 5 (1): 378–384.

105 Ruiz-Navajas, Y., Viuda-Martos, M., Sendra, E. et al. (2013). in vitro antibacterial and antioxidant properties of chitosan edible films incorporated with *Thymus moroderi* or *Thymus piperella* essential oils. *Food Control* 30 (2): 386–392.

106 Arqués, J.L., Rodríguez, E., Nuñez, M., and Medina, M. (2007, 227). Inactivation of Gram-negative pathogens in refrigerated milk by reuterin in combination with nisin or the lactoperoxidase system. *European Food Research and Technology* (1): 77–82.

107 Burt, S. (2004). Essential oils: their antibacterial properties and potential applications in foods – a review. *International Journal of Food Microbiology* 94 (3): 223–253.

108 Fasihi, H., Noshirvani, N., Hashemi, M. et al. (2019). Antioxidant and antimicrobial properties of carbohydrate-based films enriched with cinnamon essential oil by Pickering emulsion method. *Food Packaging and Shelf Life* 19: 147–154.

109 Alak, G., Guler, K., Ucar, A. et al. (2019). Quinoa as polymer in edible films with essential oil: effects on rainbow trout fillets shelf life. *Journal of Food Processing & Preservation* 43 (12): e14268.

110 Priyadarshi, R., Sauraj Kumar, B., Deeba, F. et al. (2018). Chitosan films incorporated with apricot (*Prunus armeniaca*) kernel essential oil as active food packaging material. *Food Hydrocolloids* 85: 158–166.

111 Valizadeh, S., Naseri, M., Babaei, S. et al. (2019). Development of bioactive composite films from chitosan and carboxymethyl cellulose using glutaraldehyde, cinnamon essential oil and oleic acid. *International Journal of Biological Macromolecules* 134: 604–612.

112 Roy, S. and Rhim, J.W. (2021). Fabrication of bioactive binary composite film based on gelatin/chitosan incorporated with cinnamon essential oil and rutin. *Colloids and Surfaces, B: Biointerfaces* 204: 111830.

113 Souza, V.G.L., Rodrigues, C., Ferreira, L. et al. (2019). in vitro bioactivity of novel chitosan bionanocomposites incorporated with different essential oils. *Industrial Crops and Products* 140: 111563.

114 Cardoso, L.G., Pereira Santos, J.C., Camilloto, G.P. et al. (2017). Development of active films poly(butylene adipate-*co*-terephthalate) – PBAT incorporated with oregano essential oil and application in fish fillet preservation. *Industrial Crops and Products* 108: 388–397.

115 Wang, Y., Luo, J., Hou, X. et al. (2022). Physicochemical, antibacterial, and biodegradability properties of green Sichuan pepper (*Zanthoxylum armatum* DC.) essential oil incorporated starch films. *LWT – Food Science and Technology* 161: 113392.

116 Manso, S., Cacho-Nerin, F., Becerril, R., and Nerín, C. (2013). Combined analytical and microbiological tools to study the effect on *Aspergillus flavus* of cinnamon essential oil contained in food packaging. *Food Control* 30 (2): 370–378.

117 Hemalatha, T., UmaMaheswari, T., Senthil, R. et al. (2017). Efficacy of chitosan films with basil essential oil: perspectives in food packaging. *Journal of Food Measurement and Characterization* 11 (4): 2160–2170.

118 Chein, S.H., Sadiq, M.B., and Anal, A.K. (2019). Antifungal effects of chitosan films incorporated with essential oils and control of fungal contamination in peanut kernels. *Journal of Food Processing & Preservation* 43 (12): e14235.

119 Ali, B., Al-Wabel, N.A., Shams, S. et al. (2015). Essential oils used in aromatherapy: a systemic review. *Asian Pacific Journal of Tropical Biomedicine* 5 (8): 601–611.

120 Silva, F., Caldera, F., Trotta, F. et al. (2019). Encapsulation of coriander essential oil in cyclodextrin nanosponges: a new strategy to promote its use in controlled-release active packaging. *Innovative Food Science and Emerging Technologies* 56: 102177.

121 Zhang, W., Jiang, H., Rhim, J.W. et al. (2022). Effective strategies of sustained release and retention enhancement of essential oils in active food packaging films/coatings. *Food Chemistry* 367: 130671.

122 Lammari, N., Louaer, O., Meniai, A.H., and Elaissari, A. (2020). Encapsulation of essential oils via nanoprecipitation process: overview, progress challenges and prospects. *Pharmaceutics* 12 (5): 431.

123 Gupta, S. and Variyar, P.S. (2016). Nanoencapsulation of essential oils for sustained release: application as therapeutics and antimicrobials. In: *Encapsulations* (ed. A.M. Grumezescu), 641–672. Academic Press.

124 Brockgreitens, J. and Abbas, A. (2016). Responsive food packaging: recent progress and technological prospects. *Comprehensive Reviews in Food Science and Food Safety* 15 (1): 3–15.

125 Marturano, V., Bizzarro, V., Ambrogi, V. et al. (2019). Light-responsive nanocapsule-coated polymer films for antimicrobial active packaging. *Polymers* 11 (1): 68.

4

Agro-Waste Residue-Based Food Packaging Films

Rajarathinam Nithya and Arunachalam Thirunavukkarasu

Government College of Technology, Department of Industrial Biotechnology,
Thadagam Main Rd., Coimbatore 641013, India

4.1 Introduction

Despite the petrochemicals-based synthetic food packaging materials are providing advantages such as cost-effective, scalable operation, and good resistance properties, their negative impact on the ecosystem resulted in the search of the production of such materials from alternative biodegradable sources. The primary role of any packaging material is to keep fresh, to retard deterioration, and to improve the shelf life of the food stuffs. In addition, the packaging also plays a significant role in the regulation of interactions among food ecosystem [1, 2]. On considering this fact, the awareness among the consumers has been drastically increased in the recent decades that resulted in the development of bio-based food packaging films. With such increased consumer demands and environmental threats to the conventional packing materials, enormous researches were initiated on the technological development of novel packing films from renewable and biodegradable sources [1–3]. In the search of such biological sources, agricultural by-products serves the better in terms of the availability, edibility, nontoxicity, and economy of the overall process.

Biopolymers from the agro-waste are highly suitable for the manufacturing of packaging films as it encompasses range of polysaccharides such as starch, chitosan, cellulose, pectin, and gums. Also the plant-based protein polymers including whey, gluten, gelatin, and soy and animal-based protein polymers such as chitosan or chitin, silk, and collagen and are also the primary candidate for the manufacturing of such biofilms [4]. With the advent of such demands, the global market reached around 10 billion USD in 2021 and with a steady growth rate of 17% in the period 2017–2021 [5]. Fuelling to the existing benefits, the potential use of nonedible portions of plants and animals for the manufacturing of such biofilms would not only

---

Natural Materials for Food Packaging Application, First Edition.
Edited by Jyotishkumar Parameswaranpillai, Aswathy Jayakumar,
E. K. Radhakrishnan, Suchart Siengchin, and Sabarish Radoor.
© 2023 WILEY-VCH GmbH. Published 2023 by WILEY-VCH GmbH.

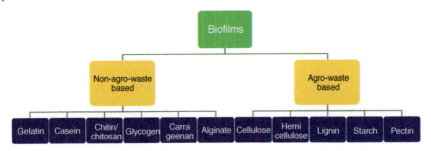

Figure 4.1 Classification of biofilms.

prevent the risk of food depletion of native communities and also cut off the overall economy of the process. The biofilms from the biological sources can be categorized as agro-waste based and non-agro-waste based as illustrated in Figure 4.1.

As the chemical complexity of these materials are susceptible to fragmentation followed by the microbial mineralization, the fate of such biofilms are not having any impact on the ecosystem. The ambient temperature, moisture content, and the enzymic action of microbial population degrade the polymer that results in the highly mineralized simple form of residues. These fragmented products are being utilized by the range of microorganisms for their food and energy requirements and thus converting the fragmented molecules into carbon dioxide, water, methane, etc. [6]. In general, the biofilms are produced with the conventional procedures such as casting, pressing, and extrusion with the follow-up of blowing process [7]. In the recent days, functionalized biofilms are being manufactured with the incorporation of new properties such as antioxidizing capacity [8, 9], antimicrobial action [10], attractive colors [11, 12], and tailor-made physical and mechanical properties. In some instances, biopolymers may be used alone or with the fusion of other biopolymeric materials, functional attributes could be attained. This chapter is intended to present a comprehensive information on the different types of agro-waste-based biofilms along with their functional properties imparted with a wide range of additives. Also, this chapter is presenting some of the research gaps which could be potentially addressed in the near future for the sustainable development in the field of food packaging and technology.

4.2 Agro-Waste-Based Biopolymers

4.2.1 Cellulose

Of the biopolymers exist in the nature, the most abundant and easily available polymer is cellulose, which has a linear rigid structure with the repeating units of D-glucose. The presence of intra and intermolecular H-bonds made the polymer with poor water solubility. Cellophane films are derived from the regenerated cellulose that has the wide application in the food packaging. It is highly transparent, mechanically rigid, and dimensionally stable. Owing to this, these films are being used in the dairy packaging, candies, and other pastries wrappings. [13] incorporated nisin, an antimicrobial natural preservative to prepare meat packagings. Similarly [14] reported the use of composite made of cellulose and lignin possess UV shielding and antimicrobial to the packed food materials. As the chemical backbone of the cellulose has too many hydroxyl moieties, it is possible to obtain various derivatives of cellulose with the suitable trainsformations. And several reported works showed that these derivatives were excellent to be used as packaging material [15]. For instance, carboxymethyl cellulose (CMC) one of the cellulose derivatives was offering an excellent mechanical properties if it has been blended with polyvinyl alcohol (PVA) in an appropriate ratio [15]. Acetate cellulose was incorporated with phages resulted in greater antimicrobial property of the packing film [16]. A similar observation was made by [17], in which the oleic acid/rosemary extract showed antifungal activity for 60 days duration at a temperature of 25 °C. A layer-by-layer coating of CMC and chitosan was made that lowered the metabolic actions and thus it retained the freshness of strawberries [18]. Cellulose cinnamate was found to be an excellent packing film for fruits such as tomatoes and berries due to its mechanical rigidity, thermal stablity, and poor permeability [19, 20].

4.2.2 Hemicellulose

Hemicellulose, a kind of branched polymer which comprises pentosans and hexosans. In general, these polymers show good solubilizing power to water and have the tendency to form films. Several reports from the literature revealed that xylans are the most abundant type in branched woods which has excellent resistance properties against oil/grease and oxygen. Likewise, galactomannan and xyloglucan were blended and films were formed by Mendes et al. [21] that showed greater resistant and thermal stabilty in packing the food materials. Despite having appreciable thermal properties, these hemicellulose films are usually susceptible to moisture and poor mechanical strength. Hence, it is imperative to chemically modify or blending with some other polymeric forms to improve their mechanical properties and also to augment range of unique properties. For instance, quarternized hemicellulose and chitosan were embedded onto montmorillonite to create a compact and robust film structures. Such composite showed improved permeation barrier to oxygen and water vapor due to the greater viscosity of chitosan. Also, it showed high mechanical stability owing to the compact nacre-like structure in its primary backbone [22]. Similarly, biofilms made of acetylated hemicellulose that was impregnated with nanocellulose followed by polycaprolactone coatings showed improved mechanical strength and reasonable hydrophobicity [23].

4.2.3 Lignin

Besides cellulose and chitin, lignin is the most abundant biopolymer on our earth as it is existing in the cell wall of all plants. Chemically, it is quite complex polymer that has cross-linked structures and usually interacted with hemicellulose through ester bonds. Primarily it has three different types of alcoholic precursors which include coniferyl, p-coumaryl, and sinapyl forms. Also, it contains different functional moieties such as hydroxyl, carbonyl, carboxyl, and benzene. Owing to its specified chemical framework and distinct functional groups, lignins are predominantly used in the food packing sectors. It can render antimicrobial properties, ultraviolet screening ability, and antioxidizing capability. Also, it is a compound with low molar mass, narrow polydispersity index (PI), less number of aliphatic hydroxyl moieties, and more number of phenolic hydroxyl moieties [24].

Hydroxy propyl methyl cellulose (HPMC)-lignin along with its chitosan composite films were evaluated as a packaging film and the study reported that the amount or concentration of lignin was significant for its antimicrobial property [25]. Likewise, an another study revealed that the PVA/gelatin mixture with lignin composite showed excellent antimicrobial properties in which the hydroxyl groups of the lignin interact with the microbial cell membrane and result in the cell lysis [26]. Several reported literatures were available on the integration of biopolymers such as PVA, PLA, chitosan, alginate, rubber latex, and gelatin with the lignin to enhance the mechanical and wettability of the biofilms [15]. Gaikwad et al. [27] proved the lignin-based biofilms can also act as an excellent antioxidation agent in the packaging of foods and grains. Of the different forms of lignin derived from diverse sources such as lignosulfonate, alkali, and kraft, alkali lignin was able to show maximum radical scavenging action as it contains free phenolic monomeric forms [28]. Owing to the augmented methoxyl groups, lignin with syringyl moieties were able to show UV-screening properties and high antioxidizing capability [14]. The biofilms made of lignin that is incorporated with PHB and PHA resulted to act as both nucleating agent and antioxidizing agent. These films were found to be rigid and hence the permeation of oxygen and carbon dioxide was minimal and can act as an excellent packing material [29].

4.2.4 Starch

Starch is a polymer that comprises linear and helical forms of amylose with the branched structure of amylopectin. As it is a semicrystalline in nature, the flexibilty is too poor and cannot be used for packaging directly. Hence, physical and/or chemical alterations are required and in some instances, plasticization and the use of enzymes were reported for improving the flexible nature of the starch. Studies reported that the use of citric acid and gelatin could improve the intrinsic properties of starch. The derivative forms such as acetylated starch was effective against water vapor and other gases and hence it was used as packaging material [15]. Further to reduce the hydrophilicity attribute of starch, plasma treatment was used to react the contact angle of 140°. Also, chemical additives such as ZnO, CNF, nanoclay, and MgO were blended with starch to improve the antimicrobial actions, UV screening ability, and better resistant properties [30]. Mendes et al. [29] recommended the incorporation of lemongrass essential oil and pectin with the starch to show good heat stability, wearability, and permeability to oxygen and water vapor.

4.2.5 Pectin

Pectin is a complex hetero form of polysachharide and it is one of the cell wall components of the plants. It is one of the widely used packing materials in the food industries. Casting or thermal-driven compression followed by the molding are the generic processes of pectin-based biofilms. As it shows greater compatibility with the various forms of biopolymers such as lipids, polysaccharides, and proteins, the polymer can be tailor-made for the specified functionalities. To improve its flexibility, glycerol was used as plasticizer in forming the films. A recent report on the use of pectin with the composite coatings of corn flour and beetroot extract modified the oxygen level, refrained the ethylene emission, and thus delayed the food decaying process [31]. A similar study was conducted by Mellinas et al. [32] with the use of pectin and corn flour to package tomatoes and the composite kept the fruit fresh for a longer period of time. Pectin films can be directly applied over cut fruits and vegetables as an encapsulating material to retard the release of phytochemically active compounds. For instance, the packaging of soybean oil using pectin films retarded the oxidation process for about 30 days. Pectin/clove essential oil composite films showed excellent resistant, mechanical, antioxidizing, and antimicrobial actions [33]. Pectin/starch/chitosan film for the coating of mint and rosemary essential oil showed high antimicrobial property, tensile strength, and thermal stabilty [34]. The tea leaves extracts were also showed to be an effective enhancer of these traits for pectin films in food packaging [35]. The augmentation of nanomeric forms of certain chemicals such as silver, gold, copper, and titanium improved the mechanical properties, UV shielding, and antimicrobial actions ([36], Mellinas et al., 2020). [37] reported the controlled release of marjoram essential oil with the use of pectin biofilms. Recently, pectin-based smart film was prepared with the use of red cabbage extract for packaging meat. As the pigment called anthocyanins was able to change its color with the change in pH value, it served as a sensor for the determination of degree of protein degradation in the meat [38].

Besides pectin as a film, it was also made into emulsions, aerogels, and hydrogels for the packaging of food materials. For instance, pectin/chitosan-based hydrogels showed good antibacterial action, prevention of enzymic browning, and prevented water losses [39]. Also, in the form of aerogels, it renders high porous volume, improved specific surface area, low bulk density, and excellent heat insulation property [40]. The rate of biodegradation of emulsion form of pectin with lemongrass essential oil was found to be very minimal and hence this pectin formulation can act as an excellent packing material for bioactive compounds [41].

4.3 Edible Coatings and Films – Classification and Properties

The two forms of edible coatings and films are not identical, edible coatings can be done by applying directly on the surface of the food products such as vegetables, fruits, and other products. The edible coating is usually done by immersing the food products into the edible material solution. But at the same time, the edible films are not in a liquid form and so it is molded as solid sheets and used to wrap and then utilized for packaging the food products [50]. In both the forms such as edible and film coatings, the components used are biodegradable and nontoxic. Edible materials are developed from biopolymer matrixes like proteins, lipids, polysaccharides, and other composite materials (Table 4.1). Both coatings and films are produced as very thin layers and it is applied on the surface of the food products or between the food components. The main aim of using edible coatings and films on the food products is to improve their shelf life and concurrently it also acts as a safety barrier. This process can slow down the migration of moisture content, the loss of volatile chemicals, also inhibit respiration, and retard the textural changes of the product. This also provides good fat and acts as oil barriers and have high selective gas permeability ratio CO_2/O_2 when compared to common synthetic films [51]. Films are also used as carriers for food additives such as antimicrobial agents [10] and antioxidants [52], as well as it increases the reliability and handling properties of the products. For certain applications, specific edible films having strong mechanical strength could be replaced with synthetic packaging films. The mechanical property and barrier property of edible films and coatings are controlled and influenced by the composition of biopolymers, viscosity, their drainage time, and other factors [4].

The control of diffusional mass transport, shear protection, and sensory appeal are the most vital properties of an edible film or coatings. Mass transfer control governs the prevention of food desiccation, gaseous transports across the food surface, and migration of ingredients and/or additrives in the packing of food materials. In some instances, the edible coatings might reduce the quantity as it directly alters the internal atmosphere of the food materials. The selection of the type of the edible film is crucial as it has direct influence on the quality of the food materials being packed. For example, in case of packing fresh fruits, edible films with moderate permeation rate to water vapor is preferable as it extends the shelf life whereas the maximum permeation rate materials will result in the moisture loss in the fruits during its storage. Likewise, the mechanical rigidness is considered to ensure the packed food items to be safely transported and not always having the packing films with high tensile strength for all the food items. Sensory appeal is the final property that need to be considered and it determines the overall acceptability and marketing of the food product among the consumers.

Table 4.1 Agro-waste-based Biofilms and their investigated functional properties.

		Functional properties studied					
Agro-waste residue	Additive	Antioxidant	Antimicrobial	Permeability	Tensile strength	Barrier	Reference
Cellulose	Nisin	—	✓	—	—	—	[13]
	Lignin	—	✓	—	—	—	[14]
	Vitamin E	✓	—	✓	✓	—	[42]
	Oleic acid containing Rosemary extract oil	—	✓	—	—	—	[17]
	Chitosan	✓	—	—	—	✓	[18]
	Bacteriophages	—	✓	—	—	—	[16]
Hemicellulose	montmorillonite (MMT)	—	—	✓	✓	✓	[22]
	Nanocellulose	—	—	✓	—	✓	[23]
Lignin	HPMC-lignin-chitosan	—	✓	—	—	—	[25]
	PVA/gelatin	—	✓	✓	✓	—	[26]
	PHA/PHB	✓	—	✓	✓	✓	[29]
Starch	Babassu	✓	—	—	—	—	[43]
	Blackberry pulp	✓	—	—	—	—	[12]
	Cranberry extract	✓	✓	—	—	—	[44]
	Curcumin	✓	—	—	—	—	[45]
	Essential oils	—	—	—	—	✓	[1]
	Tea Polyphenols	✓	✓	✓	✓	✓	[46]

(Continued)

Table 4.1 (Continued)

Agro-waste residue	Additive	Functional properties studied					Reference
		Antioxidant	Antimicrobial	Permeability	Tensile strength	Barrier	
Pectin	Transglutaminase	✓	—	—	—	—	[47]
	Ascorbic acid	✓	✓	✓	✓	✓	[48]
	Clove EO	✓	—	✓	—	✓	[33]
	Marjoram EO	✓	—	✓	—	—	[37]
	Tea extract	✓	—	—	✓	✓	[35]
	AgNPs	✓	✓	—	—	✓	[49]
	Mint and rosemary oils Nisin	✓	✓	✓	—	✓	[34]
	Red cabbage extract	—	—	✓	—	—	[38]

4.4 Conclusion and Future Prospects

The present chapter highlighted the use of different forms of agro-waste residue-based materials for the food packaging along with their imparted functional properties on the food. Also, the functional characteristics of any material to be considered for food packaging were discussed precisely. Despite the possible use of agro-waste-derived polymers for the food packaging, there are still some areas to be addressed in near future for the sustainable development of these technologies.

i) A comprehensive understanding of the physical and chemical interaction pathways among these packaging films and the food packed needs to be elucidated.
ii) For the commercial scale of production of such biofilms, one has to consider the overall economy of the production process. As the extraction, isolation, and purification of biopolymers require exhaustive collection of procedures or processes, simple techniques need to be emerged out in this context.
iii) Use of toxic solvents in the processes need to be replaced with the aqueous-based solvents to ensure the maximum level of safety.
iv) In the case of nano-based additives, the penetration of such materials into the foodstuff should be completely restricted to ensure its food safety. Hence, the futuristic studies need to concentrate or focus onto the permeation of additives onto the foodstuffs.

References

1 Mahcene, Z., Khelil, A., Hasni, S. et al. (2020). Development and characterization of sodium alginate based active edible films incorporated with essential oils of some medicinal plants. *International Journal of Biological Macromolecules* 145: 124–132.
2 Zahan, K.A., Azizul, N.M., Mustapha, M. et al. (2020). Application of bacterial cellulose film as, a biodegradable and antimicrobial packaging material. *Materials Today: Proceedings* 31: 83–88.
3 Zhao, Y., Bao, Z., Wan, Z. et al. (2019). Polystyrene microplastic exposure disturbs hepatic glycolipid metabolism at the physiological, biochemical, and transcriptomic levels in adult zebrafish. *The Science of the Total Environment* 710: 136279.
4 Sathya, A.B., Sivashankar, R., Kanimozhi, J. et al. (2018). *Biodegradable Plastics for a Green and Sustainable Environment*, 171–198. London: CRC Press.
5 Martau, G.A., Mihai, M., and Vodnar, D.C. (2019). The use of chitosan, alginate, and pectin in the biomedical and food sector—biocompatibility, bioadhesiveness, and biodegradability. *Polymers* 11: 1837.
6 Mohee, R., Unmar, G.D., Mudhoo, A., and Khadoo, P. (2008). Biodegradability of biodegradable/degradable plastic materials under aerobic and anaerobic conditions. *Waste Management* 28: 1624–1629.

7 Fakhouri, F.M., Costa, D., Yamashita, F. et al. (2013). Comparative study of processing methods for starch/gelatin films. *Carbohydrate Polymers* 2013 (95): 681–689.

8 Riaz, A., Lagnika, C., Luo, H. et al. (2020). Chitosan-based biodegradable active food packaging film containing Chinese chive (*Allium tuberosum*) root extract for food application. *International Journal of Biological Macromolecules* 150: 595–604.

9 Nogueira, G.F., Fakhouri, F.M., and de Oliveira, R.A. (2019). Incorporation of spray dried and freeze dried blackberry particles in edible films: morphology, stability to pH, sterilization and biodegradation. *Food Packaging and Shelf Lifestyles* 20: 100313.

10 Fu, Y. and Dudley, E.G. (2021). Antimicrobial-coated films as food packaging: A review. *Comprehensive Reviews in Food Science and Food Safety* 2021: 1–34. https://doi.org/10.1111/1541-4337.12769.

11 Brito, T.B., Carrajola, J.F., Gonçalves, E.C.B.A. et al. (2019). Fruit and vegetable residues flours with different granulometry range as raw material for pectin-enriched biodegradable film preparation. *Foodservice Research International* 121: 412–421.

12 Nogueira, G.F., Soares, C.T., Cavasini, R. et al. (2019). Bioactive films of arrowroot starch and blackberry pulp: physical, mechanical and barrier properties and stability to pH and sterilization. *Food Chemistry* 275: 417–425.

13 Guerra, N.P., Macias, C.L., Agrasar, A.T., and Castro, L.P. (2005). Development of a bioactive packaging cellophane using Nisaplin* as biopreservative agent. *Letters in Applied Microbiology* 40: 106–110.

14 Guo, Y., Tian, D., Shen, F. et al. (2019). Transparent cellulose/technical lignin composite films for advanced packaging. *Polymers* 11: 1455.

15 Wang, J., Euring, M., Ostendorf, K., and Zhang, K. (2021). Biobased materials for food packaging. *Journal of Bioresources and Bioproducts* https://doi.org/10.1016/j.jobab.2021.11.004.

16 Gouvêa, D.M., Mendonça, R.C.S., Soto, M.L., and Cruz, R.S. (2015). Acetate cellulose film with bacteriophages for potential antimicrobial use in food packaging. *LWT - Food Science and Technology* 63: 85–91.

17 Nasibi, S., Nargesi Khoramabadi, H., Arefian, M. et al. (2020). A review of polyvinyl alcohol/carboxiy methyl cellulose (PVA/CMC) composites for various applications. *Journal of Composites and Compounds* 2: 68–75.

18 Yan, J.W., Luo, Z.S., Ban, Z.J. et al. (2019). The effect of the layer-by-layer (LBL) edible coating on strawberry quality and metabolites during storage. *Postharvest Biology and Technology* 147: 29–38.

19 Wang, J.X., Cao, Y., Jaquet, B. et al. (2021). Self-compounded nanocomposites: toward multifunctional membranes with superior mechanical, gas/oil barrier, UV-shielding, and photothermal conversion properties. *ACS Applied Materials & Interfaces* 13: 28668–28678.

20 Wang, J.X., Emmerich, L., Wu, J.F. et al. (2021). Hydroplastic polymers as eco-friendly hydrosetting plastics. *Nature Sustainability* 4 (10): 877–883.

21 Mendes, F.R.S., Bastos, M.S.R., Mendes, L.G. et al. (2017). Preparation and evaluation of hemicellulose films and their blends. *Food Hydrocolloids* 70: 181–190.
22 Chen, G.G., Qi, X.M., Guan, Y. et al. (2016). High strength hemicellulose-based nanocomposite film for food packaging applications. *ACS Sustainable Chemistry & Engineering* 4: 1985–1993.
23 Mugwagwa, L.R. and Chimphango, A.F.A. (2020). Enhancing the functional properties of acetylated hemicellulose films for active food packaging using acetylated nanocel- lulose reinforcement and polycaprolactone coating. *Food Packaging and Shelf Life* 24: 100481.
24 Pan, X.J., Kadla, J.F., Ehara, K. et al. (2006). Organosolv ethanol lignin from hybrid poplar as a radical scavenger: relationship between lignin structure, extraction conditions, and antioxidant activity. *Journal of Agricultural and Food Chemistry* 54: 5806–5813.
25 Alzagameem, A., Klein, S.E., Bergs, M. et al. (2019). Antimicrobial activity of lignin and lignin-derived cellulose and chitosan composites against selected pathogenic and spoilage microorganisms. *Polymers* 11: 670.
26 El-Nemr, K.F., Mohamed, H.R., Ali, M.A. et al. (2020). Polyvinyl alcohol/gelatin irradiated blends filled by lignin as green filler for antimicrobial packaging materials. *International Journal of Environmental Analytical Chemistry* 100: 1578–1602.
27 Gaikwad, K.K., Singh, S., and Lee, Y.S. (2018). Oxygen scavenging films in food packaging. *Environmental Chemistry Letters* 16: 523–538.
28 Domenek, S., Louaifi, A., Guinault, A., and Baumberger, S. (2013). Potential of lignins as antioxidant additive in active biodegradable packaging materials. *Journal of Polymers and the Environment* 21: 692–701.
29 Vostrejs, P., Adamcová, D., Vaverková, M.D. et al. (2020). Active biodegradable packaging films modified with grape seeds lignin. *RSC Advances* 10: 29202–29213.
30 Nechita, P. and Roman (Iana-Roman), M. (2020). Review on polysaccharides used in coatings for food packaging papers. *Coatings* 10: 566.
31 Sucheta, Chaturvedi, K., Sharma, N., and Yadav, S.K. (2019). Composite edible coatings from commercial pectin, corn flour and beetroot powder minimize post-harvest decay, reduces ripening and improves sensory liking of tomatoes. *International Journal of Biological Macromolecules* 133: 284–293.
32 Mellinas, C., Ramos, M., Jiménez, A., and Garrigós, M.C. (2020). Recent trends in the use of pectin from agro-waste residues as a natural-based biopolymer for food packaging applications. *Materials* 13 (3): 673.
33 Nisar, T., Wang, Z.C., Yang, X. et al. (2018). Characterization of *Citrus pectin* films integrated with clove bud essential oil: physical, thermal, barrier, antioxidant and antibacterial properties. *International Journal of Biological Macromolecules* 106: 670–680.
34 Akhter, R., Masoodi, F.A., Wani, T.A., and Rather, S.A. (2019). Functional characterization of biopolymer based composite film: incorporation of natural essential oils and antimicrobial agents. *International Journal of Biological Macromolecules* 137: 1245–1255.

35 Lei, Y., Wu, H., Jiao, C. et al. (2019). Investigation of the structural and physical properties, antioxidant and antimicrobial activity of pectin-konjac glucomannan composite edible films incorporated with tea polyphenol. *Food Hydrocolloids* 94: 128–135.

36 Kumar, M., Tomar, M., Saurabh, V. et al. (2020a). Emerging trends in pectin extraction and its anti-microbial functionalization using natural bioactives for application in food packaging. *Trends in Food Science and Technology* 105: 223–237.

37 Almasi, H., Azizi, S., and Amjadi, S. (2020). Development and characterization of pectin films activated by nanoemulsion and pickering emulsion stabilized marjoram (*Origanum majorana* L.) essential oil. *Food Hydrocolloids* 99: 105338.

38 Dudnyk, I., Janeček, E.R., Vaucher-Joset, J., and Stellacci, F. (2018). Edible sensors for meat and seafood freshness. *Sensors and Actuators B: Chemical* 259: 1108–1112.

39 Torpol, K., Sriwattana, S., Sangsuwan, J. et al. (2019). Optimising chitosan–pectin hydrogel beads containing combined garlic and holy basil essential oils and their application as antimicrobial inhibitor. *International Journal of Food Science and Technology* 54: 2064–2074.

40 Nešić, A., Gordić, M., Davidović, S. et al. (2018). Pectin-based nanocomposite aerogels for potential insulated food packaging application. *Carbohydrate Polymers* 195: 128–135.

41 Mendes, J.F., Norcino, L.B., Martins, H.H.A. et al. (2020). Correlating emulsion characteristics with the properties of active starch films loaded with lemongrass essential oil. *Food Hydrocolloids* 100: 105428.

42 Mirzaei-Mohkam, A., Garavand, F., Dehnad, D. et al. (2020). Physical, mechanical, thermal and structural characteristics of nanoencapsulated vitamin E loaded carboxymethyl cellulose films. *Progress in Organic Coatings* 138: 105383.

43 Maniglia, B.C., Tessaro, L., Lucas, A.A., and Tapia-Blácido, D.R. (2017). Bioactive films based on babassu mesocarp flour and starch. *Food Hydrocolloids* 70: 383–391.

44 Wang, Z., Yan, J., Wang, T. et al. (2019). Fabrication and properties of a bio-based biodegradable thermoplastic polyurethane elastomer. *Polymers* 11 (7): 1121. https://doi.org/10.3390/polym11071121.

45 Nieto-Suaza, L., Acevedo-Guevara, L., Sánchez, L.T. et al. (2019). Characterization of *Aloe vera*-banana starch composite films reinforced with curcumin-loaded starch nanoparticles. *Food Structure* 2019 (22): 100131.

46 Wu, C.D., Huang, E., Li, W. et al. (2021). Beverages containing plant-derived polyphenols inhibit growth and biofilm formation of *Streptococcus mutans* and children's supragingival plaque bacteria. *Beverages* 7 (3): 43. https://doi.org/10.3390/beverages7030043.

47 Porta, R., Di Pierro, P., Sabbah, M. et al. (2016). Blend films of pectin and bitter vetch (*Vicia ervilia*) proteins: properties and effect of transglutaminase. *Innovative Food Science and Emerging Technologies* 36: 245–251.

48 Chiarappa, G., De'Nobili, M.D., Rojas, A.M. et al. (2018). Mathematical modeling of L-(+)-ascorbic acid delivery from pectin films (packaging) to agar hydrogels (food). *Journal of Food Engineering* 234: 73–81.

49 Lee, J.H., Jeong, D., and Kanmani, P. (2019). Study on physical and mechanical properties of the biopolymer/silver based active nanocomposite films with antimicrobial activity. *Carbohydrate Polymers* 224: 115159.

50 Aguirre-JoyaJA, M.A., Leon-Zapata, D., Alvarez-Perez, O.B. et al. (2018). Basic and applied concepts of edible packaging for foods. In: *In Handbook of Food Bioengineering, Food Packaging and Preservation*, 1–61. Academic Press.

51 Falguera, V., Quintero, J.P., Jiménez, A. et al. (2011). Edible films and coatings: structures, active functions and trends in their use. *Trends Food Science Technology* 22: 292–303. https://doi.org/10.1016/j.tifs.2011.02.004.

52 Song, T., Qian, S., Lan, T. et al. (2022). Recent advances in bio-based smart active packaging materials. *Foods* 11 (15): 2228. https://doi.org/10.3390/foods11152228.

5

Hydrogel-Based Food Packaging Films

Kunal Singha[1] and Kumar Rohit[2]

[1] *National Institute of Fashion Technology, Department of Textile Design, Salt Lake, Kolkata, West Bengal, 700098, India*
[2] *National Institute of Fashion Technology, Department of Textile Design, Patna, Bihar, 800001, India*

5.1 Introduction

Bio-based materials are being encouraged for use in food packaging in the current era. These biomaterials can readily improve the use of sustainable resources and reduce pollution. There are various biomaterials available in the food industries that can be used as the food packaging materials such as edible films, foams, and hydrogels. Among all these biomaterials, the hydrogel has earned special attention in food packaging industries due to its various advantages such as easy packaging film creation, enhancement of functional aspects like water vapor permeability, and better gas barrier properties. Recently, researchers have invented novel functional properties of hydrocolloids so that they can be used to replace synthetic polymers such as EPS (extended polystyrene). Biodegradable foams are created from a range of low-cost biopolymers such as cassava, cellulose fibers, and sunflower proteins. These hydrogel foams can easily replace EPS. Thus it can be concluded that hydrogel-based foam, film, and edible coating are becoming the popular research agendas in food industries [1].

Disposing of waste plastic packaging materials has become a serious concern worldwide due to their nonbiodegradable nature and that can inflict environmental damage day by day. Therefore, in the food industry, plastic-based food packaging is not a very suitable option. On the other hand, the hydrogel film-based food packaging aided with high water vapor transmission tests and thus allow water and air vapor inside the food packaging. Thus these types of hydrogel-based food packaging techniques are always breathable and ensure the highest level of food quality and safety. The biodegradability property testing is one of the prime characteristics any of hydrogel material. The biodegradability of these hydrogel films was evaluated using a composting environment for up to five weeks. The structural deformation of the hydrogel film over time can be shown by scanning electron microscope (SEM) analysis. The hydrogel film loses a significant amount of weight until four weeks and after which no significant hydrogel film remains in the compost bed [2].

Hydration-rich polymer gels with three-dimensional (3D) macromolecular networks that expand but do not dissolve in water are known as hydrogels. They are also useful in agriculture, such as bio and chemical sensors [2], water purification and metal ion removal [3], oil spill cleanup [4], and DNA separation [4, 5]. Hydrogels can be used in a variety of biomedical applications, including contact lens manufacturing, wound dressing, tissue engineering, and drug administration [6–8].

Hydrogel is also used in modern-day smart or intelligent food packaging [9]. Packaging may be made out of a variety of materials. Both papers and clothing are light, flexible, and recyclable. Glass and metals are widely used for packaging because they are both corrosion-resistant and robust. Polymers, particularly plastics, are in increasing demand as packaging materials. They provide transparency, softness, heat sealability, and a high strength-to-weight ratio, to name a few advantages. They are usually low-cost materials with high tear and tensile strength, as well as superior oxygen and heat barrier properties [9, 10]. On the other hand, the vast majority of plastic food packaging materials are practically nonbiodegradable and waste packaging materials have become a major global environmental concern. Biodegradable polymers have only been employed to a limited degree due to concerns about performance and processing, such as brittleness and thermal distortion temperature [11, 12]. The notion of employing starch as a biodegradable thermoplastic material initially piqued the researchers' interest. Poly-ion complex hydrogels are a unique bio-based polymer production technology for food packaging. Mixing biopolymers with bio- or synthetic polymers with differing structures and focusing on charge interactions rather than hydrogen bonding to increase polymer–polymer interactions. These innovative and revolutionary hydrogels include physical hydrogels, chemical hydrogels, and interpenetrating polymer networks (IPNs). This advanced level of the hydrogel may contain a variety of substrates, including protein chitosan/pectin, gelatin/sodium alginate, soy protein isolate/gelatin, isolate/mesquite gum, starch/cellulose fibers, gelatin/konjac glucomannan, chitosan/gelatin, gelatin/pectin, methylcellulose/whey protein, zein/starch, and so on [9]. They have several appealing features, including ease of usage, nonabrasiveness, heat seal capacity, and high weight tolerance. These hydrogel resources for food packaging are often low-cost materials with great mechanical properties such as tear and inflexibility, as well as excellent oxygen and heat barrier characteristics. On the other hand, the bulk of plastic food packaging materials are nonbiodegradable and waste packaging materials have become a well-known biological problem. The use of biodegradable hydrogel polymers has been restricted because they exhibit several concerns in their presentation and handling, such as sensitivity and temperature bending temperature independently. The recent inventions of superabsorbent polymer (SAP)-based hydrogel make better water and air absorption while making better food packaging and food preservation [10–14].

5.2 Hydrogel Nature, Definition

Hydrogels are 3D networks of polymeric chains with physical or chemical connections randomly interwoven [15, 16]. Due to their 3D network, they can have a linear or branching structure and they are distinguished by their capacity to absorb vast amounts of water or biological fluids, therefore balancing swelling and creating insoluble forms [17].

Moreover, due to the behavior of the polymeric chain bonds and the solvents involved in the system, hydrogels are colloidal gels that disperse in water and exhibit viscoelastic and structural features [18, 19].

5.2.1 Hydrogel Types and Features

Natural and synthetic hydrogels are split into two categories.

5.2.1.1 Classification According to Polymeric Composition

The preparation technique produces several different types of hydrogels. As an example, consider the following:

a) Homopolymeric hydrogels are polymer networks made up of a single monomer species, which is the basic structural unit of any polymer network. Depending on the monomer and polymerization process, homopolymers can have a cross-linked skeletal structure.

b) Two or more monomer species, each with at least one hydrophilic component, are ordered in a random, block, or alternating pattern along the polymer network's chain to form copolymeric hydrogels.

c) Multipolymer two separate cross-linked synthetic and/or natural polymer components are encased in a network structure in the interpenetrating polymeric hydrogel (IPN). A cross-linked polymer makes up one-half of a semi-IPN hydrogel, while a non-cross-linked polymer makes up the other.

5.2.1.2 Classification Based on Configuration: Classification is Done Based on the Setting

The physical structure and chemical composition of hydrogels are used to classify them.

a) An amorphous substance is neither solid nor liquid (noncrystalline).
b) Semicrystalline, a blend of amorphous and crystalline phases.
c) Crystalline.

5.2.1.3 Classification Based on the Type of Cross-Linking

Based on the chemical or physical characteristics of the cross-link junctions, hydrogels can be categorized into two categories. Temporary junctions form in chemically cross-linked networks, whereas temporary junctions form in physical networks due to polymer chain entanglements or physical interactions such as ionic contacts, hydrogen bonds, or hydrophobic interactions.

5.2.1.4 Classification Based on Physical Appearance

The appearance of hydrogels as matrix, film, or microsphere is determined by the polymerization technique used throughout the manufacturing process.

5.2.1.5 Classification According to Network Electrical Charge

Based on the presence or lack of electrical charge on the cross-linked chains, hydrogels can be divided into four groups:

a) Nonionic (neutral).
b) Ionic (including anionic or cationic).
c) Amphoteric electrolyte (ampholytic) containing both acidic and basic groups.
d) Zwitterionic (polybetaines) containing both anionic and cationic groups in each structural repeating unit.

Natural polymers that can be utilized to generate hydrogels include proteins such as collagen and gelatine, as well as polysaccharides such as starch, alginate, and agarose. Traditionally, chemical polymerization processes have been utilized to make synthetic polymers that are employed to make hydrogels [13].

5.3 Preparation of Hydrogel Film

Hydrogel film can be made in a variety of methods. Roy et al. [2] demonstrated a new method for making hydrogel films by combining 0.2% polyvinylpyrrolidone (PVP), 0.8% carboxymethyl cellulose (CMC), 1% polyethylene glycol (PEG), 2% agar, 1% glycerine, and 95% water. In a sealed glass jar, the reaction was carried out under pressure and heat (15 lb and 107 kPa) pressure and 120 °C temperature for 20 minutes. The heated polymer solution was then poured into 80 mm diameter Petri dishes and allowed to cool to ambient temperature (22–25 °C). Finally, a thin circular PVP–CMC hydrogel film was developed.

5.4 Hydrogel as Food Packaging Material

Bio-based hydrogels have recently been used in food packaging. In the food business, however, the application of SAPs as bio-based hydrogels is relatively limited. Bio-based hydrogels and new technologies that require fewer fossil fuels than non-biodegradable petroleum-based polymers are becoming increasingly popular [20–22]. In recent years, the food packaging industry has concentrated on enhancing the biodegradability, swelling properties, and tensile and thermal capabilities of polymers. Hydrogels are polymeric matrices that can absorb 100% of their weight in water or other water-compatible fluids. Superabsorbent hydrogels, on the other hand, are a type of hydrogel made up of polymers that can absorb hundreds of times their dry weight in water [23].

These materials are mostly used in food packaging to keep the inside of the package dry. Water loss caused by physicochemical changes in packaged goods, as well as water vapor penetration due to external situations, are all possible uses in the food preservation business, particularly for fresh items [15]. Hydrogel lowers water activity, which inhibits the formation of mold, yeast, and spoilage bacteria on foods like ready-to-eat meals (freshly cut vegetables) and hygroscopic products (powder food), as well as the softening of dry crispy items (fried potatoes and biscuits) [12, 24].

5.4.1 Hydrogels Functional Properties

Hydrogels' high water absorption property determines their swelling capacity and water retention feature. This is one of their most important functional features for any hydrogel material. These absorption qualities that are provided by hydrophilic groups such as hydroxyl (OH), amine (NH_2), amide ($CONH-$, $CONH_2$), and sulfate are crucial for polymeric hydrogels' potential incorporation into food packaging systems. Hydrogel properties are determined by chemical structure, solvent quality, degree of bonding between polymeric chains, and unique stimuli induced by changes in the external environment [19]. The kind of polymer, degree of crosslinking and particular elaboration process factors such as temperature and pH can all impact the water absorption capacity in this scenario [25].

Jensen [26] claims that superabsorbent hydrogels may absorb huge amounts (>100%) of certain chemicals from the environment (e.g. water) and keep them inside the polymer matrix for a long time, even when mechanical strain is applied. In addition, when polymeric chains are submerged in water, their hydrophilic groups get ionized and increase electrostatic repulsion by causing the polymer to swell more than other polymers (up to 1500% in some cases). The cross-linked network prevents water loss and allows the swelling equilibrium to be maintained [27]. The active osmotic pressure of the polyelectrolytic complex in interaction with aqueous solutions is proportional to the medium's ionic concentration [28]. Thus, the higher osmotic pressure allows the hydrogel packaging to absorb more water and able to preserve the food moisture and quality over a long time.

5.5 Classification of Hydrogel

The classification of the hydrogel and their relevant materials can be classified as shown in Table 5.1.

5.6 Hydrogels Functional Properties

High water absorption is one of the most important functional features of hydrogels, and it is closely related to swelling capacity. These qualities, which are caused by the presence of hydrophilic groups such as hydroxyl (–OH), amine (NH_2), amide (–$CONH-$, –$CONH_2$), and sulfate are extremely important in terms of the possible integration of polymeric hydrogels into food packaging systems (–SO_3H). Hydrogel characteristics can be influenced by a variety of parameters, including chemical

Table 5.1 The feature-based classification of hydrogel materials.

Main features	Classification
Nature of the polymeric matrix	Natural, semi-natural polymer, hybrid or synthetic
Crosslink, biodegradable	Chemical, physical, biodegradable, nonbiodegradable
Polymeric composition	Homopolymers, copolymers, multicomponents
Physical properties	Film, tablets, multiarticulate system
Electrical charge	Neutral/ionic or on-ionic nature
Stimulus responsiveness	Sensitive to pH, enzymes, antigens, and photoradiation

Source: Adapted from Batista et al. [13].

structure, solvent quality, the degree of bonding between polymeric chains, and specific stimuli induced by changes in the external environment [19]. The type of polymer, cross-linking degree, and specific parameters of the elaboration process, such as temperature and pH, can all influence the water absorption capacity [25].

Superabsorbent hydrogels, according to Jensen [26], may absorb high amounts (>100%) of certain compounds from the environment (e.g. water) and hold them inside the polymer matrix for some time, even when mechanical pressure is applied. When the hydrophilic groups of polymeric chains are submerged in water, they become ionized, enhancing electrostatic repulsion and resulting in a higher degree of swelling (>1500% in some situations) than other polymers. The cross-linked network inhibits water loss, allowing the swelling equilibrium to be maintained [27]. The acting osmotic pressure of the polyelectrolytic complex in interaction with aqueous solutions is proportional to the medium's ionic concentration [28]. Ions are tightly connected to the polymer network in the dry state, resulting in a higher osmotic pressure inside the hydrogel matrix. The ions separate themselves during water absorption, lowering osmotic pressure. The swelling process is restricted by the mechanical strength of the polymeric matrix, as the connections between polymeric chains weaken. As a result, the mechanical behavior of hydrogels is determined by the elasticity of the polymer during swelling. The creation of hydrogen and covalent bonds rises as the concentration of hydrophilic groups increases, increasing the polymeric matrix's hydrophilicity. Furthermore, a temperature rise may boost the degree of swelling in some circumstances, owing to increased environmental entropy and improved chain detachment (environmental entropy is defined as the disorder that can be measured in a specific system).

Because of the cross-linking effect on the solubility and spatial arrangement of the molecules, the degree of cross-linking has a direct impact on the thermal, mechanical, and swelling properties of hydrogels in various mediums. Polymer–solvent interactions, which are governed by solubility and 3D conformation, control the swelling process. As a result, the swelling potential is determined by the polymeric matrix's molecular conformation as well as environmental factors such as electrical charge, pH, temperature, and ionic force. Furthermore, the electrical charge affects the retention of absorbed water.

Particle size, drying method, and porosity, which establish the structural array of the 3D network, are further aspects to consider during hydrogel production and performance evaluation.

The swelling properties of hydrogels, particularly their kinetics, are influenced by particle size; hence, the size of hydrogel particles should be appropriate for the specific target application. Swelling happens slowly with particles larger than 80 mesh, and the liquid is maintained in the system for a long period. When employing hydrogels in agriculture, which requires swelling and delayed release to provide proper soil humidity, those sizes may be sufficient [15]. The generated vacant areas permit sufficient gas interchanges after deswelling. Particles smaller than 80 mesh, on the other hand, allow for quick fluid absorption with less retention, owing to the larger surface contact area. The particle size of hydrogels can be used in personal care (hygiene) products to aid in the quick absorption of fluids like blood or urine [25]. When using hydrogels in food packaging systems, similar qualities are required, especially when the major aim is the absorption of biological fluids (exudates) from fresh food products such as meat, chicken, fruits, and vegetables. Mechanical resistance, wettability, degradability, and thermal resistance are all significant aspects to consider when evaluating hydrogel performance.

5.7 Potential Application of Hydrogel in Food Packaging Systems

Food packaging using hydrogels has the potential to be beneficial. Moisture-absorbent food production systems based on "absorbent pads" with water-removal capabilities have lately attracted attention due to their capacity to reduce the risk of microbiological contamination while maintaining the sensory characteristics of packaged meals [29]. This active food packaging can absorb food fluids, modify package headspace, and have antibacterial action.

Hydrogels could also be employed in food packaging systems, where this sort of absorbent material could open up new options for the effective design of food packaging systems with desirable characteristics (such as shelf life, biodegradability, and mechanical resistance). Using hydrogels in food packaging systems, according to Roy et al. [30], is an innovative and promising technology for creating unique, environmentally friendly materials that can improve the shelf life of food commodities. The bulk of hydrogels used in food packaging has a film-like structure. In general, absorbent materials are used in food packaging to catch the exudate released by the food in a plastic tray or container. In this circumstance, the hydrogel in the food container absorbs the exudate while controlling the capacity of the food product to absorb it [29].

The use of absorbent materials in food packaging systems follows four essential requirements.

i) following absorption, the absorbent material should retain the exudate in the 3D structure,

Table 5.2 Potential usages of hydrogel in the food industry.

Objective	Characteristic	Hydrogel composition
Food freshness indicator	Freshness, metabolites production, signaling pH level changes	Poly(N,N-dimethyl acrylamide) copolymer
Stability and retention of volatile matters	Flavor encapsulation	Flavored nanoemulsions
Enhancement of bioactivity	Lipophilic bioactive compounds such as ß-carotene in food matric, to increase the overall bioactivity	Polysaccharides-based hydrogel
Aflatoxin B1 detection	Inspecting the network collapsing nature due to the level changing of the Aflatoxin B1	DNA hydrogel

Source: Adapted from Batista et al. [13].

ii) absorbent materials should maintain the good visual presentation and the sensory attributes of the packaged food, at a low cost,
iii) the absorbent material should present certain performance properties that ensure the structural integrity of the food packaging system during storage, and
iv) the absorbent material should increase the shelf-life of stored food products and avoid microbial growth on the food surface [31].

The usage of the hydrogel as part of an intelligent packaging system or as a carrier system incorporated directly into food matrices are two more potential applications in the food sector.

When employed as part of an intelligent packaging system, their primary function is to offer information about the freshness of the included food goods or to serve as a simple testing tool for contaminants such as aflatoxin. Other innovative hydrogel applications include serving as carriers for tastes or beneficial substances such as -carotenes, which are often integrated with nanoemulsions [32].

The various potential usages of hydrogel in the field of food industry can be classified as mentioned in Table 5.2.

5.7.1 Applications of Hydrogels In Vitro and Food Matrices

Researchers investigated the antibacterial activity of the hydrogel when coupled with additional antimicrobial components such as nanoparticles (silver nanoparticles) or bioactive compounds in vitro and food matrices. Hydrogels have certain basic difficulties, such as low mechanical and water resistance, despite their potential uses in the food sector and a range of other industries. As a result, integrating nanoparticles into the polymeric structure of the hydrogel is one of the greatest ways to overcome these obstacles.

5.7.2 Biodegradable Packaging

Because some hydrogels are made up of bio polysaccharides and are organized with glycosidic connections that can be dissolved by enzyme activity, they are biodegradable [33]. Chitosan is a biopolymer that is commonly utilized in the production of hydrogel matrices. Important connections such as glucosamine-glucosamine, glucosamine-N-acetyl-glucosamine, and N-acetyl-glucosamine-N-acetyl-glucosamine are all targeted by enzyme activity. Lysozyme and bacterial enzymes are thought to enhance chitosan biodegradation. Better plants have chitinases enzymes that operate on N-acetylglucosamine residues in a hydrogel matrix, resulting in higher biodegradability and food perseverance.

5.7.3 Biodegradability

Because they are made up of bio-polysaccharides and are arranged with glycosidic connections that can be broken by enzyme activity, some hydrogels are biodegraded [33]. Chitosan is a biopolymer that is frequently used to make hydrogel matrices. Enzyme activity targets important connections including glucosamine-glucosamine, glucosamine-N-acetyl-glucosamine, and N-acetyl-glucosamine-N-acetyl-glucosamine. Chitosan biodegradation is thought to be aided by lysozyme and bacterial enzymes. Chitinases, enzymes that act on N-acetylglucosamine residues, are found in higher plants (Figure 5.1).

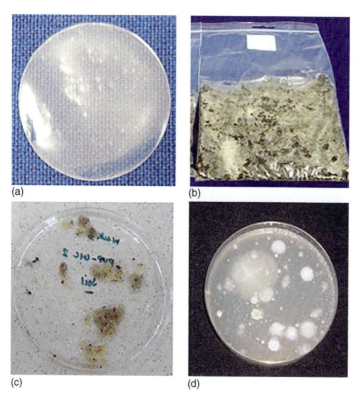

Figure 5.1 Biodegradability of PVP–CMC hydrogel film (a) hydrogel film before degradation, (b) compost bed where the film was buried for degradation, (c) hydrogel film after biodegradation (after four weeks degradation), (d) an image of microbes that are present in compost bed. Source: Roy et al. [2] / with permission from Elsevier.

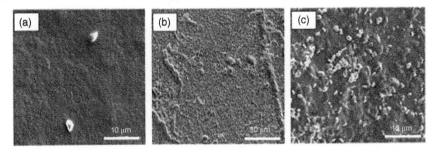

Figure 5.2 SEM micrograph of PVP–CMC hydrogel film. (a) Initial and after (b) two weeks and (c) four weeks of biodegradation. Source: Roy et al. [2] / with permission from Elsevier.

As the SEM images show, the biodegradation of PVP–CMC hydrogel sheets is extremely realistic. Even before the breakdown, the hydrogel surface was almost smooth, as seen in Figure 5.2. After two weeks, abnormalities on the hydrogel film surface may occur, indicating that the PVP–CMC hydrogel film's interior structure has begun to break down. The hydrogel film surface becomes increasingly uneven after four weeks, with occasional depositions (most probable deposition of microorganisms). SEM inspection of sample PVP–CMC hydrogel films before and after degradation was used to determine the biodegradability of the films.

5.7.4 Other Potential Applications in the Food Industry

Hydrogel might be used as part of an intelligent packaging system or as a carrier system directly in food matrices, according to other potential uses in the food business. Other innovative hydrogel applications include serving as flavor transporters or carriers for medicinal substances such as carotenes, which are commonly found in nanoemulsions [32, 34].

5.8 Latest Development in the Hydrogel in the Field of Food Packaging

The characteristics of modified graphene nanomaterials and graphene-based composite hydrogels have been thoroughly investigated. Because it is active and intelligent, this chemical has been employed in a variety of industries [13].

Hydrogels made at the nanoscale are another option for these sorts of systems (nanogels or microgels). In this case, the controlled medication release approach can be used to release active chemicals having biological activity against foodborne pathogens or microorganisms that cause food spoiling. Furthermore, the possible use of hydrogels as antimicrobial carriers in biomedicine generated a foundation of data supporting hydrogels' significant potential as active substance carriers and their likely application in food preservation when combined with antimicrobials. Finally, research in this area will expand the industry's capabilities and interest in developing novel degradable hydrogel-based goods (e.g. active food packaging) [3].

5.9 Futuristic Uses of Hydrogel in Miscellaneous Process

Synthetic hydrogels have a higher mechanical strength and stability than natural hydrogels, as well as a greater range of shape and performance, making them useful in biomedical, biosensing, and tissue engineering applications.

(a) **Unraveling the antimicrobial activity of peptide hydrogel systems: current and future perspectives**: Hydrogels have attracted interest as biomaterials and drug delivery systems for anti-infective applications. For decades, antimicrobial peptides have been hailed as a much-needed new class of antibiotics. Peptide hydrogels that self-assemble and have antibacterial properties have recently gained prominence. On the other hand, their antibacterial properties, selectivity, and mechanism of action remain mostly unknown. This review attempts to establish a link between antimicrobial and efficacy, the self-assembly process, peptide–membrane interactions, and mechanical properties by examining several reported peptide systems, such as b-hairpin/b-loop peptides, multidomain peptides, amphiphilic surfactant-like peptides, and ultrashort/low molecular weight peptides. Their role in the formation of amyloid plaques and the likelihood of an infectious etiology in Alzheimer's disease. A quick look at several cutting-edge gel characterization approaches was examined by researchers. These could be important resources for future research into this rapidly expanding field [35].

(b) **Hydrogel-based 3D bioprinting: a comprehensive review on cell-laden hydrogels, bioink formulations, and future perspectives**: Because they duplicate the physical and biological features of the natural extracellular matrix, hydrogels are vital in cell-laden 3D bioprinting (ECM). Although the ECM's complex milieu differs from the hydrogel's typical static microenvironment, 3D bioprinting has made it easier to deal with the hydrogel system's dynamic modulation and geographic heterogeneity. The hydrogel used in 3D bioprinting is influenced by the printing techniques utilized, such as micro extrusion, inkjet, laser-assisted printing, and stereolithography. Unagolla and Jayasuriya [36] developed a new 3D printable hydrogel approach that allows cells to be contained without compromising their viability. Recent research highlights for the most widely used hydrogel materials are presented in terms of hydrogel system stability, cross-linking process, support cell kinds, and post-printing cell survival. The inclusion of various organic and inorganic substances, as well as the construction of microchannels, are all strategies for increasing the mechanical and biological properties of the hydrogels. This paper also highlights recent advances in vascularized tissue constructs, as well as scaffold-free bioprinting as a potential vascularization technique. 3D bioprinting-based organ-on-chip systems, as well as four-dimensional (4D) bioprinting as a stimuli-responsive fabrication of new organs, are described [36].

(c) **A star-PEG-heparin hydrogel platform to aid cell replacement therapies for neurodegenerative diseases**: Biomolecular composition and mechanical properties of bio-functional matrices for in vivo tissue engineering techniques must be changeable. Freudenberg et al. [37] investigated a modular biohybrid hydrogel system based on covalently cross-linked heparin and star-shaped poly(ethylene glycols) (star-PEG) that allows network characteristics to be gradually changed while heparin concentrations remain constant. The degree of cross-linking of gel components was shown to be proportional to mesh size, swelling, and elastic moduli. Furthermore, subsequent heparin conversion within the biohybrid gels allowed for the covalent attachment of cell adhesion stimulating arginylglycylaspartic acid (RGD) peptides as well as the non-covalent binding of soluble mitogens such as FGF-2 (basic fibroblast growth factor. Mechanical and biomolecular inputs influenced primary nerve cells and neural stem cells, as demonstrated by biohybrid gels. The findings demonstrate how synergistic signaling events interact differently depending on the cell type, as well as biohybrid materials' potential to drive cell fate decisions selectively. These findings lead to important future implications for this chemical in cell replacement therapy for neurodegenerative diseases [37].

(d) **Hydrogel-forming microneedles: current advancements and future trends**: This condensed progress assessment covers the most recent achievements and trends in hydrogel-forming microneedles (HFMs), as well as potential future possibilities. Previously, microneedles (solid, hollow, coated, and dissolving microneedles) were primarily used to improve the efficacy of transdermal drug administration for a number of purposes, such as vaccinations and antibiotic treatment. However, due to the hydrogel's swelling nature, a recent trend in microneedle development has resulted in microneedles made of hydrogels that can deliver transdermal drugs and passively extract interstitial fluid from the skin, suggesting that they could be used for biocompatible minimally invasive monitoring devices. This review summarizes recent advancements in microneedle design, hydrogel compositions, fabrication techniques, HFM applications, and potential future opportunities for employing HFMs for personalized healthcare monitoring and therapy [38–41].

5.10 Conclusions

It has been demonstrated that hydrogel film can be both breathable and biodegradable. The film is also translucent and stretchable. Overall, this hydrogel film looks to be a viable food packaging technology with the potential to relieve some of the issues associated with plastic waste pollution.

Intelligent food packaging is the new way in the food market today, and hydrogel-based biodegradable film provides an amazing opportunity to introduce it. Fresh fruit and vegetable packing is essential for maintaining freshness and shelf life while preventing spoilage.

References

1 Regubalan, B., Pandit, P., Maiti, S. et al. (2018). Potential bio-based edible films, foams, and hydrogels for food packaging. In: *Bio-Based Materials for Food Packaging* (ed. S. Ahmed), 105–123. Singapore: Springer.

2 Roy, N., Saha, N., and Saha, P. (2011). Biodegradable hydrogel film for food packaging. In: *Recent Researches in Geography, Geology, Energy, Environment and Biomedicine*, 329–334.

3 Saha, N., Benlikaya, R., Slobodian, P., and Saha, P. (2015). Breathable and polyol-based hydrogel food packaging. *Journal of Biobased Materials and Bioenergy* 9 (2): 136–144.

4 Mathur, A.M., Moorjani, S.K., and Scranton, A.B. (1996). Methods for the synthesis of hydrogel networks: a review. *Journal of Macromolecular Science, Part C: Polymer Reviews* 36 (2): 405–430.

5 Deligkaris, K., Tadele, T.S., Olthuis, W., and van den Berg, A. (2010). Hydrogel-based devices for biomedical applications. *Sensors and Actuators B: Chemical* 147 (2): 765–774.

6 Carvalho, H.W.P., Batista, A.P., Hammer, P. et al. (2010). Removal of metal ions from aqueous solution by the chelating polymeric hydrogel. *Environmental Chemistry Letters* 8 (4): 343–348.

7 Shin, J., Braun, P.V., and Lee, W. (2010). Fast response photonic crystal pH sensor based on templated photo-polymerized hydrogel inverse opal. *Sensors and Actuators B: Chemical* 150 (1): 183–190.

8 Zheng, Y., Hua, S., and Wang, A. (2010). Adsorption behavior of Cu^{2+} from aqueous solutions onto starch-g-poly(acrylic acid)/sodium humate hydrogels. *Desalination* 263 (1–3): 170–175.

9 Farris, S., Schaich, K.M., Liu, L. et al. (2009). Development of polyion-complex hydrogels as an alternative approach for the production of bio-based polymers for food packaging applications: a review. *Trends in Food Science & Technology* 20 (8): 316–332.

10 Mahalik, N.P. and Nambiar, A.N. (2010). Trends in food packaging and manufacturing systems and technology. *Trends in Food Science & Technology* 21 (3): 117–128.

11 Aider, M. (2010). Chitosan application for active bio-based film production and potential in the food industry. *LWT – Food Science and Technology* 43 (6): 837–842.

12 De Azeredo, H.M. (2009). Nanocomposites for food packaging applications. *Food Research International* 42 (9): 1240–1253.

13 Batista, R.A., Espitia, P.J.P., Quintans, J.D.S.S. et al. (2019). Hydrogel is an alternative structure for food packaging systems. *Carbohydrate Polymers* 205: 106–116.

14 Farris, S., Schaich, K.M., Liu, L. et al. (2011). Gelatin–pectin composite films from polyion-complex hydrogels. *Food Hydrocolloids* 25 (1): 61–70.

15 Ahmed, E.M. (2015). Hydrogel: preparation, characterization, and applications: a review. *Journal of Advanced Research* 6 (2): 105–121.

16 Chang, C., Duan, B., Cai, J., and Zhang, L. (2010). Superabsorbent hydrogels based on cellulose for smart swelling and controllable delivery. *European Polymer Journal* 46 (1): 92–100.

17 Hebeish, A., Hashem, M., Abd El-Hady, M.M., and Sharaf, S. (2013). Development of CMC hydrogels loaded with silver nanoparticles for medical applications. *Carbohydrate Polymers* 92 (1): 407–413.

18 Ahmed, E.M., Aggor, F.S., Awad, A.M., and El-Aref, A.T. (2013). An innovative method for preparation of nanometal hydroxide superabsorbent hydrogel. *Carbohydrate Polymers* 91 (2): 693–698.

19 Laftah, W.A., Hashim, S., and Ibrahim, A.N. (2011). Polymer hydrogels: a review. *Polymer-Plastics Technology and Engineering* 50 (14): 1475–1486.

20 Rahman, M.M., Netravali, A.N., Tiimob, B.J. et al. (2016). Bio-inspired "green" nanocomposite using hydroxyapatite synthesized from eggshell waste and soy protein. *Journal of Applied Polymer Science* 133 (22): 43477.

21 Rhim, J.W., Park, H.M., and Ha, C.S. (2013). Bio-nanocomposites for food packaging applications. *Progress in Polymer Science* 38 (10–11): 1629–1652.

22 Santana, J.S., do Rosário, J.M., Pola, C.C. et al. (2017). Cassava starch-based nanocomposites reinforced with cellulose nanofibers extracted from sisal. *Journal of Applied Polymer Science* 134 (12): 44637.

23 Feng, E., Ma, G., Wu, Y. et al. (2014). Preparation and properties of organic-inorganic composite superabsorbent based on xanthan gum and loess. *Carbohydrate Polymers* 111: 463–468.

24 de Azeredo, H.M. (2013). Antimicrobial nanostructures in food packaging. *Trends in Food Science & Technology* 30 (1): 56–69.

25 Kabiri, K., Omidian, H., Zohuriaan-Mehr, M.J., and Doroudiani, S. (2011). Superabsorbent hydrogel composites and nanocomposites: a review. *Polymer Composites* 32 (2): 277–289.

26 Jensen, O.M. (2011). Water absorption of superabsorbent polymers in a cementitious environment. In: *International Conference on Advances in Construction Materials through Science and Engineering 2011*. Rilem Publications.

27 Bhattacharya, S.S., Shukla, S., Banerjee, S. et al. (2013). Tailored IPN hydrogel bead of sodium carboxymethyl cellulose and sodium carboxymethyl xanthan gum for controlled delivery of diclofenac sodium. *Polymer-Plastics Technology and Engineering* 52 (8): 795–805.

28 Friedrich, S. (2012). Superabsorbent polymers (SAP). In: *Application of Super Absorbent Polymers (SAP) in Concrete Construction. RILEM State of the Art Reports*, vol. 2 (ed. V. Mechtcherine and H.W. Reinhardt). Dordrecht: Springer https://doi.org/10.1007/978-94-007-2733-5_3.

29 Otoni, C.G., Espitia, P.J., Avena-Bustillos, R.J., and McHugh, T.H. (2016). Trends in antimicrobial food packaging systems: emitting sachets and absorbent pads. *Food Research International* 83: 60–73.

30 Roy, N., Saha, N., Kitano, T., and Saha, P. (2012). Biodegradation of PVP–CMC hydrogel film: a useful food packaging material. *Carbohydrate Polymers* 89 (2): 346–353.

31 Fernández, A., Soriano, E., López-Carballo, G. et al. (2009). Preservation of aseptic conditions in absorbent pads by using silver nanotechnology. *Food Research International* 42 (8): 1105–1112.

32 Park, S., Mun, S., and Kim, Y.R. (2018). Effect of xanthan gum on lipid digestion and bioaccessibility of β-carotene-loaded rice starch-based filled hydrogels. *Food Research International* 105: 440–445.

33 Croisier, F. and Jérôme, C. (2013). Chitosan-based biomaterials for tissue engineering. *European Polymer Journal* 49 (4): 780–792.

34 Kwan, A. and Davidov-Pardo, G. (2018). Controlled release of flavor oil nanoemulsions encapsulated in filled soluble hydrogels. *Food Chemistry* 250: 46–53.

35 Cross, E., Coulter, S.M., Pentlavalli, S., and Laverty, G. (2021). Unraveling the antimicrobial activity of peptide hydrogel systems: current and future perspectives. *Soft Matter* 17 (35): 8001–8021.

36 Unagolla, J.M. and Jayasuriya, A.C. (2018). Drug transport mechanisms and in vitro release kinetics of vancomycin encapsulated chitosan-alginate polyelectrolyte microparticles as a controlled drug delivery system. *European Journal of Pharmaceutical Sciences* 114: 199–209.

37 Freudenberg, U., Hermann, A., Welzel, P.B. et al. (2009). A star-PEG–heparin hydrogel platform to aid cell replacement therapies for neurodegenerative diseases. *Biomaterials* 30 (28): 5049–5060.

38 Bandiwadekar, A., Jose, J., Khayatkashani, M. et al. (2021). Emerging novel approaches for the enhanced delivery of natural products for the management of neurodegenerative diseases. *Journal of Molecular Neuroscience* 72: 653–676.

39 Fang, G., Yang, X., Wang, Q. et al. (2021). Hydrogels-based ophthalmic drug delivery systems for treatment of ocular diseases. *Materials Science and Engineering: C* 127: 112212.

40 Nguyen, T.T., Nguyen, T.T.D., and Van Vo, G. (2022). Advances of microneedles in hormone delivery. *Biomedicine & Pharmacotherapy* 145: 112393.

41 Turner, J.G., White, L.R., Estrela, P., and Leese, H.S. (2021). Hydrogel-forming microneedles: current advancements and future trends. *Macromolecular Bioscience* 21 (2): 2000307.

6

Natural Fiber-Based Food Packaging Films

G. Rajeshkumar, M. Karthick, A.K. Aseel Ahmed, T. Vikram Raj, V. Abinaya, K. Madhu Mitha, and R. Ronia Richelle

PSG Institute of Technology and Applied Research, Department of Mechanical Engineering, Neelambur, Coimbatore, Tamil Nadu, India

6.1 Introduction

Sustainable products are becoming essential in the twenty-first century. Major use of plastics slowly destroys the environment. Packaging of food is an area where plastic films and bottles are used abundantly [1]. Proper food packaging protects the food product within it against undesirable environmental conditions such as moisture, microorganisms, rancidity, and many others [2, 3]. Biodegradable films pose a lot of disadvantages due to their very own function of biodegradability. These films degrade in a short time, have poor barrier properties, and are expensive to manufacture. Hence, many of them are composites or blends of various synthetic or natural additives [4]. The lifecycle of films is depicted in Figure 6.1. Natural fiber is one attractive reinforcement due to its environment-friendly properties and acts as a load-bearing element [6, 7]. Natural fiber is widely used in automotive industry; unlike in food packaging industry, researchers and engineers are slowly moving toward natural fiber composite materials [8, 9]. Natural fibers are essentially classified into plant-based fiber, vegetable fiber, animal fiber, and mineral fiber based on their sources; Figure 6.2 depicts the classification [11, 12]. Natural fibers have excellent environmental properties, but the water properties of natural fiber are not excellent compared to polymers [13, 14]. So the additional reinforcement is added to increase the desirable properties. However, chemical treatment of natural fibers will also increase their properties significantly [15–17]. In this chapter, we discuss the important properties of some widely used fibers that are reinforced with matrix and are used in food packaging applications. This will give researchers and engineers an overview of the usage of natural fiber in food packaging applications.

Natural Materials for Food Packaging Application, First Edition.
Edited by Jyotishkumar Parameswaranpillai, Aswathy Jayakumar,
E. K. Radhakrishnan, Suchart Siengchin, and Sabarish Radoor.
© 2023 WILEY-VCH GmbH. Published 2023 by WILEY-VCH GmbH.

6 Natural Fiber-Based Food Packaging Films

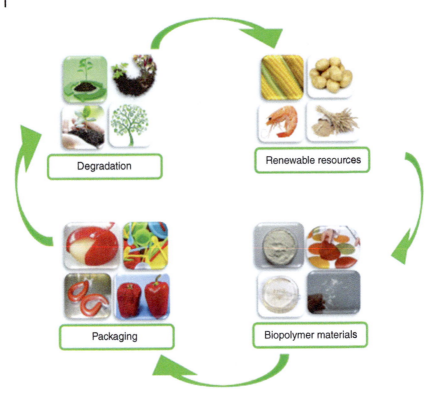

Figure 6.1 Lifecycle of biofilms. Source: Reproduced with permission from Popović et al. [5] / Reproduced with permission from Elsevier, License Number: 5352360147830.

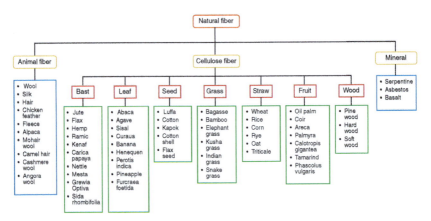

Figure 6.2 Natural fibers classification. Source: Reproduced with thanks from Elsevier, License Number: 5352420972366 [10].

6.2 Manufacturing of Fiber-Reinforced Biofilms

Manufacturing of biofilms is a tedious process. There are numerous manufacturing methods (Figure 6.3), but solution casting (Figure 6.4) is most commonly used to create biofilms with stirring and curing processes [19]. However, other methods such as electrospinning, melt mixing, and extrusion are also used [20–24]. Electronic spinning is one of the upcoming methods for manufacturing of polymers. The basic idea behind electronic spinning is melting the material and injecting it into a plate by the spinning method (Figure 6.5) [26]. It is efficient for manufacturing films for packaging meat [27]. Extrusion (Figure 6.6) is also used for manufacturing biofilms,

Figure 6.3 Manufacturing types of natural fiber.

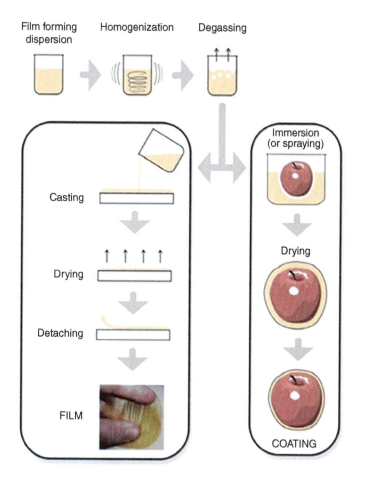

Figure 6.4 Solution casting process. Source: Reproduced with permission from Otoni et al. [18] / Reproduced with permission from John Wiley & Sons.

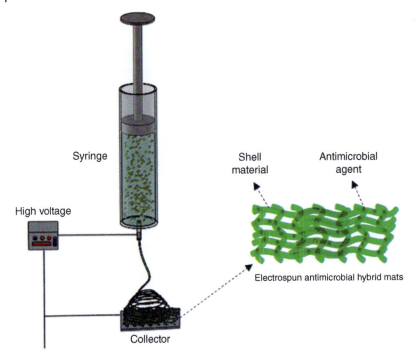

Figure 6.5 Fiber spinning. Source: Reproduced with thanks from Elsevier, License Number: 5352361225788 [25].

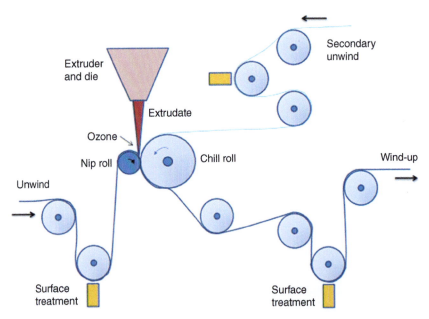

Figure 6.6 Extrusion coating and lamination. Source: Reproduced with thanks from Elsevier, License Number: 5352361412947 [28].

and also the cooling of the extrusion is also done by air, water and also the extrusion process is followed by injection molding in some cases [29, 30].

6.3 Rice Straw-Based Films

We have seen a great development and innovation in sustainable bio-composite materials, for instance, rice straw (RS), which is a good source of bioactive compounds. It is primarily utilized in making of food packaging films owing to its biodegradability, renewability, and nonpetroleum-based source. Another main advantage is low cost and easy procurability. The extracts of rice straw have shown a good antioxidant and antimicrobial properties which is good for making food packaging. Properties of rice straw-based films are listed in Table 6.1. Perumal et al. [31] present a study on bio-nanocomposite film with polyvinyl alcohol (PVA) and chitosan (CS) reinforced with cellulose nanocrystals (CNCs). Since PVA is nontoxic, chemically stable, and environmentally friendly, blending it with chitosan (CS) is essential as it further strengthens the mechanical and barrier properties. Chitosan is obtained from crab/shrimp shells, which are biocompatible, biodegradable, and are linear polysaccharide, and CNC is segregated from rice straw; the final films are prepared by solution casting method. Here, various concentrations of CNC were incorporated with PVA/CS to get optimized film for desired application. Increase in the percentage of CNC has effectively improved the thermal stability of the film. Elhussieny et al. [32] did a similar study on chitosan reinforcement with rice straw (RS) fiber and nano-rice straw (NRS). Yield strength, elastic modulus, and thermal degradation temperature of the chitosan have been enhanced. When comparing both of the above studies, in the latter one, only chitosan was employed as the matrix. For that, reinforcement content (wt%) is between 25% and 35% for both RS and NRS. Since there is no PVA involved in the study by Elhussieny et al. its tensile strength and elastic modulus values are lower when compared to the values obtained by A.B. Perumal et al. In another study conducted by Menzel et al. [33] starch films obtained from potatoes are used as the matrix. Furthermore, to provide more antimicrobial and antioxidant properties, it is blended with rice straw, which also contains phenolic compounds. Plasticizers such as glycerol are added to make the films more flexible. Melt blending and compression molding are used to manufacture films with antioxidant properties. Final resulted films showed that they are transparent with a slight red–brownish color and due to the presence of phenolic compounds, they lower the affinity for water. We can notice from the table that polymer matrices involving chitosan have better mechanical properties than starch-based polymer matrix.

6.4 Wheat Straw-Based Films

To obtain the optimized film based on polyhydroxyl-3-butyrate-*co*-3-valerate (PHBV) with wheat straw fibers (WSFs), PHBV is an encouraging polymer matrix which is made from food industry by-products; it is fully biodegradable and bio-sourced in nature.

Table 6.1 Properties of rice straw-based films.

Matrix	Reinforcement	Reinforcement content	Tensile strength (MPa)	Elastic modulus (MPa)	Thermal degradation temperature via TGA (T_{peak} °C)	Water vapor permeability (WVP) and oxygen permeability (OP)	References
Polyvinyl alcohol (PVA)/ chitosan	Cellulose nanocrystals (CNCs) from rice straw	5%	98.15	1764	367	—	[31]
Chitosan	Rice straw and nano-rice straw	25%	24	1200	252	—	[32]
Starch	Antioxidant extract (AOE) from rice straw and glycerol	3 g	11.4	374	269	WVP = 12 g mm/h m² kPa OP = 7×10¹⁴ cm³/m s Pa	[33]

Table 6.2 Properties of wheat straw-based films.

Matrix	Elongation break (%)	Stresses at break (MPa)	Elastic modulus (GPa)	Water vapor transmission rate (WVTR) (gm^{-2}/day)	Water vapor permeability (gs^{-1}/Pa m)	References
PHBV with WSF and PEG400	1.0	9	2.5	—	—	[34]
PHBV with wheat straw fibers (WSFs)	1.1	16.3	2.6	32.2	$5.2 \pm 0.4 \times 10^{-11}$	[35]
PE/PP/ NAOH treated wheat straw	119.98	—	—	51.89	—	[36]

Martino et al. [34] learned that by introducing the WSFs in various amounts into a polymer matrix led to better water vapor permeability (WVP) and could accomplish the needs of respiring food products. Hydrophobic plasticizers such as polyethylene glycol (PEG) and glycerol triacetate (GTA) were used to make the polymer matrices more stable. Solvent casting method is used to blend films at laboratory scale. Results prove that external plasticization led to improved flexibility of PHBV. Incorporation of WSF into the matrix led to an increase in elastic modulus but a decrease in both the tensile strength and elongation break. Similarly, Berthet et al. [35] also used wheat straw as a potential filler in PHBV for food packaging applications. To attain the desired property of films, WSF is used as reinforcement for PHBV. Both the mechanical and barrier properties are improved. Dixit et al. [36] made a composite film out of PE, PP, and wheat straw which was treated with sodium hydroxide (NaOH). PE and PP provide a fine mechanical property but it is not bio-sourced, so to make them biodegradable, wheat straw has been blended with the polymer matrix. The response surface methodology (RSM) process was used to create an optimal film. Mathematical and statistical data provide the improved condition by analyzing the factors affecting the food packaging film. The main purpose of RSM is to maximize the response by optimizing the process parameters. It can be seen from Table 6.2 that due to the presence of PE and PP in the matrix system, its elongation at break is higher when compared to other two cases.

6.5 Jute-Based Films

Jute is one of the abundantly available fibers and it has excellent environmental, damping properties [37]. Burrola-Núñez et al. [38] prepared a biocomposite film by reinforcing modified cut jute fibers (CJFs) in a PLA matrix by extrusion and investigated its mechanical, thermal, and water absorption properties. These properties are investigated by reinforcing with 5% untreated CJF, 10% untreated CJF, 5% and 10% NaOH treated CJF, 5% and 10% maleic anhydride functionalized CJF, 5% and 10% Gamma irradiated CJF at 10 kgy, 5% and 10% Gamma-irradiated CJF at 50 kgy with PLA matrix. The results revealed that reinforcement of 10 wt%. Gamma-irradiated CJF at 50 kgy with PLA has better mechanical properties. CJF treated with maleic anhydride, which improves thermal stability. Water absorption properties of neat PLA are better when compared to PLA reinforced with treated and untreated CJFs. The increase in mechanical properties is due to an increase in the interaction between materials [38].

Ray et al. [39] prepared a composite film by reinforcing 5% NaOH-treated jute fiber at 30 °C for a short duration of eight hours in vinylester resin and investigated its mechanical properties. These properties are investigated by reinforcing vinylester resin with 8–35 wt% of untreated and 5% of NaOH-treated jute fibers. Higher percentage of jute fiber gives better mechanical properties due to the bonding between fiber and matrix. Kumar et al. [40] made a film by reinforcing cellulose nanofibers (CNFs) in a PLA matrix, and then they tested its optical, mechanical, barrier, and thermal properties. These properties are investigated by reinforcing PLA matrix with 1, 3, and 5 wt% of CNFs. The results revealed that reinforcement of 5 wt% CNFs with PLA has better mechanical, water, thermal, and optical properties due to a change in the crystallinity of the matrix. Kumar et al. [41] prepared a film by reinforcing CNFs in Chitosan/PVP (CHP) by solvent casting method and investigated its morphology, thermal, optical, mechanical, barrier (water and oxygen transmission rate, contact angle), biodegradable, and crystallinity properties. These properties are investigated by reinforcing 1, 3, and 5 wt% of CNFs with CHP films. The results revealed that reinforcement of 5 wt% CNFs with CHP has better morphological, crystallinity, optical, barrier, mechanical, biodegradability, and thermal properties due to higher crystallinity of CNFs and dense composite structure. Baheti et al. [42] prepared a bionanocomposite film by reinforcing CNFs extracted from waste jute fibers by wet pulverization in PLA films and investigated its morphology, thermal, thermomechanical, and tensile properties. The results revealed that reinforcement of 3 wt% of treated wet-milled jute nanofibrils with PLA has better thermomechanical, thermal, and mechanical properties due to better bonding and structure of the fiber.

6.6 Pineapple-Based Films

The major producers of pineapple fibers are Asian countries and their major advantage is low cost [43]. Nikmatin et al. [44] produced a bionanocomposite film by using pineapple leaf fibers (PLFs) and tapioca starch as matrix and investigated its mechanical and optical properties. These properties are investigated by reinforcing 3%, 4%, and 5% of nanosized PLFs as fillers in tapioca starch with a glycerol plasticizer matrix. The results revealed that the reinforcement of 5% PLF with tapioca starch glycerol plasticizer matrix has better mechanical (tensile strength and elongation) and optical properties (smaller absorbance) where the bonding is better than other contents. Wahyuningsih et al. [45] produced a nanocomposite film by reinforcing nanocellulose fibrils from PLFs in a PVA matrix by the casting method and investigated its physical (moisture content, density), thermal, and barrier properties (water vapor transmission rate). These properties are investigated by reinforcing 10% nanocellulose in 90% PVA, 20% nanocellulose in 80% PVA, 30% nanocellulose in 70% PVA, 40% nanocellulose in 60% PVA, 50% nanocellulose in 50% PVA, 10% nanocellulose in 89% PVA and 1% glycerol, 20% nanocellulose in 79% PVA and 1% glycerol plasticizer, 30% nanocellulose in 69% PVA and 1% glycerol, 40% nanocellulose in 59% PVA and 1% glycerol, 50% nanocellulose in 49% PVA and 1% glycerol. The results revealed that PVA + nanocellulose films have better thermal, physical, and barrier properties. Kumar et al. [46] produced a biocomposite film by reinforcing pineapple peel extract (PPE) in polyvinyl alcohol (PVOH) and corn starch (ST) by the casting method and investigated its thermal, mechanical, barrier, and antioxidant properties. These properties are investigated by reinforcing 5%, 10%, 15%, 20% PPE in PVA and corn starch films. PVOH/ST reinforced with 5% PPE has higher elongation at break, thermal stability, and water vapor permeability. PVOH/ST reinforced with 20% PPE has the highest antioxidant activities due to the formation of hydrogen bonds between PPE and the PVOH/ST matrix. Mahardika et al. [47] produced a bionanocomposite film by reinforcing CNFs from pineapple leaf in bengkoang starch films by ultrasonication and investigated its thermal, mechanical, and barrier properties. These properties are investigated by reinforcing 0.5, 1, 1.5, and 2 wt% of CNFs in bengkaong starch (BS) films. The results revealed that the reinforcement of 2 wt% of CNFs with bengkoang starch (BS) films has better tensile properties, barrier properties, and thermal properties. Ninjiaranai [48] produced a biocomposite film by reinforcing PLF in Chitosan and Polyethylene glycol 6000 film (Chitosan/PEG 6000) and investigated its barrier and mechanical properties. Chitosan PEG 6000 film reinforced with PLFs has better barrier (less WVP) and mechanical (higher tensile strength, higher solubility) properties due to an increase in intermolecular hydrogen bonding between chitosan and the hydroxyl groups in the fiber.

6.7 Flax-Based Films

Flax is one of the most widely used natural fibers in marine, aerospace, sports, etc. [49]. Cao et al. [50] manufactured starch reinforced with flax cellulose nanocrystals (FCNs) films. Glass transition temperature increases due to FCNs restricting the mobility of starch chains. On the one hand, the tensile strength increases and Young's modulus also increases, while the elongation at break decreases. At 50% addition, the suppression of swelling occurs due to greater bonding, which leads to a decrease in water absorption. Flaxseed oil on soy protein isolate (SPI)-based emulsion-type films was manufactured by Hopkins et al. [51]. At 10% oil concentration, the tensile strength of the film increases and due to the size distribution of droplets, there is a decline in

Table 6.3 Properties of flax-based films.

Matrix reinforcement	Tensile strength (MPa)	Young's modulus (MPa)	Water vapor permeability (WVP)	References
Starch with flax cellulose nanocrystals (FCN)	11.9 ± 0.8	498.2 ± 23.4	—	[50]
Soy proties isolate with flax seed oil	5.35	—	1.950 g mm/m^2/h kPa	[51]
Flax seed gum with carvacrol	32.29 ± 1.33	—	1.10 ± 0.21 g H$_2$O/m^2/s MPa	[52, 54]
Chitosan with false flax	12.47 ± 0.20	492.09 ± 11.16	—	[53]

the tensile properties, whereas the puncture properties are not affected. The water vapor permeability was similar to the tensile strength data. The moisture content of the films decreased to 10% of oil concentration due to optimal droplet size [51].

Flaxseed gum (FG) films reinforced with carvacrol (C) by Fang et al. [52] by the film-casting method with sonication. The tensile strength of FG-C films (untreated films) decreases gradually when carvacrol percentage increases. On the other hand, the elongation at break increases by a slight margin. However, FG-C films with sonication slightly increase both tensile strength and elongation at breaks. FG-0.5C-6030 showed the highest tensile strength and elongation at break, due to an increase in film matrix bonding. The content angle increased with the increase in carvacrol content in FG film. FG-C films with sonication also increase the contact angle and the sonication effect has a good impact on the films' optical properties. False flax seed oil is reinforced in chitosan films by Gursoy et al. [53]. The water solubility of film with 2% seed oil was determined to have lower water solubility than other films. The presence of seed oil decreases the transmittance due to the bonding of materials. CS had the highest tensile strength and Young's modulus values compared to other films. The elongation at break for CSO-2 is the highest among all the films, which is 23.04 ± 3.55%. Table 6.3 depicts the optimum value of the films reinforced with flax-based materials.

6.8 Kenaf-Based Films

Single- and double-stage chemically treated kenaf fiber with PVA film developed by Ali et al. [55]. The PVA/kenaf fiber composite was manufactured by the casting method. The PVA film reinforced with chemically treated kenaf fiber has better tensile strength than the PVA film with untreated kenaf fiber due to the interfacial bonding of the treated fiber. The flexural modulus of treated fiber has greater properties than the untreated due to the same phenomena of mechanical properties. Zarina et al. [56] developed a kappa carrageenan film with kenaf's derivative CNC by solution casting. The increase in the percentage of CNCs increases the mechanical properties to an optimum level. Mechanical properties improved with the addition of CNC. It also improved the thermal properties of the material due to the uniform formation of additives. Oyekanmi et al. [57] developed Macroalgae Biopolymer with kenaf as reinforcement. The mechanical tensile strength increases compared to the neat film due to the adhesive interaction between the matrix and reinforcement. Due to the formation of bonds between CNFs and matrix polymer, the overall thermal stability of the film increases. Research based on kenaf fiber as reinforcement in corn starch (CS) film was prepared by the casting method. The tensile strength increased as the fiber loading increased (0–0.6 wt%). Similarly, Young's modulus increases with an increase in fiber content. Increasing the kenaf fiber percentage resulted in higher water solubility, but further increasing decreased water solubility. At 6% kenaf, the bonding was better than other percentiles, so the properties were better at that film, so CS/K 6% was considered as the best film which has the best bonding [58].

6.9 Hemp-Based Films

Zhang et al. [59] found a way to manufacture lignin-cellulose nanocrystals (L-CNCs) from Hemp hurd. Native lignin obtained by the ball milling method was added as a filler to the PVA matrix. The PVA is mixed with lignin of 16–24 wt%. The composite was tried with hydrolysis using sulfuric acid. The addition of hemp particles increases Young's modulus and the water vapor permeability of the film compared to the pure PVA due to increased bonding between PVA and lignin. According to the given findings in the paper, the water vapor permeability increases proportionally to the increase in weight content of hemp due to the stoppage of water molecules by lignin. Viscusi et al. [60] have manufactured natural fiber-based films with hemp and grapefruit oil with rice flour. Grapefruit oil and hemp powder were blended to prepare the film. The film consisting of rice flour and 20 wt% of hemp and grape oil shows an increase in mechanical properties and shows an increase in contact angle to the value of 98.21 ± 2.40, thereby reducing water permeability by about 99%. Hydrophobic character of the additives is one of the important reasons for these kinds of changes. Alkali-treated hemp fiber is blended with polyethylene (PE) and polypropylene (PP) films by surface methodology. The tensile properties, contact angle, and transparency of the film are increased due to interface bonding between the hemp and the matrix, proving that treated hemp fiber-based composite films are a more sustainable option for food packaging [61].

6.10 Conclusions

Food packaging films with natural fiber as reinforcement are more sustainable compared to other films. The usage of fiber increases the mechanical properties to a certain point. The fiber also influences the thermal and optical properties of the film in positive and negative ways. However, the chemically treated fiber has greater results compared to the untreated fiber. For further increases in the optical, water, and thermal properties desired, nanoparticles can be used. With the good environmental properties, the addition of good mechanical, water, and thermal properties can promote the usage of natural fiber in food packaging applications. However, further study on bacterial activity and shelf life of different food materials should be discussed in the near future.

References

1 Taylor, P. and Russell, D.A.M. (2014). Sustainable (food) packaging – an overview. *Food Additives & Contaminants: Part A* 396–401. https://doi.org/10.1080/19440049.2013.856521.
2 Gaikwad, K.K., Singh, S., and Lee, Y.S. (2018). Oxygen scavenging films in food packaging. *Environmental Chemistry Letters* 16 (2): 523–538. https://doi.org/10.1007/s10311-018-0705-z.
3 Gil, M.I., Selma, M.V., Suslow, T. et al. (2015). Pre- and postharvest preventive measures and intervention strategies to control microbial food safety hazards of fresh leafy vegetables. *Critical Reviews in Food Science and Nutrition* 55 (4): 453–468. https://doi.org/10.1080/10408398.2012.657808.
4 Siracusa, V., Rocculi, P., Romani, S., and Rosa, M.D. (2008). Biodegradable polymers for food packaging: a review. *Trends in Food Science and Technology* 19 (12): 634–643. https://doi.org/10.1016/j.tifs.2008.07.003.
5 Popović, S.Z., Lazić, V.L., Hromiš, N.M. et al. (2018). Chapter 8 – Biopolymer packaging materials for food shelf-life prolongation. In: *Biopolymers for Food Design* (ed. A.M. Grumezescu and A.M. Holban), 223–277. Academic Press https://doi.org/10.1016/B978-0-12-811449-0.00008-6.
6 Taj, S., Munawar, M.A., and Khan, S. (2007). Natural fiber-reinforced polymer composites. *Pakistan Academy of Sciences* 2007: 129–144.
7 Mitra, B.C. (2014). Environment friendly composite materials: biocomposites and green composites. *Defence Science Journal* 64 (3): 244–261. https://doi.org/10.14429/dsj.64.7323.
8 Faruk, O., Bledzki, A.K., Fink, H.P., and Sain, M. (2014). Progress report on natural fiber reinforced composites. *Macromolecular Materials and Engineering* 299 (1): 9–26. https://doi.org/10.1002/mame.201300008.
9 Rajeshkumar, G., Devnani, G.L., Maran, J.P. et al. (2021). Characterization of novel natural cellulosic fibers from purple bauhinia for potential reinforcement in polymer composites. *Cellulose* 28 (9): 5373–5385. https://doi.org/10.1007/s10570-021-03919-2.
10 Rajeshkumar, G., Seshadri, S.A., Devnani, G.L. et al. (2021). Environment friendly, renewable and sustainable poly lactic acid (PLA) based natural fiber reinforced composites – a comprehensive review. *Journal of Cleaner Production* 310: https://doi.org/10.1016/j.jclepro.2021.127483.
11 Kozłowski, R.M. and Mackiewicz-Talarczyk, M. (2012). *Introduction to Natural Textile Fibres*. Woodhead Publishing Limited https://doi.org/10.1533/9780857095503.1.

12 Rajeshkumar, G. (2021). Characterization of surface modified *Phoenix* sp. fibers for composite reinforcement. *Journal of Natural Fibers* 18 (12): 2033–2044. https://doi.org/10.1080/15440478.2019.1711284.

13 Khalid, M.Y., Al Rashid, A., Arif, Z.U. et al. (2021). Natural fiber reinforced composites: sustainable materials for emerging applications. *Results in Engineering* 11: 100263. https://doi.org/10.1016/j.rineng.2021.100263.

14 Zhong, Y., Godwin, P., Jin, Y., and Xiao, H. (2020). Biodegradable polymers and green-based antimicrobial packaging materials: a mini-review. *Advanced Industrial and Engineering Polymer Research* 3 (1): 27–35. https://doi.org/10.1016/j.aiepr.2019.11.002.

15 Li, X., Tabil, L.G., and Panigrahi, S. (2007). Chemical treatments of natural fiber for use in natural fiber-reinforced composites: a review. *Journal of Polymers and the Environment* 15 (1): 25–33. https://doi.org/10.1007/s10924-006-0042-3.

16 Kabir, M.M., Wang, H., Lau, K.T., and Cardona, F. (2012). Chemical treatments on plant-based natural fibre reinforced polymer composites: an overview. *Composites. Part B, Engineering* 43 (7): 2883–2892. https://doi.org/10.1016/j.compositesb.2012.04.053.

17 Rajeshkumar, G., Hariharan, V., Devnani, G.L. et al. (2021). Cellulose fiber from date palm petioles as potential reinforcement for polymer composites: physicochemical and structural properties. *Polymer Composites* 1–11. https://doi.org/10.1002/pc.26106.

18 Otoni, C.G., Avena-Bustillos, R., De Azeredo, H.M.C. et al. (2017). Recent advances on edible films based on fruits and vegetables – a review. *Comprehensive Reviews in Food Science and Food Safety* 16 (5): 1151–1169. https://doi.org/10.1111/1541-4337.12281.

19 Rathinavel, S. and Saravanakumar, S.S. (2021). Development and analysis of poly vinyl alcohol/orange peel powder biocomposite films. *Journal of Natural Fibers* 18 (12): 2045–2054. https://doi.org/10.1080/15440478.2019.1711285.

20 Tomé, L.C., Fernandes, S.C.M., Sadocco, P. et al. (2012). Antibacterial thermoplastic starch-chitosan based materials prepared by melt-mixing. *BioResources* 7 (3): 3398–3409.

21 Guimarães, M., Botaro, V.R., Novack, K.M. et al. (2015). Starch/PVA-based nanocomposites reinforced with bamboo nanofibrils. *Industrial Crops and Products* 70: 72–83. https://doi.org/10.1016/j.indcrop.2015.03.014.

22 Van Hai, L., Choi, E.S., Zhai, L. et al. (2020). Green nanocomposite made with chitin and bamboo nanofibers and its mechanical, thermal and biodegradable properties for food packaging. *International Journal of Biological Macromolecules* 144: 491–499. https://doi.org/10.1016/j.ijbiomac.2019.12.124.

23 Amjadi, S., Almasi, H., Ghorbani, M., and Ramazani, S. (2020). Reinforced ZnONPs/rosemary essential oil-incorporated zein electrospun nanofibers by κ-carrageenan. *Carbohydrate Polymers* 232: https://doi.org/10.1016/j.carbpol.2019.115800.

24 Priya, B., Gupta, V.K., Pathania, D., and Singha, A.S. (2014). Synthesis, characterization and antibacterial activity of biodegradable starch/PVA composite films reinforced with cellulosic fibre. *Carbohydrate Polymers* 109: 171–179. https://doi.org/10.1016/j.carbpol.2014.03.044.

25 Hemmati, F., Bahrami, A., Esfanjani, A.F. et al. (2021). Electrospun antimicrobial materials: advanced packaging materials for food applications. *Trends in Food Science and Technology* 111: 520–533. https://doi.org/10.1016/j.tifs.2021.03.014.

26 Zhao, L., Duan, G., Zhang, G. et al. (2020). Electrospun functional materials toward food packaging applications: a review. *Nanomaterials* 10 (1): https://doi.org/10.3390/nano10010150.

27 Amna, T., Yang, J., Ryu, K.S., and Hwang, I.H. (2015). Electrospun antimicrobial hybrid mats: innovative packaging material for meat and meat-products. *Journal of Food Science and Technology* 52 (7): 4600–4606. https://doi.org/10.1007/s13197-014-1508-2.

28 Morris, B.A. (2017). Converting processes. *The Science and Technology of Flexible Packaging* 25–49. https://doi.org/10.1016/b978-0-323-24273-8.00002-2.

29 Xu, J., Manepalli, P.H., Zhu, L. et al. (2019). Morphological, barrier and mechanical properties of films from poly(butylene succinate) reinforced with nanocrystalline cellulose and chitin whiskers using melt extrusion. *Journal of Polymer Research* 26 (8): https://doi.org/10.1007/s10965-019-1783-8.

30 Yang, S., Bai, S., and Wang, Q. (2018). Sustainable packaging biocomposites from polylactic acid and wheat straw: enhanced physical performance by solid state shear milling process. *Composites Science and Technology* 158: 34–42. https://doi.org/10.1016/j.compscitech.2017.12.026.

31 Perumal, A.B., Sellamuthu, P.S., Nambiar, R.B., and Sadiku, E.R. (2018). Development of polyvinyl alcohol/chitosan bio-nanocomposite films reinforced with cellulose nanocrystals isolated from rice straw. *Applied Surface Science* 449: 591–602. https://doi.org/10.1016/j.apsusc.2018.01.022.

32 Elhussieny, A., Faisal, M., D'Angelo, G. et al. (2020). Valorisation of shrimp and rice straw waste into food packaging applications. *Ain Shams Engineering Journal* 11 (4): 1219–1226. https://doi.org/10.1016/j.asej.2020.01.008.

33 Menzel, C., González-Martínez, C., Vilaplana, F. et al. (2020). Incorporation of natural antioxidants from rice straw into renewable starch films. *International Journal of Biological Macromolecules* 146: 976–986. https://doi.org/10.1016/j.ijbiomac.2019.09.222.

34 Martino, L., Berthet, M.A., Angellier-Coussy, H., and Gontard, N. (2015). Understanding external plasticization of melt extruded PHBV-wheat straw fibers biodegradable composites for food packaging. *Journal of Applied Polymer Science* 132 (10): 1–11: https://doi.org/10.1002/app.41611.

35 Berthet, M.A., Angellier-Coussy, H., Machado, D. et al. (2015). Exploring the potentialities of using lignocellulosic fibres derived from three food by-products as constituents of biocomposites for food packaging. *Industrial Crops and Products* 69: 110–122. https://doi.org/10.1016/j.indcrop.2015.01.028.

36 Dixit, S. and Yadav, V.L. (2020). Comparative study of polystyrene/chemically modified wheat straw composite for green packaging application. *Polymer Bulletin* 77 (3): 1307–1326. https://doi.org/10.1007/s00289-019-02804-0.

37 Duan, L. and Yu, W. (2006). Review of recent research in nano cellulose preparation and application from jute fibers. In: *Proceedings of the 2016 3rd International Conference on Materials Engineering, Manufacturing Technology and Control,*

Taiyuan, China (27-28 February 2016), 742–748. https://doi.org/10.2991/icmemtc-16.2016.148. Atlantis Press.

38 Burrola-Núñez, H., Herrera-Franco, P., Soto-Valdez, H. et al. (2021). Production of biocomposites using different pre-treated cut jute fibre and polylactic acid matrix and their properties. *Journal of Natural Fibers* 18 (11): 1604–1617. https://doi.org/10.1080/15440478.2019.1693473.

39 Ray, D., Sarkar, B.K., Rana, A.K., and Bose, N.R. (2001). Effect of alkali treated jute fibres on composite properties. *Bulletin of Materials Science* 24 (2): 129–135. https://doi.org/10.1007/BF02710089.

40 Kumar, R., Kumari, S., Rai, B. et al. (2019). Effect of nano-cellulosic fiber on mechanical and barrier properties of polylactic acid (PLA) green nanocomposite film. *Materials Research Express* 6 (12): 125108. https://doi.org/10.1088/2053-1591/ab5755.

41 Kumar, R., Rai, B., and Kumar, G. (2019). A simple approach for the synthesis of cellulose nanofiber reinforced chitosan/PVP bio nanocomposite film for packaging. *Journal of Polymers and the Environment* 27 (12): 2963–2973. https://doi.org/10.1007/s10924-019-01588-8.

42 Baheti, V., Mishra, R., Militky, J., and Behera, B.K. (2014). Influence of noncellulosic contents on nano scale refinement of waste jute fibers for reinforcement in polylactic acid films. *Fibers and Polymers* 15 (7): 1500–1506. https://doi.org/10.1007/s12221-014-1500-5.

43 Mishra, S., Mohanty, A.K., Drzal, L.T. et al. (2004). A review on pineapple leaf fibers, sisal fibers and their biocomposites. *Macromolecular Materials and Engineering* 289 (11): 955–974. https://doi.org/10.1002/mame.200400132.

44 Nikmatin, S., Rudwiyanti, J.R., Prasetyo, K.W., and Yedi, D.A. (2015). Mechanical and optical characterization of bio-nanocomposite from pineapple leaf fiber material for food packaging. *International Seminar on Photonics, Optics, and Its Applications* 9444 (ISPhOA 2014): 94440R. https://doi.org/10.1117/12.2081112.

45 Wahyuningsih, K., Iriani, E.S., and Fahma, F. (2016). Utilization of cellulose from pineapple leaf fibers as nanofiller in polyvinyl alcohol-based film. *Indonesian Journal of Chemistry* 16 (2): 181–189. https://doi.org/10.22146/IJC.21162.

46 Kumar, P., Tanwar, R., Gupta, V. et al. (2021). Pineapple peel extract incorporated poly(vinyl alcohol)-corn starch film for active food packaging: preparation, characterization and antioxidant activity. *International Journal of Biological Macromolecules* 187: 223–231. https://doi.org/10.1016/j.ijbiomac.2021.07.136.

47 Mahardika, M., Abral, H., Kasim, A. et al. (2019). Properties of cellulose nanofiber/bengkoang starch bionanocomposites: effect of fiber loading. *LWT – Food Science and Technology* 116: https://doi.org/10.1016/j.lwt.2019.108554.

48 Ninjiaranai, P. (2015). Biopolymer films based on chitosan and polyethylene glycol with pineapple leaf fiber for food packaging applications. *Macromolecular Symposia* 354 (1): 294–298. https://doi.org/10.1002/masy.201400090.

49 Chauhan, V., Kärki, T., and Varis, J. (2019). Review of natural fiber-reinforced engineering plastic composites, their applications in the transportation sector and processing techniques. *Journal of Thermoplastic Composite Materials* https://doi.org/10.1177/0892705719889095.

50 Cao, X., Chen, Y., Chang, P.R. et al. (2008). Starch-based nanocomposites reinforced with flax cellulose nanocrystals. *Express Polymer Letters* 2 (7): 502–510. https://doi.org/10.3144/expresspolymlett.2008.60.

51 Hopkins, E.J., Chang, C., Lam, R.S.H., and Nickerson, M.T. (2015). Effects of flaxseed oil concentration on the performance of a soy protein isolate-based emulsion-type film. *Food Research International* 67: 418–425. https://doi.org/10.1016/j.foodres.2014.11.040.

52 Fang, S., Qiu, W., Mei, J., and Xie, J. Effect of sonication on the properties of flaxseed gum films incorporated with carvacrol. *International Journal of Molecular Sciences* 21 (5): 1637.

53 Gursoy, M., Sargin, I., Mujtaba, M. et al. (2018). False flax (*Camelina sativa*) seed oil as suitable ingredient for the enhancement of physicochemical and biological properties of chitosan films. *International Journal of Biological Macromolecules* 114: 1224–1232. https://doi.org/10.1016/j.ijbiomac.2018.04.029.

54 Yang, C., Tang, H., Wang, Y. et al. (2019). Development of PLA-PBSA based biodegradable active film and its application to salmon slices. *Food Packaging and Shelf Life* 22: https://doi.org/10.1016/j.fpsl.2019.100393.

55 Ali, M.E., Yong, C.K., Ching, Y.C. et al. (2015). Effect of single and double stage chemically treated kenaf fibers on mechanical properties of polyvinyl alcohol film. *BioResources* 10: 822–838.

56 Zarina, S. and Ahmad, I. (2015). Biodegradable composite films based on κ-carrageenan reinforced by cellulose nanocrystal from kenaf fibers. *BioResources* 10 (1): 256–271. https://doi.org/10.15376/biores.10.1.256-271.

57 Oyekanmi, A.A., Saharudin, N.I., Hazwan, C.M. et al. (2021). Improved hydrophobicity of macroalgae biopolymer film incorporated with kenaf derived cnf using silane coupling agent. *Molecules* 26 (8): https://doi.org/10.3390/molecules26082254.

58 Hazrol, M.D., Sapuan, S.M., Zainudin, E.S. et al. (2022). Effect of kenaf fibre as reinforcing fillers in corn starch-based biocomposite film. *Polymers (Basel)* 14 (8): https://doi.org/10.3390/polym14081590.

59 Zhang, Y., Haque, A.N.M.A., and Naebe, M. (2021). Lignin–cellulose nanocrystals from hemp hurd as light-coloured ultraviolet (UV) functional filler for enhanced performance of polyvinyl alcohol nanocomposite films. *Nanomaterials* 11 (12): https://doi.org/10.3390/nano11123425.

60 Viscusi, G., Adami, R., and Gorrasi, G. (2021). Fabrication of rice flour films reinforced with hemp hurd and loaded with grapefruit seed oil: a simple way to valorize agro-waste resources toward low cost materials with added value. *Industrial Crops and Products* 170: 113785. https://doi.org/10.1016/j.indcrop.2021.113785.

61 Dixit, S., Mishra, G., and Yadav, V.L. (2022). Optimization of novel bio-composite packaging film based on alkali-treated hemp fiber/polyethylene/polypropylene using response surface methodology approach. *Polymer Bulletin* 79 (4): 2559–2583. https://doi.org/10.1007/s00289-021-03646-5.

7

Natural Clay-Based Food Packaging Films

Ram Kumar Deshmukh[1], Dakuri Ramakanth[2], Konala Akhila[1], and Kirtiraj K. Gaikwad[1]

[1] Indian Institute of Technology Roorkee, Department of Paper Technology, Haridwar Highway, Roorkee, 247667, Uttarakhand, India
[2] Indian Institute of Technology Roorkee, Department of Polymer and Process Engineering, Haridwar Highway, Roorkee, 247667, Uttarakhand, India

7.1 Introduction

Consumer demands change quickly, so food packaging has dramatically evolved over the past few decades. Packaging's essential purpose in transportation and marketing has grown into a valuable role for preservation, identification, and information. Packaging has undergone a significant paradigm shift to improve customer health and safety. An active packaging feature has been designed to prolong the shelf life by incorporating elements that emit or draw substances into or from the packaged food. In contrast, smart packaging methods are those packaging systems that monitor packaged foods' conditions to access packaged food quality information. With all of these purposes, packaging has recently emerged to become the world's third-largest sector, representing around 2% of the gross national product in developed nations [1, 2]. The packaging industry is quickly expanding due to the growing customer demand for processed and packed food. Recent attempts have fulfilled consumers' expectations while ensuring product quality and safety during distribution through flexible packaging, high barrier material inclusion, sustainable packing, and active and intelligent packaging. Changing food packaging and its environmental and socioeconomic consequences will impact the entire food processing sector. "The packaging industry has grown exceptionally at worth USD 700 billion at the global level. The food packaging business is at USD 277.9 billion and is expected to grow to USD 441.3 billion by 2025, at a CAGR of 5.1percent." [3]

The packing and packaging material has evolved from a simple container to an integral aspect of product design in the previous two centuries. Packaging has the

---

Natural Materials for Food Packaging Application, First Edition.
Edited by Jyotishkumar Parameswaranpillai, Aswathy Jayakumar,
E. K. Radhakrishnan, Suchart Siengchin, and Sabarish Radoor.
© 2023 WILEY-VCH GmbH. Published 2023 by WILEY-VCH GmbH.

objective of serving high-quality, appealing, and inexpensive products to the customers. The packaging industry undertakes significant efforts for industrial and sustainable solutions to optimize packaging material that will minimize the impact on the environment through reducing, reusing, and recycling [4]. Plastic, glass, and metal are primarily nonbiodegradable raw materials in packaging applications today, which has prompted concerns about environmental pollution, indicating a severe problem for the global environment. Therefore, some biodegradable nanocomposites (NCs) having different functional characteristics for food packaging can be developed. The biodegradable notion in packaging is known as green packaging, and it includes plant, animal, and metal nanoparticles. They can replace traditional sources of packaging material, hence reducing the negative impact on the environment [5]. Nanoparticles have been proven to increase packaging features such as mechanical, physical, barrier, and antimicrobial to improve the safe storage period. Low nanoparticle loading in conventional packaging materials can improve the polymer composite's initial properties, manufacturing technique, and process parameters. Nanocomposite packaging materials may contain a variety of polymers and nanoparticles with the advancement in the features of food packaging. Nanotechnology has proven its potential reliability to improved food packaging functions from basic to advanced level, i.e. containment, storage, preservation, distribution, communication, and marketing of the food packages [6, 7]. A nanotechnology is a new approach that includes the characterization, production, and manipulation of structures, devices, or materials with a length of at least 1–100 nm [8–10]. Nanotechnology has diverse applications in food packaging, which has completely transformed the food sector in recent years by facilitating packaging functions such as active and intelligent packaging [4]. Nanotechnology has the potential to serve as the primary factor for food packaging development, such as nanocomposite, nanosensor, nanofiber, a nanoplate, nanoantimicrobial, nanoceuticals, and nanocochleate [11, 12]. As a result of these efforts, the search for acceptable materials for food packaging has been intensified. In recent developments, clay materials reinforced polymers known as polymer–clay nanocomposites (PCNs) have demonstrated the potential to supplement the limitations of traditional food packaging solutions, especially cost-effectiveness, environmental sustainability, and consumer safety [13]. Crystallites are a type of nanostructured material composed of organic and inorganic additions with specific geometrical properties. It is immobilized and restrained by its crystal lattice and incorporated into bulk material such as polymers from synthesis and biopolymeric sources such as polyamide (PA), polystyrene (PS), polypropylene (PP), high-density polyethylene (HDPE), linear low-density polyethylene (LLDPE), polyethylene (PE), nylons, polyvinyl chloride (PVC), polyurethane, and polyethylene terephthalate (PET), polylactic acid (PLA), poly (glycolic acid) (PGA), poly (caprolactone) (PCL), poly (butylene succinate) (PBS), poly (vinyl alcohol) (PVOH) [6, 9, 10].

    Clay-based packaging is one of the sustainable nanotechnology applications in food packaging, where the polymers are reinforced with organic and synthetic nanoclay materials. The term clay has several senses that refer to the relevant subject and its application. The clay materials are utilized in different commercial applications. It is used in geology for weather report prediction, petroleum industry, civil construction, agriculture, ceramic goods, plastic, rubber industry, and others, as shown in Figure 7.1. Clay has received attention in industries, laboratories, and academics due to its abundance in nature and affordable cost. It also possesses a unique crystal structure, larger surface space, surface electric charge, and cation energy capacity (CEC). CEC promotes hydration, plasticity, swelling, and thixotropy in water [12]. The concept of PCNs emerged in the 1980s. Toyota was the first enterprise to experiment with PCN in the vehicle models for several years [14]. PCN is a hybrid material made from nanoscale particles such as sheets of silicates; for example, the Nanotechnology Product Database (NPD, 2015) collected and analyzed nanotechnology products available in the global market. More than 8694 nanotechnology-based products from 2194 companies in more than 60 countries worldwide were registered in NPD. By August 2020, the NPD has recorded 222 nanoclay-based products available in the market from 125 companies in 25 countries. These products are categorized into 70 classes, including electronics medicine, construction, cosmetics, environment, automotive, food and its packaging, agriculture, sporting and toys, paper and printings, soil treatment, and water treatment. The United States of America, China, India, Japan, and the United Kingdom contribute 29.7, 17.6, 7.7, 7.7, and 5.0%, respectively, of the total global production of nanoclay-based products [15, 16]. Table 7.1 shows the commercially available nanoclay-based food packaging materials in the current global market.

Figure 7.1 Commercial applications of nanoclay minerals.

Table 7.1 Commercially available polymer nanoclay composites available in the current global market along with the manufacturer, brand name, application, and observation.

Polymer nanoclay composites	Manufacturer	Brand name	Food application	Observation	Film/container
Polyethylene terephthalate/organoclay	Mitsubishi and Nanocor Ltd., Japan	Imperm®	Alcoholic and nonalcoholic beverage	High barrier and improved oxygen transmission rate	
Nanoclay	Debbie Meyer Bread Bags, United States of America	Bread Bags™	Bakery products	High moisture barrier	
Polyethylene terephthalate/organoclay	Honeywell, United States of America	Aegis™ OX	Beer	Improved clarity and transparency	

Polymer nanoclay composites	Manufacturer	Brand name	Food application	Observation	Film/container
Montmorillonite-based polymer	FUJIGEL SANGYO LTD, Japan	NaturaSorb® SN series	Food package barrier film	Chemical inert, economic	
Polyethylene terephthalate/nanoclay particles N-coat	Multifilm Packaging Corporation, United States of America	FOOTHILL FARMS™	Cereal based Snacks and Ready-to-eat food	High gas barrier	
Kaolin	Imerys SA, France	BARRISURF™ FX	Fast foods	Gas barrier, i.e. Oxygen, Nitrogen, Carbon dioxide, or microorganism barrier	
Polyethylene, ethylene vinyl alcohol, polyamide, polypropylene, and organoclay	Mondi Uralplastic, Russia	Poliplen®	Meat, fish, and dairy products	The barrier to gases and odor	
Polyethylene/polypropylene/organoclay	TERM Mondi Uralplastic, Russia	Polyelf®	Meat and cheese package	Reduced moisture migration	

(Continued)

Table 7.1 (Continued)

Polymer nanoclay composites	Manufacturer	Brand name	Food application	Observation	Film/container
Nylon resin/organoclay	Honeywell, United States of America	Aegis BarrierPro2™	Beverages	Excellent barrier to CO_2 and O_2 clarity and recyclability	
Organoclay	Laviosa Chimica Mineraria, Italy	DELLITE™	Fruits and vegetables	Less weight, greater barrier to gases	
Polyethylene/polyamide/organoclay	BYBK Materials Technology, China	Suhou™	Soups	Antibacterial activity	
Kaolinite	NanoBioMatters Industries S.L, Spain	O_2Block™	Meat	O_2 scavenging activity	

Source of images in Table 7.1: Transcontinental Inc.; david_franklin / Adobe Stock; Mondi Uralplastic; Honeywell International Inc.; fullempty / Adobe Stock; BYBK Materials Technology.

Clay materials can improve the different properties and characteristics that are essential for the successful application of clay material in polymer composite matrices, such as dispersion tendency, high surface area, hydrophilic nature, and controlled polymer interaction in the matrix [12]. The application of clay materials in food packaging is now getting attention due to their mechanical and chemical barrier (against oxygen, carbon dioxide, ultraviolet, moisture, and volatiles), thermal properties, and biodegradable nature. There are several benefits of clay-based food packaging over conventional neat polymer packaging. These include improved transparency, toughness, gaseous, moisture and odor barrier, puncture resistance, abrasion and flex cracking, heat stability, and neutral to fat, grease, and oil. This chapter focuses on the naturally occurring clay materials for the food packaging application for a sustainable environment and a cost-effective source of biodegradable NCs for the improved food packaging application. It emphasizes the classification, properties, and application of different clay materials used in packaging applications. We also provide insight into the safety concerns of the customers overusing nanoclay-based food packaging.

7.2 Clay Materials Classification

Clay materials are composed of silicates in a layered structure which is termed aluminum phyllosilicate. It contains ions of oxides as earth metals, alkali metals, organic metals, etc. The clay materials have been extensively explored as filler material in polymer matrix over the available nanofillers like carbon nanotube, graphene, nanocellulose, and nano-silica due to their reinforcing ability and suitable with polymer matrices [17–22]. The stack of sheets is the primary form of raw clay materials. The sheets are composed of piles of tetrahedral $[SiO_4]^{4-}$ and octahedral $[AlO_3(OH)_3]^{6-}$ structures that render the whole clay materials assembly. The tetrahedral are linked at corners as hexagons, whereas octahedrons are connected by one side [12, 23]. The tetrahedrons and octahedrons are abbreviated as "T" and "O," respectively. The classification of clay materials is based on the number and ratio of the layer in fundamental structure, total charge, existing valency in the O and T. According to number and layer ratio, clay materials have three classes as below.

7.2.1 TO or 1:1 Type (One-One Tetra-octahedral Layer)

It is a typical two sheets of tetrahedrons and octahedrons with an equal distance (inner layer distance) of 0.7 nm. All the elements are distributed in an arrangement where the total electric charge between two sheets is neutral (quasineutral). These are also known as the kaolinite and serpentine groups. Sheets are layered so that oxygen is present in front of the OH group of adjacent layer, resulting in a neutral network by interlayer hydrogen bonds. Pile of sheets is rich in platelets with a few hundred nanometers of lateral extension and tens of nanometers of thickness. Clays such as serpentine, halloysite, and kaolinite are prominent examples in this group.

7.2.2 TOT or 2:1 Type (One-Octahedral in Between Two Tetrahedral Layers)

Two tetrahedral layers (side-wise) linked to one octahedral sheet form a three-layer materials type, designated as 2:1 or TOT type of clay materials such as mica, smectite, and vermiculite groups. This materials group has an exchangeable cation, which can be substituted easily with other available cations that form absorption capacity on the surface of the sheets.

7.2.3 2:1:1 or TOTO Type (Two Tetrahedral with Two Octahedral)

In the 2:1:1 structure, each octahedral is connected to another tetrahedral layer, and the net electric charge between these four sheets is 1.1 to 3.3. This clay contains mica-type sheets with a negative charge compensated by the positively charged octahedron. Chlorites, donbassite, and chamosite are essential in this group of clay materials.

The classification of clay is ambiguous due to its diverse microcrystalline structured species that is prone to variation in the arrangement with many possibilities of replacements. Hence the classification of clay materials has advanced in a different perspective and commercial application. "The nomenclature committee of the International Association for the study of clays (AIPEA) relies on the crucial structural data which depict there are different possible classifications of clay materials" based on various characteristics and properties [12]. Following are some categories:

- The combination of layers "(T/O or 1/1; T/O/T or 2/1; T/O/T/O or 2/1/1)"
- The position of cation in the octahedron
- The cationic filler interlayer and
- The interlayer components, such as cations and water molecules, are present.

7.3 Preparation of Natural Clay Nanocomposites

The preparation of clay material starts with mining and collecting raw clay, followed by crushing, drying, and screening of materials. Later clay materials are classified based on the polymer's degree of dispersion/intercalation, followed by different nanocomposite preparation methods and their applications in food packaging, as summarized in Figure 7.2. Dispersion of clay is essential in polymer matrices to develop the polymer composite packaging material. Covalent bonds are a matter of concern in between clay layers that causes clay dispersion in polymer matrices. To address this problem, clay particles are modified before dispersion in the polymer matrix. Changing clay materials involves expanding the space between clay sheets with functional moieties surfactant or hydrophobic grafting. This variation leads to ease in clay incorporation into the polymer composites, even in the hydrophobic polymer nanocomposites (PNCs) [24]. Several studies have been conducted in the late 1990s on manufacturing characterization for food packaging applications. The polymer nanocomposite (PNC) development involves different clay materials such as montmorillonite (MMT), kaolinite, hectorite, and sepiolite. A wide range of synthetic (PE, PVC, nylon) and

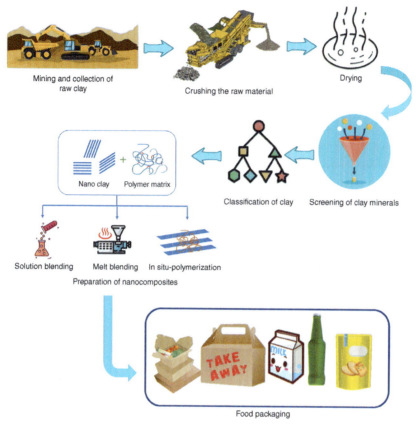

Figure 7.2 Schematic representation of nanoclay material life cycle beginning with the raw material collection, processing, and preparation of nanocomposites and their applications in food packaging.

biobased polymers (starch, cellulose, PLA) have been investigated for the development with varying amounts of nanoclay, i.e. usually 1 to 5 wt% [13, 25].

The physical and chemical methods are the two most common methods for modifying clay particles. The physical alteration technique involves the adsorption of modifying materials on the clay surface. There is no change in the clay structure, which results in a minor improvement in the final polymer composites. On the other hand, the chemical procedure is that the polymer is coupled with specific functional groups or organo-silane molecules onto the surface of the clay materials [22]. Furthermore, alteration is performed through an ion-exchange mechanism caused by cationic and anionic functional groups. As a result, improved intercalation between the clay particles and modifying agents was achieved with chemical modification. Modifying agents improvised the dispersion of clay particles in polymer composite matrix when clay materials are spread in polymer matrices having a higher aspect ratio (length/width) and wide surface area, resulting in better performance features such as increased mechanical, barrier, thermal, and optical properties of composite matrix [22]. There are several methods for modifying clay materials for better incorporation in a polymer composite; however, in situ polymerization, solution-induced intercalation, and melt processing are widely accepted procedures.

7.3.1 In situ Polymerization Method

The polymerization method involves polymer insertion between the clay layers then expanding and dispersing the layers into the uniform fine matrix. This process enhances the suitability of polymer and clay particles. Then clay particles are infused with a fine layer to the bulk matrix, which later undergoes various processes like extrusion, compression, and casting [22, 24].

7.3.2 Solution-Induced Intercalation

Solution-induced intercalation involves dissolving clays in a polymer solution to expand and distribute them. Initially, the clay material is mixed separately in a similar solvent/solution. Then this mixture is allowed to combine with polymer-solvent/solution and homogenize for a short duration before casting it on a flat surface and removing the solvent/solution by evaporation [24]. Due to the high cost of solvents, this process was not economically viable for commercial NCs for polymers. There are specific health and safety concerns with this approach. However, this process is suited for water-soluble polymers. It is economical for the commercial manufacture of NCs because of the low cost of employing water as a solvent and the minimum health and safety issues [22].

7.3.3 Melt Processing

Clay and polymers are intercalated during the melting process in this approach. When compared to the solution blending approach, this method provides better mixing of polymer and clay materials. The melt-integration method directly reinforces clay particles in polymer matrices [25, 26]. The efficiency of this approach may not be as excellent as in situ polymerization, and the composites formed by this process frequently have a partially exfoliated layered structure. On the other hand, the polymer processing industry can use the methodology to create NCs using the traditional method. This approach has a critical role in increasing the commercial production of clay/polymer NCs [22].

7.4 Properties of Natural Clay-Based Nanocomposite Polymer

Nanoclay is used to develop nanocomposite-reinforced polymer due to its high mechanical, thermal, barrier, high benignity, stability, low cost, availability, and sustainability [26–28]. The type of polymer, nanoclay, and degree of dispersion are vital aspects in improving nanocomposite properties [29].

7.4.1 Mechanical Properties

Mechanical property is one of the utmost requirements to develop a packaging material to sustain wear and tear during different operations throughout the complete food processing chain until the product reaches the customers. Tensile strength (TS), elongation at break (EB), and Young's modulus are significant mechanical properties in food packaging. A polymer such as Nylon-6 demonstrates substantial improvement in mechanical properties, such as a 103% and 49% increase in Young's modulus and TS, respectively, when using nanoclay as a filler even at low loading (1–5 wt%.) [30]. An improvement was observed in the TS of gelatin-based biofilm with the increase in the halloysite nanoclay, with low filler loading of 5% nanoclay. TS is an essential requirement for food packaging materials as it allows them to withstand the applied stress during food processing, storage, and shipping [31]. A decreasing value of EB was noticed; however, it was not significant with three percentages of nanoclay addition. The researchers noticed that decreasing value of EB is correlated with the potential of biodegradability of the film [32]. The elastic/Young's modulus is associated with the content of nanoclay in the polymer. Earlier, at 3% addition of nanoclay, Young's modulus was not noticeable; however with the increasing the concentration of nanoclay, the value of Young's modulus improved by ~60%, which can significantly improve the food packaging application [32]. One of the studies depicted that the enhancement of TS of starch/MMT film with decreasing EB corresponded with increasing concentration of MMT [33]. A recent study was involved a potato starch mixed with nanoclay MMT and a comparison of mechanical properties was made to control film, where they found that elevation in the level of nanoclay leads to improved Young's modulus (1.32–2.99 MPa%) and reduced EB (80.80–55.72%). Reduced EB value attributed to intercalation of silicate sheets, which cause a decrease in the flexibility and rigidity of film [34]. Similar results have been recited of mechanical properties like EB and TS in an earlier study where the nanoclay halloysite was added as a plasticizer in different volumes (0, 2, 4, 6, and 8 wt%). The halloysite at concentration of two percent significantly affects the mechanical properties as the lower filler loading level shows maximum TS and EB. When the filler is more than two wt%, both values TS and EB are reduced. This improvement is due to the accumulation of the nanoclay; however, when glycerol was added without any nanoclay, it exhibited maximum TS, and the film shows maximum EB value with eight percentage of halloysite nanoclay addition to the film [35]. In a recent study with plasticizers and 2:1 Phyllosilicate smectite, hectorite was found to be applied in the preparation of film using a casting process. The mechanical parameters of the produced film support prior research that shows that nanoclay increases the film's mechanical properties, such as TS, EB, and Young's modulus. Young's modulus and toughness of the packaging film improved with increasing nanofiller hectorite content, but it increases dramatically beyond 10% weight hectorite [36].

7.4.2 Barrier Properties

Barrier property is generally defined by the water vapor permeability and oxygen permeability of nanocomposite polymer or film. Clay-based NCs have shown excellent barrier properties compared to neat polymers. The diffusing molecules are restricted by well-dispersed, randomly orientated sheets of clay materials. The distribution of clay materials in the composite improves the aspect ratio and delay/hamper in the shortest pathway (tortuous pathway) for the migration of diffusive molecules such as gases, water vapor, and volatile compounds, as illustrated in Figure 7.3. This arrangement of clay particles restricts the passage of gas and liquid molecules through the polymer matrix, resulting in a decrease in permeability. Barrier improvement in the clay/polymer composites suggested an attractive application of these materials infiltration for different industrial solutions [37]. Various models have been developed earlier, assuming that random dispersion of diffusing molecules depends on the adjacent parallel layers of clay materials placed vertically to the diffusion orientation [38, 39]. The aspect ratio of the clay particle is responsible for reducing the permeability of the diffusing molecule through the polymer composites. They found that organoclay/nanoclay alone cannot support the developed NCs. It has a limited driving force in the PET polymerization, and degradation occurred at polymerization temperature. These disadvantages of organoclay can be overcome using clay-supported catalysts. The intercalation of stimulus into the clay layer improves the barrier property of the surface. The interlayer also creates a tortuous path that delays the migration of molecules from packaging material to food [40].

PET is extensively utilized for food packaging development. The exfoliation was stimulated at one percent OC (N-methyl diamine ester with apidic acid), which causes a two-fold reduction in the permeability [41]. When a catalyst such as chlorotitanium (tri-isopropoxide) was provided on the MMT, the oxygen permeability of film was reduced by 10–15 times with 1–5 wt percent clay [40]. The addition of two percent dodecyl ammonium-MMT lowered the water vapor permeability coefficient by 50% [42]. Clay materials with low levels decreased significant absorption of O_2, H_2O, N_2, and other diffusing substances. Several findings suggest that incorporating

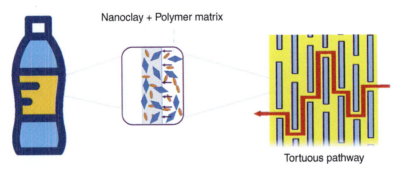

Figure 7.3 Improvement of the tortuous pathway with nanoclay composite.

catalysts improves barrier properties of the clay-based nanocomposite polymer for different applications. The clay materials and natural rubber (NR) latex NCs show excellent barrier properties for film or lamination in packaging film. The addition of 1–3phr (per hundred resin) reduced O_2 permeability by 66% [43], use of 20phr of Na^+-MMT led to a reduction of 50 % of N_2 permeation as compared to neat polymer matrix [44], 40% toluene, and 50% oxygen reduction observed with 3phr of dodecyl ammonium-MMT in the NR-clay based NCs matrix [45]. Different industrial companies, such as Nanocor and Mitsubishi Gas chemicals (New York), have produced nanocomposite nylon Imperim® with better barrier properties in film and bottle manufacture for food packaging industries. Barrier property is essential for liquid food packaging such as fruit juices, dairy products, and carbonated drinks. [46]

7.4.3 Thermal Stability of Clay-Based Polymer Composites

Clay materials improve the dimensional stability (linear dimensional) of the nanocomposite polymer. The coefficient of thermal expansion (CTE) is high for the neat polymers, which is responsible for the elongation of the material on heating. Clay materials enable these polymers to sustain in high temperatures during the different processes with the preparation of PCNs. Higher CTE is responsible for dimensional changes during molding and hampers in the automotive process in food and pharmaceutical packaging [47]. High-temperature processing causes degradation of neat polymer-based packaging in different industrial application. So, it is imperative to understand the degradation mechanism of the polymer at high temperature, which will lead to modification with clay reinforcing into the neat polymer that provides heat stability to the clay/polymer NCs [48, 49]. It has been reported that some NCs combining with clay materials with the polymers HDPE [50], LDPE [51], LLDPE [52], PA [53], PS [54], PET [55], PVC [56], PBT [57], PPT [58], nylon 6 [59], epoxy resins [60] showed significant improvement in thermal stability as well as structure of the clay/PNC. The researchers depict that the clay layer's degree of dispersion and intercalation into polymers matrix led to high thermal stability.

7.4.4 Oxygen and Ethylene Scavenging Activity of Nano-Clay Polymer Composite

The presence of oxygen inside packaged food can lead to product deterioration through different reactions such as browning, rancidity, and loss of essential components. Likewise, the excess of ethylene gas in the climacteric fruit and vegetable's storage atmosphere leads to an uncontrolled ripening process, ultimately resulting in product quality loss. Other undesirable compounds produced by fruits and vegetables during storage can also result in unwanted product quality changes [61]. Therefore, it is crucial to develop a packaging system to eliminate or reduce the released quantity of oxygen and ethylene in packaged foods. The scavenging property is attributed to trapping the target component and creating a tortuous path for diffusion, facilitating the mobility of elements across the food package [62]. One of the emerging technologies in this area is developing nanocomposite material to improve the properties of packaging film [63]. The conventional way to adsorbent oxygen and ethylene with heavy metal adsorbents is its mesoporous structure. Clay is now an emerging innovative way to scavenge the oxygen and ethylene from the food package. Porous clay heterostructure (PCH) is widely applied to prepare material having intercalation of a surfactant within the galleries of the clay layers. The aim of adding a functional group on the surface of the clay is to improve ethylene adsorption. PCH is modified with MTS to bind the methyl group for increasing ethylene adsorption in ethylene scavenger packaging. Two layers of phyllosilicate where tetrahedral sheets contain areas of si^{4+} or Al^{3+} and octahedral layers consist of Mg^{2+} or Al^{3+}. Ethylene may be removed with surface adsorption active edges on the material such as kaolinite, cristobalite, and clinoptilolite. These clays can be incorporated into ethylene permeable sachet by extrusion process [64]. Ethylene removal using zeolite as an adsorbent material has received significant attention in pharmaceutical and food industry applications. It is mentioned that advantageous properties of the hollow porous area with cation exchange, adsorption, and molecular separating attributes. So, it is utilized as ethylene scavenging material into packaging films. There are several publications available in which authors have been demonstrated zeolite application for ethylene scavenging activity [65–67]. A sachet containing a $KMnO_4$-loaded sepiolite was evaluated for ethylene scavenging activity and found helpful in reducing the release of ethylene content in the produce [68]. Ethylene scavenger system developed with $KMnO_4$-sepiolite and combined with the modified atmosphere packaging (MAP). Results showed a significant reduction in product weight loss and delayed titratable acidity (TA) loss [69]. Halloysite nanotubes (HNT) are widely available green material from natural deposits. A high aspect ratio permits them to utilize as nanocarrier in polymeric structure which was used as a discharge of active agents in NCs [70].

The ethylene scavenger activity of HNT has been examined and it was found that the material significantly improved the ethylene scavenging activity by one g of alkali-HNT from the produce within 24 h of 49 Microliter from packed produce. HNTs are categorized as generally recognized as safe (GRAS) for application in food packaging [64]. HDPE film was modified with clay kaolinite to create oxygen scavenging film, and they observed the improved O_2 scavenging activity of 43 ml O_2/g at 100% relative humidity (RH) and 37 ml O_2/g at 50% RH [4, 71]. A polymeric film with organo-modified montmorillonite (OMMT) and iron nanoparticles affects the physicochemical properties of the film with 2% of OMMT to PP film, the oxygen scavenging ability of the film increased by 77% [72]. Oxygen and ethylene scavenging properties of clay-based NCs promise feature in food packaging, enabling safe storage of fresh and packed foods. The application of clay-based composite for oxygen and ethylene scavenging activity depends on the approval and cost-effectiveness of the food grower and processing units. Thus, extra efforts should be taken to tackle the limitations and reduce the production cost of these technologies.

7.5 Application of Natural Clay in Food Packaging Film

Clay materials are widely available in nature, so they are inexpensive and rational to be utilized as a functional agent for food packaging. They significantly enhance the physicochemical and degradable properties of composite even at a lower filler load (<10 wt%). Different surface modification of nanoclay offers desirable divergent properties, making it a suitable choice for the numerous commercial applications [73]. Extensive research on nanoclay has valuable functions in food packaging applications, such as active and intelligent activity. Other factors are responsible for incorporating nanoclay into polymer matrices like polymer types, nanoclay, filler loading percentage, processing techniques, desirable properties, and end-product applications. In all inorganic nanoadditives, nanoclay materials are one of the most extensive studied in recent times. Although phyllosilicates are commonly found in clay, they may also contain additional elements that give them strength and flexibility [74]. The majority of phyllosilicates used as nanoclays in food packaging are layered because they are made up of layers that can be separated into individuals when needed. These layers range in thickness from 0.7 nanometers to several nanometers in length, resulting in nanoparticles with a high aspect ratio. The scientific literature on polymer-layered NCs (biodegradable and synthetic polymer), both in general and in the context of packaging, shown in Tables 7.2 and 7.3 had some excellent assessments in the field published in recent years [110, 111]. MMT, hectorite, saponite, bentonite, vermiculite, rectorite, and asbestos (chrysolite) are primary clay materials, which are most suitable for food packaging application with some other clay-like halloysite, sepiolite, and palygorskite [112, 113].

Table 7.2 Examples of synthetic polymer–clay nanocomposites and their properties.

Polymers	Filler types	Quantity (wt %)	Observation	References
Ethylene vinyl alcohol	Montmorillonite		Improved barrier properties	[75]
polypropylene (PP)	Clay and hollow glass microspheres		Oxygen permeability performance improved	[76]
Polypropylene	Talc nanocomposite		Lightweight and strong	[77]
Low-density polyethylene	Montmorillonite modified with copper	4%	Antimicrobial effect increases up to 94%	[78]

(Continued)

Table 7.2 (Continued)

Polymers	Filler types	Quantity (wt %)	Observation	References
Low-density polyethylene	Montmorillonite	—	Increases compatibility and function in polymers	[79]
Low-density polyethylene	Montmorillonite	3%	Improved thermal stability	[80]
	Montmorillonite	10%	TS improved; O_2 and water vapor permeability decreases	
Low-density polyethylene	Montmorillonite	0.5–5.0%	Increases crystallinity and crystallization temperature; Tensile modulus and strength of filament increase	[81]
LDPE, maleic anhydride grafted polyethylene (MAPE) and ethylene-vinyl acetate (EVA)	Montmorillonite	4.0%	Improvement in TS, tear strength, and oxygen barrier properties	[82]
Low-density polyethylene	Montmorillonite	—	Tensile and barrier properties significantly	[83]
Low-density polyethylene	Silver-montmorillonite	5%	Improved antibacterial activity against E. coli (70% reduction)	[84]
High-density polyethylene (HDPE)	Cloisite® 15A	3%	Maximum diffusion rate reduction	[85]
PE	Closite 20A	5%	Selectable barrier property	[86]
Polyvinyl alcohol (PVA) low-density polyethylene film (LDPE)	Vermiculite	—	Oxygen barrier properties were	[87]
Polypropylene (PP)	Montmorillonite	—	Reduced oxygen permeability	[88]

Polymers	Filler types	Quantity (wt %)	Observation	References
Polyamide (PA)	Cloisite 30B (C30B)	—	Increment in stiffness and oxygen barrier properties	[89]
Polyamide (PA)	Dellite 43B (D43B)	—	Lower oxygen transmission rate	
Polyethylene (PE)	Montmorillonite	0.3%	Lowering in O_2 and water vapor permeabilities about 55% and 70%	[90]
Polyethylene (PE)	Montmorillonite	2phr	Thermal stability increases	[91]
Linear low-density polyethylene (LLDPE)	Organo-montmorillonite	2–4%	TS improved by 45%; 48% increase in transverse stiffness; dart impact improves by 20%; the tear resistance rose 33%; and the creep resistance by 20%.	[92]
Polyethylene (PE)	Cloisite 20A	2%	Reduced water vapor transmission rate	[93]
Polyethylene (PE)	Organo-montmorillonite	—	Clay dispersion and barrier property improved	[94]
Ethylene-vinyl acetate (EVA)	Organo-montmorillonite	5%	Enhanced barrier by 30%; and tensile modulus by 37%	[95]
Low-density polyethylene, High-densitypoly ethylene	Cloisite 20A	5% 5%		
Low-density polyethylene/ Maleic anhydride-grafted polyethylene (MAPE)	Organo-montmorillonite	7%	TS improved significantly	[96]
Low-density polyethylene/Dimethyl dodecyl ammonium (DDA)	Organo-montmorillonite	0.5%	CO_2 and O_2 barrier properties increase significantly	[97]
Low-density polyethylene/ salt and octadecyl trimethyl ammonium (OTA)		2.0%	WVP permeability decrease 2.5 times	

Table 7.3 Examples of biodegradable polymer–clay nanocomposites and the changes in their properties.

Polymers	Filler types	Quantity (wt %)	Observation	References
Pectin	Halloysite	5–30	Improved antioxidant activity, thermal stability, high hydrophobicity	[73]
Chitosan	Montmorillonite		Better antibacterial, better oxygen barrier	[98]
Polylactic acid	Cloisite 20A	5%	Water vapor barrier properties improved	[99]
Soy protein isolate (SPI)/polyvinyl alcohol (PVA) blends	Montmorillonite (MMT)	—	The improved TS and Young's; exhibit subtle reinforcing effect.	[100]
Poly (lactic acid) (PLA)	Cloisite 30B, Cloisite 15A, and Dellite 43B	3–5%	Enhanced thermal stability	[101]
Polyvinyl alcohol (PVA)	Saponite	5%	Thermal properties and oxygen transmission rate (OTR)	[102]
	Bentonite	5%	Improved thermal stability	
Hemicellulose from oil palm	Montmorillonite (MMT)	—	Improved thermogravimetry analysis (TGA), mechanical properties, and decreased water vapor permeability (WVP)	[103]
Chitosan	Montmorillonite (MM)	5%	Improved barrier properties Inhibitory effect on *Staphylococcus aureus* and *Escherichia coli*	[104]
Agar-carboxymethyl cellulose (agar-CMC) bio nanocomposites film	Montmorillonite	3–10%	Antibacterial activity against Gram-positive and Gram-negative bacteria	[105]
Cassava starch	Laponite	1–6%	Textural properties of film improved	[106]
Collagen	Laponite	—	Thermal stability increased	[107]

Polymers	Filler types	Quantity (wt %)	Observation	References
Pectin	Laponite	1–7%	Moisture migration reduced antimicrobial effect against *E. coli* and *S. aureus*	[108]
Carrageenan	Laponite	1–7	Adhesion property improved; barrier against O_2 water vapor reduced	[109]

7.5.1 Montmorillonite (MMT)-Based Nanocomposite

Among these clays, MMT is the most widely utilized in NCs preparation with the chemical formula $(Na, Ca)_{0.33} (Al, Mg)_2 (Si_4O_{10}) (OH)_2 \cdot nH_2O$. MMT is widely available, and it has a high surface area and activity with high acceptance. These properties enable MMT a suitable clay for exfoliation and intercalation in the host polymer matrices. Closite is a modified MMT clay material that contains quaternary ammonium surfactant [114]. The structure of MMT is made up of two tetrahedral sheets linked to an octahedral sheet in the center. The modification of MMT with silicon, aluminum, iron, and cation was by replacing oxide anion from the tip of the tetrahedral subunit [115, 116]. Organophilic chemicals make up the majority of polymers. The layered silicates are miscible with nonpolar polymers, and a cationic-organic surfactant must replace the alkali counter-ions. Other "onium" salts, such as sulfonium and phosphonium, can be employed instead of alkylammonium ions. Surfactants can be utilized to enhance the clay's dispersibility. So, it is used to separate the layers (d-spacing) at different degrees according to the number of polar units in polymer, and it is termed as organo-modified layers silicates (OMLS) or OMMT [117, 118]. MMT and OMMT have several advantages in food packaging application such as:

- MMT is compatible with hydrophilic polymers such PVOH, PLA, and biopolymers like starch, chitosan, and proteins to improve mechanical, thermomechanical, and oxygen and water vapor barrier characteristics in packaging films. [119–121].
- MMT is modifiable with Ag^+ or Cu^{2+} nanoparticles due to its strong ion exchange capacity.
- OMMT can be easily combined with most hydrophobic and biopolymers to increase mechanical characteristics and water vapor and oxygen barrier capabilities.
- Antioxidants and antibacterial components are utilized to maintain the high surface area of MMT to develop nanocarriers for intelligent packaging.
- Enzymes can be embedded to the MMT surface to develop smart packaging, and metal ions can incorporate into the interlayer of MMT to create nanosensors for food packaging.

The polymer can be divided into three classes such as intercalated, flocculated, and exfoliated NCs, as depicted in Figure 7.4. In the first case, polymer chains are placed within equal distances of clay sheets in a crystallographical manner despite the ratio structure. Hydroxylated edge-to-edge involvement can be seen in the second interacted NCs and, finally, the third class of NCs, where clay sheets are randomly dispersed with polymer chains to make an advanced exfoliated structure. This clay and polymer chain orientation improved significant properties such as tensile and flexural strength and barrier properties [120]. A long list of polymers is utilized in developing food packaging like PE, PP, PET, PS, and PVC [122]. This problem can be overcome with nanoclay involvement in developing PCNs.

The diffusion of gases, water vapor, and volatile molecule has a critical effect on the performance of the food packaging material. So, clay-based NCs are getting profound attention in recent times. The structures of the clay nanoparticles (CNPs) have a higher aspect ratio which improves the distribution of the clay sheets in the polymer matrix that create a tortuous path with better barrier property [120].

PE has been widely used in food packaging due to its mechanical and physical properties. The main types of PE are HDPE, LLDPE, very low-density polyethylene. Neat PE is branched plastic having a long molecular chain that causes low crystallinity value and possesses low tensile and compressive strength [122]. Low-density maleic anhydride-grafted (LDMAPE) is reinforced with organo-modified MMT with dimethyl tallow benzyl ammonium ion to prepare PE/OMMT nanocomposite film. It has been observed that the addition of OMMT improved the clay nanoplatelet's dispersion in the polymer matrix, which led to an increasingly significant barrier property [95]. Polypropylene properties were evaluated with LDPE, PLA,

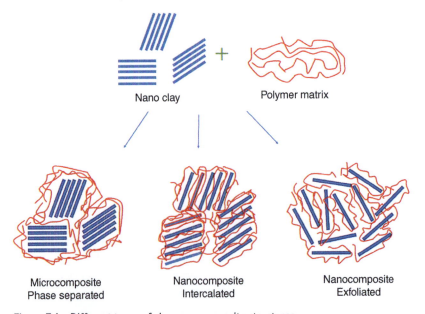

Figure 7.4 Different types of clay nanocomposite structures.

and MMT incorporation. PP/LDPE at 80 : 20 w/w blended with PLA and MMT. PP/LDPE/PLA/MMT composite showed improved EB and TS. The morphological improved properties were found in the scanning electron microscope (SEM), transmission electron microscope (TEM), and x-ray diffraction (XRD) analysis which depicts the blend as biodegradable. The developed film exhibited holes and cracks on the surface after 60, 90, and 150 days of soil burial [123]. PS/OMMT NCs were prepared via solution blending procedure using $CHCl_3$ and CCl_4 as a solvent modified with hexadecyl trimethyl-ammonium bromide (CTAB) and improved thermal stability and water barrier properties observed [124]. Another study on the PS/MMT changed with dimethyl dehydrogenated tallow (DMDT) NCs prepared using a novel spray casting method, and they noticed a significant improvement in the barrier properties [125]. PET/clay nanocomposite prepared using OMMT cloisite 30B and cloisite Na^+, which exhibited enhanced mechanical and barrier properties [126]. Several other studies on the synthesis of biopolymers based on the MMT clay nanocomposite have shown tremendous improvement in nanocomposite properties.

7.5.2 Laponite-Reinforced Polymer Nanocomposite

Laponite is classified as 2:1 clay with a chemical formula $Si_8(Mg_{5.45}Li_{0.4}) H_4O_{24}Na_{0.75}$, and it is hydrophilic and biocompatible. Its well-defined nanosize layered form has been employed as a model in fundamental studies of PNCs to improve their physicomechanical properties [127]. The alteration makes laponite clay materials with different silane coupling agents such as aminopropyl trimethoxy silane (APTS), dimethyl-octyl methoxy silane (DMOMS), and aminopropyl dimethyl ethoxy silane (APDES). It is used to reinforce laponite in polymers and bind to the clay's surface; the hydroxyl groups were replaced by silane groups [128]. Apart from the silane group, some surfactants were also used to modify laponite like CTAB [129]. Inorganic compounds carry clay materials such as silica and iron that render the clear and transparent colloidal suspension of clay in solution. Because of this reason, laponite is suitable for the manufacturing of household products, personal care, paper, and polymer film for food packaging by casting method [130]. Laponite (1, 3, 5, and 10 wt%) was reinforced into Kafirin film, and it was found that the addition of laponite into the film makes it more robust and less pliable, suitable for various packaging and coating applications [131]. Laponite and carboxymethylcellulose (CMC) composites exhibit improved mechanical properties, water vapor barrier function (lowered 42%), and increased degradation temperature (about 65 °C) compared to the virgin CMC [132]. Cellulose nanofiber (CNF) was reinforced with the laponite in the 3.5:1 by mass and found to have higher thermal stability and improved water vapor permeability (157% for CNF/Laponite 1:1 by mass) [133].

7.5.3 Sepiolite-Reinforced PNC

Sepiolite is a member of the 2:1 phyllosilicate family known as hydrated magnesium silicate, and its chemical formula is $Si_{12}O_{30}Mg_8 (OH)_4(OH_2)_4 \cdot (H_2O)_8$. It is similar to the MMT in structural base; the only difference is the insufficient octahedral layer in the sepiolite structure [134–136]. This clay shows nanoscale tunnels structure in micro-fibrous morphological with particle size range 2–10 μm. This structure possesses a higher specific surface area ($>300\,m^2/g$) as well as porous volume ($0.4\,cm^3/g$). The silica layer has a discontinuity in series at the tunnel's margins due to the presence of silane groups (Si-OH) [137]. It could be utilized as an adsorbent, catalyst, or catalyst carrier and thermal protector due to its unique structural orientation. A large interface area can be applied to polymer modification, which leads to solid intercalation and dispersion between the polymer matrix [25, 138]. When clay is used as an additive, it is generally known that it improves the characteristics significantly when it is nanometric in size, so it is applied as a nanofiller in NCs. These clay materials show a lower bonding tendency to the hydrophobic organic polymer, limiting its commercial application [139]. Sepiolite was reinforced with polyamide6 (PA6) and trimethyl hydrogenated tallow quaternary ammonium (TMTH) modification, and the catalytic modifier effects were evaluated. They observed that the elastic rigidity and heat deflection temperature in PA6/sepiolite nanocomposite increased ~2.5 times compared to neat PA6 [140]. Sepiolite was used as a reinforcing element to prepare a 90/10 (w/w) nanocomposite of PLA/styrene-ethylene-butylene styrene-g-maleic anhydride copolymer/sepiolite (SEBS-g-MA/Sep). They discovered that adding 0.5 and 2.5 weight percent sepiolite to the mix enhanced the tensile modulus by 36.0 and 17.0%, respectively [141]. The solution casting method developed an alginate/sepiolite nanocomposite modified with myrtle berries extract (MBE) rich in polyphenol. The physicochemical, mechanical, and antioxidant properties were evaluated and improved EB, TS, WVP, and UV barrier properties. The antioxidant property was significantly enhanced with the MBE concentration and raised the potential applicability of sepiolite-incorporated NCs in the food packaging [142]. A recent study of PET/Sepiolite nanocomposite and tray has been developed and evaluated the different properties. And it is found that there was improved permeability of 30% even with less nanomaterial loading (1.37%) in the polymer matrix. Maximum mechanical strength has been noticed at 1.88% sepiolite loading. The antimicrobial property of the nanocomposite was found to be significantly improved with lowering the colony-forming units of mesophilic bacteria in the tray from (log colony-forming unit (CFU)/g 6.57–5.25) as compared to neat PET tray with (log CFU/g 3.83–2.98) [143].

7.5.4 Bentonite-Reinforced Polymer Nanocomposite

Bentonite clay is composed of a 2:1 silicate layer containing calcium and sodium ions. The presence of impurities in naturally occurring bentonite clay components, such as quartz and mica, reduced their thermal stability. As a result, bentonite clay components must be cleansed before undergoing the modification operation [144]. Purification of bentonite clay components is thus required before integration and modification of the nanocomposite material for various applications. "Numerous modification methods improve the physicochemical and mechanical properties of bentonite clay, including CTAB, stearyl dimethyl ammonium chloride (SDAC), ammonium polyphosphate (APP), dimethyl dioctadecyl ammonium chloride (DDAC), and tributyl hexadecyl phosphonium bromide (THPB), tetradecyl ammonium bromide (TDAB), and benzyl triphenylphosphonium bromide (BTPPB)." Clay modification results in enhanced dispersion, more robust thermal stability, TS, Young's modulus, EB, increased char yield, and polymer compatibility [145–149]. Bentonite clay is a low-cost filler that can be utilized in polymer matrices to improve desired qualities. Bentonite clay-based biodegradable food packaging films development shows significant potential benefits in sustainable food application to reduce the environmental pollution load due to synthetic and conventional packaging material. Active zein biofilm with different Zataria multiflore boiss and essential oil mixed with the sodium bentonite (2% and 4%)exhibits improved physicochemical and mechanical of active films by 2% sodium bentonite clay addition.

Additionally, antibacterial activity improved against *Listeria monocytogenes*, *Escherichia coli* [150]. "Cassava starch/sodium bentonite/cinnamon oil (0.75% sodium bentonite, 2% glycerol, and 2.5% cinnamon oil w/w basis) based film showed significant antimicrobial potential against *Escherichia coli, Salmonella typhimurium, and Staphylococcus aureus.*" The meatball microbiological quality was evaluated at ambient temperature with the developed clay-based film. It has significantly reduced the bacterial proliferation to 96 hours below the standard limit compared to the control film with 48 hours [151]. Starch/glycerol/bentonite coating showed a higher barrier to water vapor as the water vapor transmission rate (WVTR) of the layer reduced significant level with 780 to $340 \pm 20\,\text{g}\,\text{m}^2$ per day with starch alone; this was further reduced 48 to $66\,\text{g}\,\text{m}^2$ per day with bentonite added to the coating formulation [152]. A recent study of bentonite effects on the cellulose nanofiber for food packaging application was evaluated. Bentonite at different loads (15, 30, and 45 wt %) was incorporated into the CNF matrix, and observed physicochemical and barrier properties of the matrix lowered the CNF degradation temperature and was monitored and reduced the WVTR of the nanocomposite. The prepared nanocomposite with bentonite and CNF shows the potential solution as an eco-friendly alternative packaging material [153].

7.5.5 Hectorite-Reinforced Polymer Nanocomposite

The 2:1 phyllosilicate is the most widely utilized layered clay material for PNCs, particularly smectite. Hectorite is plain and whitish in color smectite with the chemical formula $Na_{0.3}(Mg, Li)_3Si_4O_{10}(OH)_2$ [154]. It has a soft greasy texture and is an expensive clay with unique thixotropic nature. Natural hectorite has varied crystallinity, numerous impurities, and different geological and environmental conditions, limiting its application. Hectorite is a clay material easily produced in the laboratory and the industry using a hydrothermal method. These active edges react with one another; hence, modification is necessary. Acid treatment, ion exchange, grafting, and pillaring are significant improvement methods [155]. PA 66/hectorite nanocomposite has been observed under bending strain using a custom-built bending fatigue test setup in the controlled condition of the laboratory. Dynamic mechanical analysis (DMA) data showed a considerable rise in loss of modulus rigidity in frequency from 1 to 0.1 Hz and lowering the value of cyclic softening in the 2 to 0.5 Hz range of fatigue test frequency [156]. A biopolymer based on potato polysaccharide reinforced with hectorite starch–clay nanocomposite (SC-NC) exhibited higher mechanical characteristics and reduced crystallinity with an increasing percentage of hectorite clay in the composite. This improved the biodegradability of the nanocomposite SC-NC while the swelling behavior decreased, an essential property of food packaging films [36]. "Fluorohectorite is a mixture of silicate, oxides, fluorides of lithium, and magnesium. It has a high aspect ratio, surface area, and good ion exchange capacity." EVA nanocomposite reinforced with nanofiller demonstrated high reinforcing capacity in the produced film [157]. Hectorite offers a heat stability advantage over montmorillonite despite being smectite because it lacks acidic sites that could promote polymer breakdown during melt-mixing procedures. Nanocomposite PVA has been used in packaging to strengthen the barrier (water, volatile chemical, gaseous) with hectorite [158].

7.5.6 Rectorite-Reinforced Polymer Nanocomposite

Rectorite clay possesses an alternative arrangement; a dioctahedral mica-like layer of (non-expandable) and a dioctahedral montmorillonite-like layer (expandable) in 1:1 ratio has a structure of both MMT and mica. Rectorite layer thickness is about 2nm and can extend up to several microns. Rectorite has utilization in nanocomposite filler to develop the food packaging films and coating; however, the reported work is limited. In general, the rectorite is hydrophobic, which is incompatible to reinforced with hydrophilic matrices [157]. Modified asphalt's physical and aging properties with waste polypropylene packaging (WPP) and rectorite were evaluated. The result showed that the developed composite had excellent flexibility and plasticity when four WPP and 1.5 wt% of rectorite content were added. The composite improves softening, deformation, and high-temperature stability than base asphalt without rectorite content [159]. In another study, lysozyme (LY)/recorite incorporated into chitosan film exhibited enhanced antibacterial properties with higher mechanical strength. The composite of the film reveals that the dispersion of the LY and rectorite was homogeneous in the chitosan film. The addition of LY and rectorite enhanced the hydrophobic characteristics of the chitosan films, as measured by the water contact angle. After adding LY-rectorite, the mechanical parameters of the composite films were reduced by 27.58% compared to chitosan films, yet they still had high TS [160]. Epoxy acrylate (EA) coating was incorporated with modified rectorite through ultraviolet curing technique. Rectorite was modified using octadecyl trimethyl ammonium chloride (OTAC) and [2-(methacryloyloxy)ethyl] trimethylammonium chloride (MAOTMA) rectorite. The nanocomposite morphology was investigated with the SEM and TEM, and it exhibited better dispersion in the composite. The nanocomposite with three percent rectorite showed better thermal stability; flexibility decreased with clay percentage [161].

7.5.7 Other Nanoclay Materials-Based Nanocomposites

Kaolinite is a simple clay found on the earth's surface that is widely applied as a raw material in various household and industrial utilities. Kaolinite is a 1:1 phyllosilicate with an octahedral aluminum hydroxide with a tetrahedral silicon oxide sheet. This asymmetric structure allows strong hydrogen bonds between layers, resulting in high cohesive energy. The introduction of polymer chains between the kaolinite platelets is considerably hampered due to the high layer-to-layer contacts [162]. Surface modification of kaolinite employing a silane coupling agent improved the mechanical characteristics of a virgin polymer mix significantly [163]. Cassava starch/kaolinite composite film prepared with dimethyl sulfoxide (DMSO) modification. The developed film exhibited improved transparency, water uptake in the matrix, and UV light transmission reduced [164]. A similar study was conducted for nanocomposite of EVOH/kaolinite by melting process. The nanocomposite exhibited higher mechanical and thermal performance, the water diffusion coefficient in the developed matrix decreased by 50%, and the oxygen barrier increased by 50% [165]. Sago starch/kaolinite/polyolefin and PCL/Ag-kaolinite with iron modified developed for active food packaging [71, 166, 167]. The application of vermiculite in food packaging has been reported; however, its application is not as widespread as other clay materials. Butyl rubber/vermiculite composite and polylactic/vermiculite blended by in situ intercalation polymerization. Polyethylene/organo-modified vermiculite nanocomposite showed blend has high mechanical, physical, barrier, and antibacterial properties [112, 168, 169]. Some phyllosilicates, such as asbestos (chrysotile), halloysite, and sepiolite/palygorskite, are fibrillar rather than laminar. Both halloysite and palygorskite have a lot of applications in food packaging.

7.6 Challenges of Using Clay in Food Packaging Applications

Nanoclays have attracted much attention due to their excellent physical and mechanical strength, antibacterial property, and high barrier. Hence that allow them to be included in nanocomposite production with various syntheses. Despite all the advantageous features, the application of nanoclay is still a matter of concern for the safety and well-being of customers. The major problem is nanoparticle migration from food packaging to food or beverages. According to studies, the smaller the nanoparticles are and the lower their density, the more likely they are to be transferred to food and cause health concerns for the consumer [170, 171]. The migration of nanomaterial from nanocomposite to food depends on the exposure level. The development of nanoclay-enabled NCs for commercial food packaging applications prioritizes consumer safety. Nanoparticles may have varying toxicity due to differences in physical, chemical, optical, and magnetic properties, but their impact on human health cannot be underestimated. There is no evidence of nanoparticles causing acute toxicity, but the long-term effect of accumulation in the food chain remains unanswered [172]. The liver, kidneys, spleen, heart, lungs, and brain receive bio-distribution through systemic circulation. It is well understood that the liver, kidneys, and colons are principally responsible for nanoparticle and potential metabolite excretion [173]. There is no standard regulation in Europe or elsewhere to monitor the application of nanoclay in food packaging formulation. As a result, "nanotechnologies in food packaging covered by existing legislation comply with the provisions of the European framework regulation EC 1935/2004 (European Commission, 2004) that sets the general standard to ensure that migration of the substance from the packaging into the food can be prevented." [174] Food and Drug Administration (FDA) has also not established any regulatory definitions for "nanomaterial," "nanotechnology," and "nanoscale." [175, 176]. The European Food Safety Authority's (EFSA) strategic line is based on a proper risk assessment technique and the potential risk associated with PNC utilized as food contact material. In particular, in nanomaterials, especially new technologies create substances in particle sizes that exhibit physicochemical properties considerably divergent from those at a larger scale. Nanoparticle risk assessment is conducted on individual events until further understanding about the innovative technology is obtained [176].

As a result, the risk assessment associated with nanoclay in food packaging applications is far broader than the nanoscale paradigm, which entails two unique challenges:

- Migration of nanoparticles and their constituents form nanocomposite into the food.
- The impact of nanoparticle incorporation on other components (monomers, food additives, or processing aids) with potential toxicity.

The inclusive evaluation of any polymer NCs for direct contact must necessarily include these above two investigations [177].

7.6.1 Migration and Exposure of Nanoclay Materials to Humans and the Environment

Nanoclay is natural; however, nanoclay-based PNCs for food packaging necessitates careful consideration of the harmful effects on consumer and environmental health [178]. Nanoparticles can enter into food after interacting with components during processing, storage, and distribution. Clay-based nanocomposite materials must demonstrate the safety of the nanoparticles during regular use, disposal, and recycling. Nanoparticles can enter the biological system via various routes or migrate to food from contact material and eventually interact with body cells, as shown in Figure 7.5. As a result of recycling and disposal, they may contact plants, animals, and humans [179]. Migration is generally defined as the mass transfer of low molecular particles into the host product. Since packaging material cannot be chemically inert, direct contact can substantially migrate into the product. It becomes vital for the non-intended transfer that has been occurring because of undesirable packaging constituents that may affect the consumer's safety [180]. Nanomaterials can migrate into food from composite, which can change the quality of food products

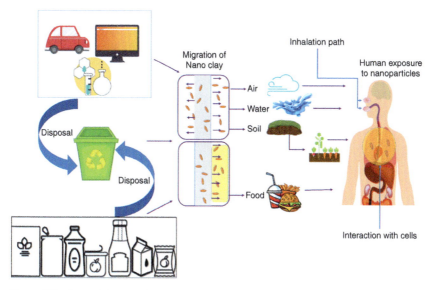

Figure 7.5 Potential routes for the migration of nanomaterials to humans and the environment.

such as TiO_2 that causes rancidity in high-fat foods [181]. Several researchers have been interested in gaining attention to the potential and possible conditions for the migration of nanoparticles from the PNCs in recent years. Evidence is available associated with nanoparticle release during machining, weathering, washing, contact, and incineration. The available publications show the migration of nanomaterials from the food packaging materials like Ag-nanoparticle, TiO_2, TiO_2, and ZnO [182, 183].

The researchers classified the migration of nanoparticles based on the particles' quantity, type, and size. However, the informed incidents of nanoclay migration from the nanoclay NCs are scarce. In a study, a theoretical approach to MMT diffusion is considered negligible from the packaging material due to its slow transfer rate because of larger size and morphological differences between platelet-like nanoclay and spherical silver, iron, zinc, or titanium nanoparticles [184]. There are some investigations available that demonstrate the MMT release by measuring the essential component (mainly Si) from 3% to 5% (w/w) of nanoclay loading into the PNCs [185–187]. The migration value of MMT measured from the PNCs in various food simulants is very low, which did not exceed the considered level frequently seen in food materials. In this regard, consumer exposure and safety risks PCNs were minimal, even though recent research indicated migration of the nanoclay MMT from commercial packaging during shredding [188]. A study observed that PP/MMT composite has lower migration of airborne particles than the neat PP during shredding.

The nanoclay particles remain attached to the PNCs, preventing the nanoclay from migrating into the atmosphere. According to the findings, recycling nanocomposite should not cause any health concerns to workers or the general public than recycling neat polymer [189]. Similar results were obtained during mechanical drilling with PA 6/silica and PA 6/MMT NCs, implying that the presence of nanoclay can lower the concentration of generated particles during drilling and particle deposition [190, 191]. Mechanically driven processes such as desorption, dissolution, and degradation of the matrix with clay also influence nanoclay release. Weak hydrogen bond led to the desorption of the nanoclay into the migrants, and concentration gradient also induces migration of the nanoparticles from the interior of the NCs. Another critical factor is high temperature or UV exposure of nanoparticles, leading to polymer matrix degradation during processing [192]. Time is also a significant factor because nanoclays have a large surface contact area with the polymer; it may take longer for food simulants to erode the polymer and dislodge a nanoplatelet than a spherical NP. Although nanoclay is embedded into the polymer matrix with different organo modifications to enhance the overall nanocomposite performance's compatibility and properties, these particles are still released during wear and tear. According to EU regulations, the total limit for migrating plastic-based food packaging material or particles is 10 mg/dm^2 of packaging material surface area. However, for bigger containers, such as 500 ml–10 l food containers or filled products, the measurement surface area of the stopper, caps, gaskets, or closing area is problematic; the restriction mentioned above is increased to 60 mg of particle emitted per kilogram of foodstuffs [193, 194]. The existing migration studies suggest that the migration from clay composites is modest and acceptable to be used for food packaging. There is a need for more extensive research that focuses on each food simulant individually; on the other hand; it could help with more effective evaluation and regulation.

7.6.2 Toxicity of Nanoclay

The wide application of nanoparticles has gained in recent times with the advancement in the material performance it has brought attention toward safety-related issues. The environmental emission via air, groundwater, and soil has reached the internal part of organs which might be sensitive to this nanoparticle to some extent. The migration of nanoparticles into the environment harms resources like water, soil, plant, and living organisms. Moreover, for food packaging, these nanoparticles can bind nutrients or interact in an undesired way, which could be the source of harmful reactions of nanoparticles to the consumers [27]. Nanoparticles can enter the environment through different mechanisms, but their stability duration in the atmosphere is still an unanswered question. Boxall et al. estimated nanoparticle concentration in the air, soil, and water in the range of ng/l to μg/l. They observed the level of nanoparticles in the environment is lower than lethal and sublethal effects with low indication level of risk [195]. There are three possible ways of nanoparticles entering the body: inhalation, penetration through the skin, and ingestion through an oral path. Still, there is doubt in the customer's mind that indirect exposure due to migration can ultimately affect the health and safety of individuals. The workers in the nanomaterial-producing factories are prone to inhalation, and penetration of nanoparticles through the skin is entirely a safety issue. It is recommended them to protect using gloves, eyeglasses, and mask with highly efficient filters. It is vital first to investigate the migration extent of nanoparticles from the packaging material to food for food packages related to migration. If there is any migration, the duration of ingestion and penetration of nanoparticles in the human body from mouth to the gastrointestinal tract is essential to investigate its effect on the body. It is imperative to understand the accumulation and excretion mechanism of nanoparticles in human organs [27]. The penetration of nanomaterials depends upon the skin type and the properties of the nanomaterial. Available data suggested that nanoparticles with a size of more than 10 nm are not penetrating through the skin. If any event is observed, toxicity depends on the penetration site's barrier function and clearance mechanism.

Additionally, the translocation of these nanoparticles from the penetration site depends upon the nanoparticles and organism interaction [196–198]. There are proofs available on the toxicity of nanoclay on a living organism. Bentonite, MMT, cloisite 30B, and organ-modified nanoclay did show some cytotoxicity [101, 199–201]. Researchers found that platelets structured nanoclay are more toxic than tubular ones, so the toxicity depends upon the nanoclay materials [202]. According to EU regulation, the implementation of food packaging is considered to be a safe practice. The size of the host polymer and nanoclay sheets in the NCs is a crucial characteristic from the EU standpoint. The surfactant employed for organic modification of nanoclay must also be approved for food contact by EU law or the FDA [203–206].

7.7 Future Outlook and Conclusion

Understanding the physicochemical properties of food and the packaging material plays a vital role in food packaging development practices. There is no absolute inert material that has zero migration at all from packaging to food. Nanoclay has shown the potential for emerging filler materials for the host composite matrix. Synthetic polymers hamper environmental health. Hence, biopolymers are getting priority over synthetic polymers due to their biodegradability and compatibility. Nanoclay provides a remarkable improvement in the resulting material's physicochemical, mechanical, barrier properties, and degradation capacity. Compatibility and excellent dispersion qualities are essential in the preparation of PCNs. The application of nanoclay has broadened its paradigm with different promising functions such as antimicrobial activity, active processes, colorimetric indicator, and additive partitioning. Environmental protection has become a priority, and food packaging has a considerable contribution in terms of conventional synthetic polymer-based material which, is nonbiodegradable. It is an urgent requirement to protect the environment from replacing synthetic material with biobased or degradable material for food packaging applications in the future. Again, understanding and knowledge of the available biodegradable resources for commercial application implementation remains an obvious challenge. The lack of consumer and environmental safety legislation for nanoparticles appears restrictive in applying nanoclay in food packaging development. The advantages and limitations of the nanoclay have become evident now. But there is still a long way to go to attain and tune this attribute. Research and development in these areas will benefit not only current uses but also lead to new markets in developing new materials for commercial applications in the future. The construction of PNCs with clay materials expanded for extended desired properties of packaged food is a recent concept with recognized future potential.

References

1 Han, J.H. (2005). *Innovations in Food Packaging*, 517. Elsevier.
2 Robertson, J.M.C., Robertson, P.K.J., and Lawton, L.A. (2005 Sep). A comparison of the effectiveness of TiO_2 photocatalysis and UVA photolysis for the destruction of three pathogenic micro-organisms. *Journal of Photochemistry and Photobiology A: Chemistry* 175 (1): 51–56.
3 Khedkar, D. and Khedkar, R. (2020). New innovations in food packaging in food industry. In: *Emerging Technologies in Food Science* (ed. M. Thakur and V. Modi), 165–185. Singapore: Springer.
4 Thakur, M. and Modi, V.K. *Emerging Technologies in Food Science*, vol. 273. Singapore: Springer.
5 Kuswandi, B. (2017). Environmental friendly food nano-packaging. *Environmental Chemistry Letters* 15 (2): 205–221.
6 He, X., Deng, H., and Hwang, H. (2019). The current application of nanotechnology in food and agriculture. *Journal of Food and Drug Analysis* 27 (1): 1–21.
7 Nourizadeh, H. and Bakhshayesh, A. (2020). Nanoclay-based products across global markets. In: *StatNano Applied and Industrial Series: Nanoclay-Based Products Across Global Markets: Applications and Properties*, 3–33.
8 Duncan, T.V. (2011). Applications of nanotechnology in food packaging and food safety: barrier materials, antimicrobials and sensors. *Journal of Colloid and Interface Science* 363 (1): 1–24.
9 Anirudhan, T.S., Athira, V.S., and Sekhar, V.C. (2018). Electrochemical sensing and nano molar level detection of Bisphenol-A with molecularly imprinted polymer tailored on multiwalled carbon nanotubes. *Polymer (Guildf)*. 146: 312–320.
10 Rovera, C., Ghaani, M., and Farris, S. (2020). Nano-inspired oxygen barrier coatings for food packaging applications: an overview. *Trends in Food Science & Technology* 97: 210–220.
11 Kargozar, S. and Mozafari, M. (2018). Nanotechnology and Nanomedicine: start small, think big. *Materials Today Proceedings* 5 (7): 15492–15500.
12 Pal, M. (2017). Nanotechnology: a new approach in food packaging. *Journal of Food Microbiology Safety & Hygiene* 2: 121.

13 Collister, J. (2002). *Commercialization of Polymer Nanocomposites*. ACS Publications.
14 Konta, J. (1995). Clay and man: clay raw materials in the service of man. *Applied Clay Science* 10 (4): 275–335.
15 NPD. Nanotechnology in Food Industry | NPD. Retrieved September 4, 2021, from https://product.statnano.com/industry/food.
16 Ayhan, Z., Cimmino, S., Esturk, O. et al. (2015). Development of films of novel polypropylene based nanomaterials for food packaging application. *Packaging Technology and Science* 28 (7): 589–602.
17 Ray, S., Quek, S.Y., Easteal, A., and Chen, X.D. (2006). The potential use of polymer-clay nanocomposites in food packaging. *International Journal of Food Engineering* 2 (4).
18 Sahoo, N.G., Rana, S., Cho, J.W. et al. (2010). Polymer nanocomposites based on functionalized carbon nanotubes. *Progress in Polymer Science* 35 (7): 837–867.
19 Potts, J.R., Dreyer, D.R., Bielawski, C.W., and Ruoff, R.S. (2011). Graphene-based polymer nanocomposites. *Polymer (Guildf)*. 52 (1): 5–25.
20 Lee, K.-Y., Aitomäki, Y., Berglund, L.A. et al. (2014). On the use of nanocellulose as reinforcement in polymer matrix composites. *Composites Science and Technology* 105: 15–27.
21 Barus, S., Zanetti, M., Lazzari, M., and Costa, L. (2009). Preparation of polymeric hybrid nanocomposites based on PE and nanosilica. *Polymer (Guildf)*. 50 (12): 2595–2600.
22 Zhang, J., Manias, E., and Wilkie, C.A. (2008). Polymerically modified layered silicates: an effective route to nanocomposites. *Journal of Nanoscience and Nanotechnology* 8 (4): 1597–1615.
23 Jlassi, K., Krupa, I., and Chehimi, M.M. (2017). Chapter 1: Overview: clay preparation, properties, modification. In: *Polymer Nanocomposites* (ed. K. Jlassi, M.M. Chehimi, and S. Thomas), 1, 1–28, 28. https://doi.org/10.1016/B978-0-323-46153-5.00001-X.
24 Murray, H.H. (1991). Overview — clay mineral applications. *Applied Clay Science* 5 (5–6): 379–395.
25 Valapa, R.B., Loganathan, S., Pugazhenthi, G. et al. (2017). An overview of polymer–clay nanocomposites. *Clay-Polymer Nanocomposites* 29–81.
26 Najafi, N., Heuzey, M.C., and Carreau, P.J. (2012). Polylactide (PLA)-clay nanocomposites prepared by melt compounding in the presence of a chain extender. *Composites Science and Technology* 72 (5): 608–615.
27 Silvestre, C., Duraccio, D., and Cimmino, S. (2011). Food packaging based on polymer nanomaterials. *Progress in Polymer Science* 36 (12): 1766–1782.
28 Radfar, R., Hosseini, H., Farhoodi, M. et al. (2020). Optimization of antibacterial and mechanical properties of an active LDPE/starch/nanoclay nanocomposite film incorporated with date palm seed extract using D-optimal mixture design approach. *International Journal of Biological Macromolecules* 158: 790–799.
29 Bai, C., Ke, Y., Hu, X. et al. (2020). Preparation and properties of amphiphilic hydrophobically associative polymer/montmorillonite nanocomposites. *Royal Society Open Science* 7 (5): 200199.

30 Kojima, Y., Usuki, A., Kawasumi, M. et al. (1993). One-pot synthesis of nylon 6–clay hybrid. *Journal of Polymer Science Part A: Polymer Chemistry* 31 (7): 1755–1758.

31 Rivero, S., Garcia, M.A., and Pinotti, A. (2009). Composite and bi-layer films based on gelatin and chitosan. *Journal of Food Engineering* 90 (4): 531–539.

32 Voon, H.C., Bhat, R., Easa, A.M. et al. (2012). Effect of addition of halloysite nanoclay and SiO_2 nanoparticles on barrier and mechanical properties of bovine gelatin films. *Food and Bioprocess Technology* 5 (5): 1766–1774.

33 Tang, X., Alavi, S., and Herald, T.J. (2008). Barrier and mechanical properties of starch-clay nanocomposite films. *Cereal Chemistry* 85 (3): 433–439.

34 de Jara, E.M., García-Hernández, E., Quequezana-Bedregal, M.J. et al. (2020). Potato starch-based films: Effects of glycerol and montmorillonite nanoclay concentration. *Revista Mexicana de Ingeniería Química* 19 (2): 627–637.

35 Risyon, N.P., Othman, S.H., Basha, R.K., and Talib, R.A. (2016). Effect of halloysite nanoclay concentration and addition of glycerol on mechanical properties of bionanocomposite films. *Polymers and Polymer Composites* 24 (9): 795–802.

36 Islam, H.B.M.Z., Susan, M.A.B.H., and Bin, I.A. (2021). High-strength potato starch/hectorite clay-based nanocomposite film: synthesis and characterization. *Iranian Polymer Journal* 30 (5): 513–521.

37 Penaloza, D.P. Jr. (2019). Enhanced mechanical, thermal and barrier properties of clay-based polymer nanocomposite systems. *Építőanyag (Online)* 3: 74–79.

38 Paul, D.R. and Robeson, L.M. (2008). Polymer nanotechnology: nanocomposites. *Polymer (Guildf)* 49 (15): 3187–3204.

39 Bharadwaj, R.K. (2001). Modeling the barrier properties of polymer-layered silicate nanocomposites. *Macromolecules* 34 (26): 9189–9192.

40 Joon Choi, W., Kim, H., Han Yoon, K. et al. (2006). Preparation and barrier property of poly (ethylene terephthalate)/clay nanocomposite using clay-supported catalyst. *Journal of Applied Polymer Science* 100 (6): 4875–4879.

41 Kang, D., Kim, D., Yoon, S. et al. (2007). Properties and dispersion of EPDM/modified-organoclay nanocomposites. *Macromolecular Materials and Engineering* 292 (3): 329–338.

42 Yano, K., Usuki, A., Okada, A. et al. (1993). Synthesis and properties of polyimide–clay hybrid. *Journal of Polymer Science Part A: Polymer Chemistry* 31 (10): 2493–2498.

43 Jacob, A., Kurian, P., and Aprem, A.S. (2008). Transport properties of natural rubber latex layered clay nanocomposites. *Journal of Applied Polymer Science* 108 (4): 2623–2629.

44 Wu, Y.-P., Wang, Y.-Q., Zhang, H.-F. et al. (2005). Rubber–pristine clay nanocomposites prepared by co-coagulating rubber latex and clay aqueous suspension. *Composites Science and Technology* 65 (7–8): 1195–1202.

45 Li, P., Wang, L., Song, G. et al. (2008). Characterization of high-performance exfoliated natural rubber/organoclay nanocomposites. *Journal of Applied Polymer Science* 109 (6): 3831–3838.

46 Brody, A.L. (2006). Nano and food packaging technologies converge. *Food Technology* 60 (3): 92–94.

47 Galimberti, M., Cipolletti, V.R., and Coombs, M. (2013). Applications of clay–polymer nanocomposites. In: *Developments in Clay Science* (ed. F. Bergaya and G. Lagaly), 539–586. Elsevier https://doi.org/10.1016/B978-0-08-098259-5.00020-2.

48 Pielichowski, K. and Njuguna, J. (2005). *Thermal Degradation of Polymeric Materials*. iSmithers Rapra Publishing.

49 Bikiaris, D. (2011). Can nanoparticles really enhance thermal stability of polymers? Part II: an overview on thermal decomposition of polycondensation polymers. *Thermochimica Acta* 523 (1–2): 25–45.

50 Zhai, H., Xu, W., Guo, H. et al. (2004). Preparation and characterization of PE and PE-g-MAH/montmorillonite nanocomposites. *European Polymer Journal* 40 (11): 2539–2545.

51 Zhang, J. and Wilkie, C.A. (2003). Preparation and flammability properties of polyethylene–clay nanocomposites. *Polymer Degradation and Stability* 80 (1): 163–169.

52 Qiu, L., Chen, W., and Qu, B. (2006). Morphology and thermal stabilization mechanism of LLDPE/MMT and LLDPE/LDH nanocomposites. *Polymer (Guildf)* 47 (3): 922–930.

53 Pramoda, K.P., Liu, T., Liu, Z. et al. (2003). Thermal degradation behavior of polyamide 6/clay nanocomposites. *Polymer Degradation and Stability* 81 (1): 47–56.

54 Chigwada, G., Jiang, D.D., and Wilkie, C.A. (2005). Polystyrene nanocomposites based on carbazole-containing surfactants. *Thermochimica Acta* 436 (1–2): 113–121.

55 Davis, C.H., Mathias, L.J., Gilman, J.W. et al. (2002). Effects of melt-processing conditions on the quality of poly (ethylene terephthalate) montmorillonite clay nanocomposites. *Journal of Polymer Science Part B Polymer Physics* 40 (23): 2661–2666.

56 Liang, Z., Wan, C., Zhang, Y. et al. (2004). PVC/montmorillonite nanocomposites based on a thermally stable, rigid-rod aromatic amine modifier. *Journal of Applied Polymer Science* 92 (1): 567–575.

57 Xiao, J., Hu, Y., Wang, Z. et al. (2005). Preparation and characterization of poly (butylene terephthalate) nanocomposites from thermally stable organic-modified montmorillonite. *European Polymer Journal* 41 (5): 1030–1035.

58 Chang, J., Mun, M.K., and Kim, J. (2006). Poly (trimethylene terephthalate) nanocomposite fibers comprising different organoclays: thermomechanical properties and morphology. *Journal of Applied Polymer Science* 102 (5): 4535–4545.

59 Mohanty, S. and Nayak, S.K. (2007). Mechanical, thermal and viscoelastic behavior of nylon 6/clay nanocomposites with cotreated montmorillonites. *Polymer - Plastics Technology and Engineering* 46 (4): 367–376.

60 Wang, K., Chen, L., Kotaki, M., and He, C. (2007). Preparation, microstructure and thermal mechanical properties of epoxy/crude clay nanocomposites. *Composites. Part A, Applied Science and Manufacturing* 38 (1): 192–197.

61 Sharma, C., Dhiman, R., Rokana, N., and Panwar, H. (2017). Nanotechnology: an untapped resource for food packaging. *Frontiers in Microbiology* 8: 1735.

62 Mihindukulasuriya, S.D.F. and Lim, L.T. (2014). Nanotechnology development in food packaging: a review. *Trends in Food Science and Technology* 40 (2): 149–167.

63 Srithammaraj, K., Magaraphan, R., and Manuspiya, H. (2012). Modified porous clay heterostructures by organic–inorganic hybrids for nanocomposite ethylene scavenging/sensor packaging film. *Packaging Technology and Science* 25 (2): 63–72.

64 Gaikwad, K.K., Singh, S., and Negi, Y.S. (2020). Ethylene scavengers for active packaging of fresh food produce. *Environmental Chemistry Letters* 18 (2): 269–284.

65 Kumar, A., Gupta, V., Singh, S. et al. (2021). Pine needles lignocellulosic ethylene scavenging paper impregnated with nanozeolite for active packaging applications. *Industrial Crops and Products* 170: 113752.

66 Coloma, A., Rodríguez, F.J., Bruna, J.E. et al. (2014). Development of an active film with natural zeolite as ethylene scavenger. *Journal of the Chilean Chemical Society* 59 (2): 2409–2414.

67 de Bruijn, J., Gómez, A., Loyola, C. et al. (2020). Use of a copper-and zinc-modified natural zeolite to improve ethylene removal and postharvest quality of tomato fruit. *Crystals* 10 (6): 471.

68 Álvarez-Hernández, M.H., Martínez-Hernández, G.B., Castillejo, N. et al. (2021). Development of an antifungal active packaging containing thymol and an ethylene scavenger. Validation during storage of cherry tomatoes. *Food Packaging and Shelf Life* 29: 100734.

69 Álvarez-Hernández, M.H., Martínez-Hernández, G.B., Avalos-Belmontes, F. et al. (2020). Postharvest quality retention of apricots by using a novel sepiolite–loaded potassium permanganate ethylene scavenger. *Postharvest Biology and Technology* 160: 111061.

70 Yuan, P., Tan, D., and Annabi-Bergaya, F. (2015). Properties and applications of halloysite nanotubes: recent research advances and future prospects. *Applied Clay Science* 112: 75–93.

71 Busolo, M.A. and Lagaron, J.M. (2012). Oxygen scavenging polyolefin nanocomposite films containing an iron modified kaolinite of interest in active food packaging applications. *Innovative Food Science and Emerging Technologies* 16: 211–217.

72 Khalaj, M.J., Ahmadi, H., Lesankhosh, R., and Khalaj, G. (2016). Study of physical and mechanical properties of polypropylene nanocomposites for food packaging application: nano-clay modified with iron nanoparticles. *Trends in Food Science and Technology* 51: 41–48. https://doi.org/10.1016/j.tifs.2016.03.007.

73 Makaremi, M., Pasbakhsh, P., Cavallaro, G. et al. (2017). Effect of morphology and size of halloysite nanotubes on functional pectin bionanocomposites for food packaging applications. *ACS Applied Materials & Interfaces* 9 (20): 17476–17488.

74 Jiménez, A., Peltzer, M., and Ruseckaite, R. (2014). *Poly (lactic acid) Science and Technology: Processing, Properties, Additives and Applications*. Royal Society of Chemistry.

75 Kim, S.W. and Cha, S. (2014). Thermal, mechanical, and gas barrier properties of ethylene–vinyl alcohol copolymer-based nanocomposites for food packaging films: effects of nanoclay loading. *Journal of Applied Polymer Science* 131 (11).

76 Jung, B.N., Kang, D., Cheon, S. et al. (2019). The addition effect of hollow glass microsphere on the dispersion behavior and physical properties of polypropylene/clay nanocomposites. *Journal of Applied Polymer Science* 136 (14): 47476.

77 Wang, G., Zhao, G., Dong, G. et al. (2018). Lightweight and strong microcellular injection molded PP/talc nanocomposite. *Composites Science and Technology* 168: 38–46.

78 Bruna, J.E., Peñaloza, A., Guarda, A. et al. (2012). Development of MtCu^{2+}/LDPE nanocomposites with antimicrobial activity for potential use in food packaging. *Applied Clay Science* 58: 79–87.

79 Muñoz-Shugulí, C., Rodríguez, F.J., Bruna, J.E. et al. (2019). Cetylpyridinium bromide-modified montmorillonite as filler in low density polyethylene nanocomposite films. *Applied Clay Science* 168: 203–210.

80 Bumbudsanpharoke, N., Lee, W., Choi, J.C. et al. (2017). Influence of montmorillonite nanoclay content on the optical, thermal, mechanical, and barrier properties of low-density polyethylene. *Clays and Clay Minerals* 65 (6): 387–397.

81 Beesetty, P., Patil, B., and Doddamani, M. (2020). Mechanical behavior of additively manufactured nanoclay/HDPE nanocomposites. *Composite Structures* 247: 112442.

82 Majeed, K., Arjmandi, R., and Hassan, A. (2018). Mechanical and oxygen barrier properties of LDPE/MMT/MAPE and LDPE/MMT/EVA nanocomposite films: a comparison study. *Journal of Physical Science* 29 (1).

83 Majeed, K., Arjmandi, R., and Hassan, A. (2018). LDPE/RH/MAPE/MMT nanocomposite films for packaging applications. In: *Bionanocomposites for Packaging Applications* (ed. M. Jawaid and S. Swain), 209–225. Springer, Cham https://doi.org/10.1007/978-3-319-67319-6_11.

84 Savas, L.A. and Hancer, M. (2015). Montmorillonite reinforced polymer nanocomposite antibacterial film. *Applied Clay Science* 108: 40–44.

85 Ait Cherif, G., Kerkour, A., Baouz, T. et al. (2018). Investigating the diffusional behaviour of Irganox® 1076 antioxidant in HDPE/Cloisite® 15A nanocomposite-based food contact packaging films: effect of nanoclay loading. *Packaging Technology and Science* 31 (9): 621–629.

86 Ebrahimi, H., Abedi, B., Bodaghi, H. et al. (2018). Investigation of developed clay-nanocomposite packaging film on quality of peach fruit (*Prunus persica* Cv. Alberta) during cold storage. *Journal of Food Processing & Preservation* 42 (2): e13466.

87 Kim, J.M., Lee, M.H., Ko, J.A. et al. (2018). Influence of food with high moisture content on oxygen barrier property of polyvinyl alcohol (PVA)/vermiculite nanocomposite coated multilayer packaging film. *Journal of Food Science* 83 (2): 349–357.

88 Yussuf, A.A., Al-Saleh, M.A., Al-Samhan, M.M. et al. (2018). Investigation of polypropylene-montmorillonite clay nanocomposite films containing a pro-degradant additive. *Journal of Polymers and the Environment* 26 (1): 275–290.

89 Garofalo, E., Scarfato, P., Di Maio, L., and Incarnato, L. (2018). Tuning of co-extrusion processing conditions and film layout to optimize the performances of PA/PE multilayer nanocomposite films for food packaging. *Polymer Composites* 39 (9): 3157–3167.

90 Chaiko, D.J. (2006). Activation of organoclays and preparation of polyethylene nanocomposites. *e-Polymers* 6 (1): https://doi.org/10.1515/epoly.2006.6.1.242.

91 Zhao, C., Qin, H., Gong, F. et al. (2005). Mechanical, thermal and flammability properties of polyethylene/clay nanocomposites. *Polymer Degradation and Stability* 87 (1): 183–189.

92 Said, M., Challita, G., and Seif, S. (2020). Development of blown film linear low-density polyethylene-clay nanocomposites: Part B: mechanical and rheological characterization. *Journal of Applied Polymer Science* 137 (16): 48590.

93 Said, M., Seif, S., and Challita, G. (2020). Development of blown film linear low-density polyethylene–clay nanocomposites: Part A: manufacturing process and morphology. *Journal of Applied Polymer Science* 137 (16): 48589.

94 Jacquelot, E., Espuche, E., Gérard, J. et al. (2006). Morphology and gas barrier properties of polyethylene-based nanocomposites. *Journal of Polymer Science Part B: Polymer Physics* 44 (2): 431–440.

95 Zhong, Y., Janes, D., Zheng, Y. et al. (2007). Mechanical and oxygen barrier properties of organoclay-polyethylene nanocomposite films. *Polymer Engineering and Science* 47 (7): 1101–1107.

96 Arunvisut, S., Phummanee, S., and Somwangthanaroj, A. (2007). Effect of clay on mechanical and gas barrier properties of blown film LDPE/clay nanocomposites. *Journal of Applied Polymer Science* 106 (4): 2210–2217.

97 Xie, L., Lv, X.-Y., Han, Z.-J. et al. (2012). Preparation and performance of high-barrier low density polyethylene/organic montmorillonite nanocomposite. *Polymer - Plastics Technology and Engineering* 51 (12): 1251–1257.

98 Cesur, S., Köroğlu, C., and Yalçın, H.T. (2018). Antimicrobial and biodegradable food packaging applications of polycaprolactone/organo nanoclay/chitosan polymeric composite films. *Journal of Vinyl & Additive Technology* 24 (4): 376–387.

99 Rhim, J.-W., Hong, S.-I., and Ha, C.-S. (2009). Tensile, water vapor barrier and antimicrobial properties of PLA/nanoclay composite films. *LWT-Food Science and Technology* 42 (2): 612–617.

100 Guo, G., Tian, H., and Wu, Q. (2019). Nanoclay incorporation into soy protein/polyvinyl alcohol blends to enhance the mechanical and barrier properties. *Polymer Composites* 40 (9): 3768–3776.

101 Araújo, A., Botelho, G., Oliveira, M., and Machado, A.V. (2014). Influence of clay organic modifier on the thermal-stability of PLA based nanocomposites. *Applied Clay Science* 88: 144–150.

102 Chang, J.-H., Ham, M., and Kim, J.-C. (2014). Comparison of properties of poly (vinyl alcohol) nanocomposites containing two different clays. *Journal of Nanoscience and Nanotechnology* 14 (11): 8783–8791.

103 Haafiz, K.M., Taiwo, O.F.A., Razak, N. et al. (2019). Development of green MMT-modified hemicelluloses based nanocomposite film with enhanced functional and barrier properties. *BioResources.* 14 (4): 8029–8047.

104 Cui, R., Yan, J., Cao, J. et al. (2021). Release properties of cinnamaldehyde loaded by montmorillonite in chitosan-based antibacterial food packaging. *International Journal of Food Science and Technology* 56 (8): 3670–3681.

105 Makwana, D., Castaño, J., Somani, R.S., and Bajaj, H.C. (2020). Characterization of Agar-CMC/Ag-MMT nanocomposite and evaluation of antibacterial and mechanical properties for packaging applications. *Arabian Journal of Chemistry* 13 (1): 3092–3099.

106 Valencia, G.A., Luciano, C.G., Lourenco, R.V., and do Amaral Sobral, P.J. (2018). Microstructure and physical properties of nano-biocomposite films based on cassava starch and laponite. *International Journal of Biological Macromolecules* 107: 1576–1583.

107 Shi, J., Zhang, R., Yang, N. et al. (2020). Hierarchical incorporation of surface-functionalized laponite clay nanoplatelets with Type I collagen matrix. *Biomacromolecules* 22 (2): 504–513.

108 Vishnuvarthanan, M. and Rajeswari, N. (2019). Food packaging: pectin–laponite–Ag nanoparticle bionanocomposite coated on polypropylene shows low O_2 transmission, low Ag migration and high antimicrobial activity. *Environmental Chemistry Letters* 17 (1): 439–445.

109 Vishnuvarthanan, M. and Rajeswari, N. (2019). Preparation and characterization of carrageenan/silver nanoparticles/Laponite nanocomposite coating on oxygen plasma surface modified polypropylene for food packaging. *Journal of Food Science and Technology* 56 (5): 2545–2552.

110 Vasile, C. (2018). Polymeric nanocomposites and nanocoatings for food packaging: a review. *Materials (Basel)* 11 (10): 1834.

111 Hassan, T., Salam, A., Khan, A. et al. (2021). Functional nanocomposites and their potential applications: a review. *Journal of Polymer Research* 28 (2): 1–22.

112 Hundáková, M., Tokarský, J., Valášková, M. et al. (2015). Structure and antibacterial properties of polyethylene/organo-vermiculite composites. *Solid State Sciences* 48: 197–204.

113 Wang, X., Du, Y., Luo, J. et al. (2007). Chitosan/organic rectorite nanocomposite films: structure, characteristic and drug delivery behaviour. *Carbohydrate Polymers* 69 (1): 41–49.

114 Bharadwaj, R.K., Mehrabi, A.R., Hamilton, C. et al. (2002). Structure–property relationships in cross-linked polyester–clay nanocomposites. *Polymer (Guildf)* 43 (13): 3699–3705.

115 Bergaya, F. and Lagaly, G. (2006). General introduction: clays, clay minerals, and clay science. *Developments in Clay Science* 1: 1–18.

116 Bergaya, F., Lagaly, G., and Vayer, M. (2006). Cation and anion exchange. *Developments in Clay Science* 1: 979–1001.

117 Ghanbarzadeh, B., Oleyaei, S.A., and Almasi, H. (2015). Nanostructured materials utilized in biopolymer-based plastics for food packaging applications. *Critical Reviews in Food Science and Nutrition* 55 (12): 1699–1723.

118 De Paiva, L.B., Morales, A.R., and Díaz, F.R.V. (2008). Organoclays: properties, preparation and applications. *Applied Clay Science* 42 (1–2): 8–24.

119 Youssef, A.M. (2013). Polymer nanocomposites as a new trend for packaging applications. *Polymer - Plastics Technology and Engineering* 52 (7): 635–660.

120 Campos-Requena, V.H., Rivas, B.L., Pérez, M.A. et al. (2015). Polymer/clay nanocomposite films as active packaging material: modeling of antimicrobial release. *European Polymer Journal* 71: 461–475.

121 Rhim, J.-W., Park, H.-M., and Ha, C.-S. (2013). Bio-nanocomposites for food packaging applications. *Progress in Polymer Science* 38 (10–11): 1629–1652.

122 Khanam, P.N. and AlMaadeed, M.A.A. (2015). Processing and characterization of polyethylene-based composites. *Advanced Manufacturing: Polymer & Composites Science* 1 (2): 63–79.

123 Mooninta, S., Poompradub, S., and Prasassarakich, P. (2020). Packaging film of PP/LDPE/PLA/clay composite: physical, barrier and degradable properties. *Journal of Polymers and the Environment* 28 (12): 3116–3128.

124 Giannakas, A., Spanos, C.G., Kourkoumelis, N. et al. (2008). Preparation, characterization and water barrier properties of PS/organo-montmorillonite nanocomposites. *European Polymer Journal* 44 (12): 3915–3921.

125 Dunkerley, E. and Schmidt, D. (2010). Effects of composition, orientation and temperature on the O_2 permeability of model polymer/clay nanocomposites. *Macromolecules* 43 (24): 10536–10544.

126 Dini, M., Mousavand, T., Carreau, P.J. et al. (2014). Effect of water-assisted extrusion and solid-state polymerization on the microstructure of PET/Clay nanocomposites. *Polymer Engineering and Science* 54 (8): 1723–1736.

127 Chouhan, D.K., Rath, S.K., Kumar, A. et al. (2015). Structure-reinforcement correlation and chain dynamics in graphene oxide and Laponite-filled epoxy nanocomposites. *Journal of Materials Science* 50 (22): 7458–7472.

128 Daniel, L.M., Frost, R.L., and Zhu, H.Y. (2008). Edge-modification of laponite with dimethyl-octylmethoxysilane. *Journal of Colloid and Interface Science* 321 (2): 302–309.

129 Wheeler, P.A., Wang, J., and Mathias, L.J. (2006). Poly(methyl methacrylate)/laponite nanocomposites: exploring covalent and ionic clay modifications. *Chemistry of Materials* 18 (17): 3937–3945.

130 Perotti, G.F., Barud, H.S., Messaddeq, Y. et al. (2011). Bacterial cellulose–laponite clay nanocomposites. *Polymer (Guildf)* 52 (1): 157–163.

131 Olivera, N., Rouf, T.B., Bonilla, J.C. et al. (2019). Effect of LAPONITE® addition on the mechanical, barrier and surface properties of novel biodegradable kafirin nanocomposite films. *Journal of Food Engineering* 245: 24–32.

132 Silva, J.M., Barud, H.S., Meneguin, A.B. et al. (2019). Inorganic-organic bio-nanocomposite films based on Laponite and Cellulose Nanofibers (CNF). *Applied Clay Science* 168: 428–435.

133 de Oliveira, R.L., da Silva, B.H., De Salvi, D.T.B. et al. (2015). Transparent organic–inorganic nanocomposites membranes based on carboxymethylcellulose and synthetic clay. *Industrial Crops and Products* 69: 415–423.

134 Suárez, M. and García-Romero, E. (2011). Advances in the crystal chemistry of sepiolite and palygorskite. In: *Developments in Clay Science*, vol. 3 (ed. E. Galàn and A. Singer), 33–65. Elsevier, Elsevier.

135 Zheng, Y. and Zheng, Y. (2006). Study on sepiolite-reinforced polymeric nanocomposites. *Journal of Applied Polymer Science* 99 (5): 2163–2166.

136 Ma, J., Bilotti, E., Peijs, T., and Darr, J.A. (2007). Preparation of polypropylene/sepiolite nanocomposites using supercritical CO_2 assisted mixing. *European Polymer Journal* 43 (12): 4931–4939.

137 Ovarlez, S., Giulieri, F., Chaze, A. et al. (2009). The incorporation of indigo molecules in sepiolite tunnels. *Chemistry A European Journal* 15 (42): 11326–11332.

138 Mohd Zaini, N.A., Ismail, H., and Rusli, A. (2017). Short review on sepiolite-filled polymer nanocomposites. *Polymer - Plastics Technology and Engineering* 56 (15): 1665–1679.

139 Alvarez, A., Santaren, J., Esteban-Cubillo, A., and Aparicio, P. (2011). Current industrial applications of palygorskite and sepiolite. In: *Developments in Clay Science*, vol. 3 (ed. E. Galàn and A. Singer), 281–298. Elsevier.

140 García-López, D., Fernández, J.F., Merino, J.C. et al. (2010). Effect of organic modification of sepiolite for PA 6 polymer/organoclay nanocomposites. *Composites Science and Technology* 70 (10): 1429–1436.

141 Nehra, R., Maiti, S.N., and Jacob, J. (2018). Poly (lactic acid)/(styrene-ethylene-butylene-styrene)-g-maleic anhydride copolymer/sepiolite nanocomposites: Investigation of thermo-mechanical and morphological properties. *Polymers for Advanced Technologies* 29 (1): 234–243.

142 Cheikh, D., Martín-Sampedro, R., Majdoub, H., and Darder, M. (2020). Alginate bionanocomposite films containing sepiolite modified with polyphenols from myrtle berries extract. *International Journal of Biological Macromolecules* 165: 2079–2088.

143 Fernández-Menéndez, T., García-López, D., Argüelles, A. et al. (2021). Application of PET/sepiolite nanocomposite trays to improve food quality. *Foods* 10 (6): 1188.

144 Abdallah, W. and Yilmazer, U. (2011). Novel thermally stable organo-montmorillonites from phosphonium and imidazolium surfactants. *Thermochimica Acta* 525 (1–2): 129–140.

145 Ramos Filho, F.G., Mélo, T.J.A., Rabello, M.S., and Silva, S.M.L. (2005). Thermal stability of nanocomposites based on polypropylene and bentonite. *Polymer Degradation and Stability* 89 (3): 383–392.

146 Dahiya, J.B., Muller-Hagedorn, M., Bockhorn, H., and Kandola, B.K. (2008). Synthesis and thermal behaviour of polyamide 6/bentonite/ammonium polyphosphate composites. *Polymer Degradation and Stability* 93 (11): 2038–2041.

147 Barbosa, R., Alves, T.S., Araújo, E.M. et al. (2014). Flammability and morphology of HDPE/clay nanocomposites. *Journal of Thermal Analysis and Calorimetry* 115 (1): 627–634.

148 Seyidoglu, T. and Yilmazer, U. (2012). Use of purified and modified bentonites in linear low-density polyethylene/organoclay/compatibilizer nanocomposites. *Journal of Applied Polymer Science* 124 (3): 2430–2440.

149 Abdallah, W. and Yilmazer, U. (2013). Polyamide 66 nanocomposites based on organoclays treated with thermally stable phosphonium salts. *Journal of Applied Polymer Science* 127 (1): 772–783.

150 Kashiri, M., Maghsoudlo, Y., and Khomeiri, M. (2017). Incorporating *Zataria multiflora* Boiss. essential oil and sodium bentonite nano-clay open a new perspective to use zein films as bioactive packaging materials. *Food Science and Technology International* 23 (7): 582–596.

151 Iamareerat, B., Singh, M., Sadiq, M.B., and Anal, A.K. (2018). Reinforced cassava starch based edible film incorporated with essential oil and sodium bentonite nanoclay as food packaging material. *Journal of Food Science and Technology* 55 (5): 1953–1959.

152 Breen, C., Clegg, F., Thompson, S. et al. (2019). Exploring the interactions between starches, bentonites and plasticizers in sustainable barrier coatings for paper and board. *Applied Clay Science* 183: 105272.

153 Zheng, M., Tajvidi, M., Tayeb, A.H., and Stark, N.M. (2019). Effects of bentonite on physical, mechanical and barrier properties of cellulose nanofibril hybrid films for packaging applications. *Cellulose* 26 (9): 5363–5379.

154 Wang, D., Jang, B.N., Su, S. et al. (2005). *Fire Retardancy of Polystyrene-Hectorite Nanocomposites*. Cambridge, UK: Royal Society of Chemistry.

155 Zhang, J., Zhou, C.H., Petit, S., and Zhang, H. (2019). Hectorite: synthesis, modification, assembly and applications. *Applied Clay Science* 177: 114–138.

156 Venkata Timmaraju, M., Gnanamoorthy, R., Kannan, K., and Sriharsha, G. (2018). Experimental and numerical prediction of effect of frequency on bending fatigue performance of polyamide 66/hectorite nanocomposite. *Plastics Rubber and Composites* 47 (6): 282–295.

157 Cabedo, L. and Gamez-Pérez, J. (2018). Inorganic-based nanostructures and their use in food packaging. In: *Nanomaterials for Food Packaging* (ed. M.Â.P.R. Cerqueira, J.M. Lagaron, L.M.P. Castro, and A.A.M. de Oliveira Soares Vicente), 13–45. Elsevier https://doi.org/10.1016/B978-0-323-51271-8.00002-4.

158 Hansen, T., Barber, P., Ma, J. et al. (2006). Layered oxide polymer nanocomposites: synthesis, characterization, and strategies for achieving enhanced barrier property. In: *NSTI-Nanotech*, 845–848.

159 Cheng, Y., Fu, Q., Fang, C. et al. (2019). Preparation, structure, and properties of modified asphalt with waste packaging polypropylene and organic rectorite. *Advances in Materials Science and Engineering* 2019: https://doi.org/10.1155/2019/5362795.

160 Li, X., Tu, H., Huang, M. et al. (2017). Incorporation of lysozyme-rectorite composites into chitosan films for antibacterial properties enhancement. *International Journal of Biological Macromolecules* 102: 789–795.

161 Wang, Y., Cao, Z., Liu, F., and Xue, X. (2017). Synthesis and characterization of UV-curing epoxy acrylate coatings modified with organically modified rectorite. *Journal of Coatings Technology and Research* 14 (1): 107–115.

162 Gardolinski, J.E., Carrera, L.C.M., Cantao, M.P., and Wypych, F. (2000). Layered polymer-kaolinite nanocomposites. *Journal of Materials Science* 35 (12): 3113–3119.

163 Wierer, K.A. and Dobiáš, B. (1988). Exchange enthalpies of H^+ and OH^- adsorption on minerals with different characters of potential-determining ions. *Journal of Colloid and Interface Science* 122 (1): 171–177.

164 Mbey, J.-A., Hoppe, S., and Thomas, F. (2012). Cassava starch–kaolinite composite film. Effect of clay content and clay modification on film properties. *Carbohydrate Polymers* 88 (1): 213–222.

165 Cabedo, L., Villanueva, M.P., Lagarón, J.M., and Giménez, E. (2017). Development and characterization of unmodified kaolinite/EVOH nanocomposites by melt compounding. *Applied Clay Science* 135: 300–306.

166 Ruamcharoen, J., Munlee, R., and Ruamcharoen, P. (2020). Improvement of water vapor barrier and mechanical properties of sago starch-kaolinite nanocomposites. *Polymer Composites* 41 (1): 201–209.

167 Benhacine, F., Ouargli, A., and Hadj-Hamou, A.S. (2019). Preparation and characterization of novel food packaging materials based on biodegradable PCL/Ag-kaolinite nanocomposites with controlled release properties. *Polym Technol Mater* 58 (3): 328–340.

168 Zhang, J.H., Zhuang, W., Zhang, Q. et al. (2007). Novel polylactide/vermiculite nanocomposites by in situ intercalative polymerization. I. Preparation, characterization, and properties. *Polymer Composites* 28 (4): 545–550.

169 Takahashi, S., Goldberg, H.A., Feeney, C.A. et al. (2006). Gas barrier properties of butyl rubber/vermiculite nanocomposite coatings. *Polymer (Guildf)* 47 (9): 3083–3093.

170 Han, W., Yu, Y., Li, N., and Wang, L. (2011). Application and safety assessment for nano-composite materials in food packaging. *Chinese Science Bulletin* 56 (12): 1216–1225.

171 Daneshniya, M., Maleki, M.H., Amini, F. et al. (2020). Positive and negative aspects of nanocomposites utilization in food packaging. In: *3rd International Congress of Science*. Engineering and Technology.

172 Tiede, K., Boxall, A.B.A., Tear, S.P. et al. (2008). Detection and characterization of engineered nanoparticles in food and the environment. *Food Additives and Contaminants* 25 (7): 795–821.

173 Bertrand, N. and Leroux, J.-C. (2012). The journey of a drug-carrier in the body: an anatomo-physiological perspective. *Journal of Controlled Release* 161 (2): 152–163.

174 Commission E (2004). Regulation (EC) No 1935/2004 of the European Parliament and of the Council of 27 October 2004 on materials and articles intended to come into contact with food and repealing Directives 80/590/EEC and 89/109/EEC. *Official Journal of the European Union* L338 (47): 4–17.

175 Guidance, D. (2011). *Considering Whether an FDA-Regulated Product Involves the Application of Nanotechnology*. FDA.

176 Potočnik, J. (2011). Commission recommendation of 18 October 2011 on the definition of nanomaterial (2011/696/EU). *Official Journal of the European Union* 275: 38–40.

177 Dudefoi, W., Villares, A., Peyron, S. et al. (2018). Nanoscience and nanotechnologies for biobased materials, packaging and food applications: new opportunities and concerns. *Innovative Food Science and Emerging Technologies* 46: 107–121.

178 Ray, S.S. (2013). *Environmentally Friendly Polymer Nanocomposites: Types, Processing and Properties*. Elsevier.

179 Froggett, S.J., Clancy, S.F., Boverhof, D.R., and Canady, R.A. (2014). A review and perspective of existing research on the release of nanomaterials from solid nanocomposites. *Particle and Fibre Toxicology* 11 (1): 1–28.

180 Torres, A., Guarda, A., Moraga, N. et al. (2012). Experimental and theoretical study of thermodynamics and transport properties of multilayer polymeric food packaging. *European Food Research and Technology* 234 (4): 713–722.

181 de Azeredo, H.M.C. (2013). Antimicrobial nanostructures in food packaging. *Trends in Food Science and Technology* 30 (1): 56–69.

182 von Goetz, N., Fabricius, L., Glaus, R. et al. (2013). Migration of silver from commercial plastic food containers and implications for consumer exposure assessment. *Food Additives & Contaminants: Part A* 30 (3): 612–620.

183 Šimon, P., Chaudhry, Q., and Bakoš, D. (2008). Migration of engineered nanoparticles from polymer packaging to food--a physicochemical view. *Journal of Food and Nutrition Research* 47 (3).

184 Bott, J., Störmer, A., and Franz, R. (2014). A comprehensive study into the migration potential of nano silver particles from food contact polyolefins. In: *Chemistry of Food, Food Supplements, and Food Contact Materials: From Production to Plate*, vol. 1159, 51–70. ACS Publications, American Chemical Society.

185 Farhoodi, M., Mousavi, S.M., Sotudeh-Gharebagh, R. et al. (2014). Migration of aluminum and silicon from PET/clay nanocomposite bottles into acidic food simulant. *Packaging Technology and Science* 27 (2): 161–168.

186 Maisanaba, S., Pichardo, S., Jordá-Beneyto, M. et al. (2014). Cytotoxicity and mutagenicity studies on migration extracts from nanocomposites with potential use in food packaging. *Food and Chemical Toxicology* 66: 366–372.

187 Schmidt, B., Katiyar, V., Plackett, D. et al. (2011). Migration of nanosized layered double hydroxide platelets from polylactide nanocomposite films. *Food Additives & Contaminants: Part A* 28 (7): 956–966.

188 Echegoyen, Y. and Nerín, C. (2013). Nanoparticle release from nano-silver antimicrobial food containers. *Food and Chemical Toxicology* 62: 16–22.

189 Raynor, P.C., Cebula, J.I., Spangenberger, J.S. et al. (2012). Assessing potential nanoparticle release during nanocomposite shredding using direct-reading instruments. *Journal of Occupational and Environmental Hygiene* 9 (1): 1–13.

190 Sachse, S., Silva, F., Zhu, H. et al. (2012). The effect of nanoclay on dust generation during drilling of PA6 nanocomposites. *Journal of Nanomaterials* 2012: 26.

191 Sachse, S., Silva, F., Irfan, A. et al. (2012). Physical characteristics of nanoparticles emitted during drilling of silica based polyamide 6 nanocomposites. In: *IOP Conference Series: Materials Science and Engineering*, vol. 40, 012012. IOP Publishing.

192 Huang, J.-Y., Li, X., and Zhou, W. (2015). Safety assessment of nanocomposite for food packaging application. *Trends in Food Science and Technology* 45 (2): 187–199.

193 European Commission Commission Directive 90/128/EEC of 23 February 1990 relating to plastics materials and articles intended to come into contact with foodstuffs. *Official Journal of the European Communities* 1990: 75.

194 Simoneau, C. (2016). Guidelines on testing conditions for articles in contact with foodstuffs. Eur. Communities(JRC Sci. Tech. Reports) Retrieved January 2009;14.

195 Boxall ABA, Tiede K, Chaudhry Q. Engineered nanomaterials in soils and water: how do they behave and could they pose a risk to human health? 2007 919-927

196 Souza, P.M.S., Morales, A.R., Marin-Morales, M.A., and Mei, L.H.I. (2013). PLA and montmorillonite nanocomposites: properties, biodegradation and potential toxicity. *Journal of Polymers and the Environment* 21 (3): 738–759.

197 National Research Council (2012). A Research Strategy for Environmental, Health, and Safety Aspects of Engineered Nanomaterials.

198 Stern, S.T. and McNeil, S.E. (2008). Nanotechnology safety concerns revisited. *Toxicological Sciences* 101 (1): 4–21.

199 Nam, J.Y., Sinha Ray, S., and Okamoto, M. (2003). Crystallization behavior and morphology of biodegradable polylactide/layered silicate nanocomposite. *Macromolecules* 36 (19): 7126–7131.

200 Lordan, S., Kennedy, J.E., and Higginbotham, C.L. (2011). Cytotoxic effects induced by unmodified and organically modified nanoclays in the human hepatic HepG$_2$ cell line. *Journal of Applied Toxicology* 31 (1): 27–35.

201 Maisanaba, S., Llana-Ruíz-Cabello, M., Pichardo, S. et al. (2017). Toxicological assessment of two silane-modified clay minerals with potential use as food contact materials in human hepatoma cells and *Salmonella typhimurium* strains. *Applied Clay Science* 150: 98–105.

202 Maisanaba, S., Pichardo, S., Puerto, M. et al. (2015). Toxicological evaluation of clay minerals and derived nanocomposites: a review. *Environmental Research* 138: 233–254.

203 Störmer, A., Bott, J., Kemmer, D., and Franz, R. (2017). Critical review of the migration potential of nanoparticles in food contact plastics. *Trends in Food Science and Technology* 63: 39–50.

204 Food and Drug Administration (2011). List of indirect additives used in food contact substances.

205 Directive, C (2022). 72/EC Commission Directive 2002/72/EC of 6 August 2002 relating to plastic materials and articles intended to come into contact with foodstuffs. *Official Journal of the European Communities* 220: 18–58.

206 Inventory of Effective Food Contact Substance (FCS) Notifications | FDA (2010). Retrieved july, 2022, from. https://www.fda.gov/food/packaging-food-contact-substances-fcs/inventory-effective-food-contact-substance-fcs-notifications.

8

Curcumin-Based Food Packaging Material

Leidy T. Sanchez[1,2], Andres F. Cañon-Ibarra[3], J. Alejandro Arboleda-Murillo[1], and Cristian C. Villa[3]

[1] Universidad del Quindío, Facultad de Ciencias Agroindustriales, Programa de Ingeniería de Alimentos, Carrera 15 Calle 12N, Armenia, Quindío, Colombia
[2] Universidad del Quindío, Facultad de Ciencias Básicas y Tecnologías, Doctorado en Ciencias – Química, Armenia, Quindío, Colombia
[3] Universidad del Quindío, Facultad de Ciencias Básicas y Tecnologías, Programa de Química, Armenia, Quindío, Colombia

Curcumin, known as diferuloylmethane (1,7-bis(4-hydroxy-3-ethoxyphenyl)-1,6-heptadiene-3,5-dione), is a highly hydrophobic, yellow polyphenol extracted from turmeric (*Curcuma longa*) [1]. Over centuries, turmeric has been used in food products as a spice and natural colorant ([2]). Likewise, turmeric has been commonly used in India and other parts of the world for the treatment of rheumatoid arthritis, conjunctivitis, wound healing, urinary tract infections, liver ailments, stomach pains, and fever, among other ailments [3]. Furthermore, curcumin has shown great potential as an anticancer, antimicrobial, antifungal, antioxidant, and anti-inflammatory agent [4]. Due to its great benefits, curcumin is commonly available in several forms such as capsules, tablets, ointments, energy drinks, soaps, and cosmetics [5]. Curcumin is considered "Generally Recognized As Safe" (GRAS) by the US Food and Drug Administration (FDA) and was approved for food applications by the European Union Scientific Committee on Food in 1975 [6]. Although the beneficial properties of curcumin are well documented, several drawbacks limit its application in the pharmaceutical, cosmetic, and food industries, such as its low water solubility (WS), photo instability, and low bioavailability, among others. However, these problems can be overcome by several strategies such as micro and nanoencapsulation [7].

8.1 Structural Characteristics of Curcumin

Most of the commercially available extracts prepared from turmeric crude or even refined contain three major compounds: curcumin (60–70%), demethoxycurcumin

Natural Materials for Food Packaging Application, First Edition.
Edited by Jyotishkumar Parameswaranpillai, Aswathy Jayakumar,
E. K. Radhakrishnan, Suchart Siengchin, and Sabarish Radoor.
© 2023 WILEY-VCH GmbH. Published 2023 by WILEY-VCH GmbH.

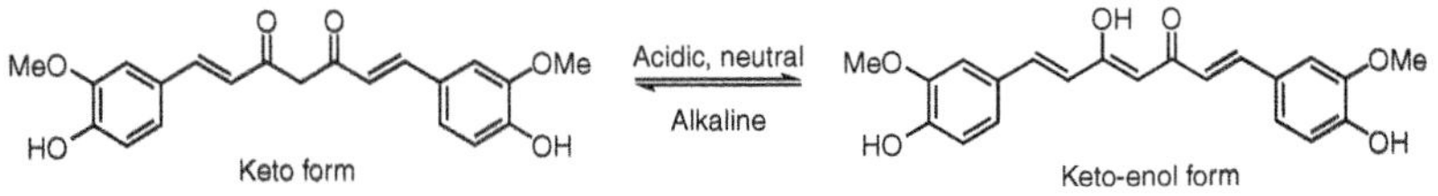

Figure 8.1 Tautomerism of curcumin under different pH values. Source: Cheng et al. [9] / MDPI / Licensed under CC BY 4.0.

(20–27%), and bisdemethoxycurcumin (10–15%) [8]. The chemical structure of all three compounds is shown in Figure 8.1. As shown in Figure 8.1, curcumin is a bis-α,β-unsaturated β-diketone allowing keto–enol tautomerism and cis–trans isomerism related to the solution's pH [10]. At low and neutral pH values the keto form is dominant while solutions maintain a bright yellow color, on the other hand, ad high pH values the enol form predominates, and the solution changes to an orange-red color [11].

One of the main characteristics of curcumin is that it has a very low WS at acidic and neutral pH reaching a maximum solubility value of 8 µg/ml [12]. However, curcumin is highly soluble in polar and nonpolar organic solvents, as well as in alkaline water solutions [13]. Likewise, curcumin is highly sensitive to environmental conditions, such as high temperatures, extreme pH, light, and oxygen [14]. It has been reported that curcumin experience hydrolysis at basic pH values and even in phosphate buffer conditions. Furthermore, photodegradation of curcumin occurs both in solution and in solid form, as well as in aerated and deaerated organic solutions and aqueous micellar solutions [15].

8.2 Antimicrobial, Antifungal, and Antioxidant Properties of Curcumin

Curcumin is known for its broad-spectrum antibacterial, antifungal, antiviral and antiprotozoal activity [16, 17]. The antimicrobial activity of curcumin takes place through the interaction with the essential cell division-initiating protein FtsZ [18]. Likewise, curcumin has shown the capability to inhibit bacterial growth by targeting the bacterial cell membrane, cell wall, and other cellular structures [16]. Furthermore, curcumin can be used as a photosensitizer for the photoinactivation of several bacteria by using blue light [19].

Despite its poor solubility in water and low bioavailability, the antimicrobial activity of curcumin has been widely studied [20]. Curcumin presents activity against a wide group of bacteria, however, it is known that the antimicrobial activity of curcumin is heavily dependent on the bacteria type and strain [21]. Over the last decades, the efficiency of curcumin against gram-positive bacteria has been well documented, especially *Staphylococcus aureus, Staphylococcus epidermidis, Streptococcus pyogenes, Micrococcus luteus* [17, 22, 23]. Likewise, some reports of its antimicrobial activity against gram-negative bacteria such as *Escherichia coli* and *Pseudomonas aeruginosa* [24, 25].

The antifungal activity of curcumin has been widely known and has been attributed to a reduction in proteinase secretion and several alterations of membrane-associated properties of ATPase activity caused by the presence of the polyphenolic molecule [26–28]. Curcumin has shown great antifungal activity against *Candida albicans*, especially when used in the photoinactivation process [19, 29]. Likewise, curcumin has shown antibiofilm activity against biofilms created by *C. albicans* and other bacteria [30]. Curcumin has also activity against fungi from the *Aspergillus* genus, such as *Aspergillus niger, Aspergillus flavus,* and *Aspergillus parasitucus* [31–33].

Curcumin is a well-known antioxidant molecule [34–36]. Two mechanisms have been proposed for the antioxidant mechanism of curcumin. The first one result from the formation of phenoxy radicals, involving an electron transfer to the free radical, forming a radical cation, which in turn produces a phenoxy radical by a proton loss. The second mechanism involves the direct abstraction of hydrogen, while hydroxyl groups of curcumin are vulnerable to radical attack [37, 38].

8.3 Nanoencapsulation of Curcumin

One way to overcome curcumin's low WS and increase its biological activities and stability is through nanoencapsulation or the inclusion of curcumin molecules into different types of nanocarriers that can be formed by different types of materials, such as polymers, carbohydrates, proteins, and lipids [7]. Nanoencapsulation not only increases the curcumin's stability, but also allows a controlled release of the bioactive molecule, which has been used in the development of active food packaging.

Nanoemulsions are among the most common nanocarriers used for curcumin nanoencapsulation, as their lipidic nature allows high encapsulation efficiency and they are relatively easy to produce [39]. Nanoemulsions are defined as biphasic dispersion of two immiscible phases (water and oil) with one dispersed in the other, droplets stabilized by an amphiphilic surfactant [40]. Due to their small droplet size (below 100 nm), nanoemulsions are kinetically stable systems, with longer shelf life [41]. In oil-in-water nanoemulsions, curcumin molecules are located in the hydrophobic interior of the droplets, although some molecules may also be located close to the oil–water interface due to the polar groups on the curcumin [42]. Solid lipid nanoparticles (SLNps) are another lipidic nanocarrier that has been increasingly used for curcumin nanoencapsulation [43–45]. SNLps are formed by a solid-lipid core stabilized by surfactant molecules [46, 47]. Finally, among lipid-based nanocarriers, liposomes are among the most well-known. Liposomes are spherical systems formed by a bilayer of lipids (phospholipids) that can be used to encapsulate both hydrophobic and hydrophilic molecules ([48–50]). Several studies have shown that liposomes can be used to increase the bioavailability and stability of curcumin [51–53].

Carbohydrate-based nanocarriers include starch nanoparticles (SNPs) and chitosan nanoparticles (ChNPs), among others. SNPs are amorphous particles formed mostly through the controlled precipitation of gelatinized starch. SNPs are biodegradable and biocompatible and their physicochemical properties can be changed through the chemical, physical and enzymatic modifications of the different native starches [54–56]. They have been used to increase the curcumin's WS and photostability [57–59]. ChNPs are formed mostly through ionic gelation of chitosan and a cross-linking molecule [60, 61]. They have been used to increase the antimicrobial and anticancer properties of curcumin [62, 63].

8.4 Curcumin-Based Food Packaging

Over the last decades, curcumin has been used in the development of active food packaging [64]. Due to the aforementioned biological activities of curcumin, it can be used to increase the shelf life of food products by reducing bacterial and fungal spoilage, and lipid oxidation. Furthermore, the inclusion of curcumin can lead to several desired changes in the mechanical and barrier properties of the films due to the intermolecular bonds formed between the polyphenol and the polymeric matrix [64]. Several studies have shown that curcumin concentrations of up to 10 mg/g of polysaccharide, lead to increasing values in the tensile strength (TS) of films made from chitosan. This can be attributed to the H-bonding between the hydroxyl of curcumin and the amino groups of chitosan molecules [65]. However, at high curcumin concentrations, TS values tend to decrease due to the formation of curcumin aggregates, which lead to a nonuniform distribution through the polymeric matrix [66, 67]. Similar behaviors have been reported in films made from k-carrageenan, cellulose and its derivates, and native and modified starch [68–72]. Furthermore, it has been reported that the inclusion of curcumin creates a plasticizing that increases elongation at break (EB) of the different types of polysaccharide-based polymeric films [64]. Different behaviors have been reported for the mechanical properties of protein-based films as curcumin was incorporated. Both TS and EB values of gelatin films tend to decrease as curcumin concentration increases [11]. A similar behavior was reported for composite films made from chicken skin gelatin/ rice starch [73]. The decrease of TS and EB values in the curcumin concentration increased in protein-based films has been attributed to a disruption of the protein–protein interactions in the film network due to specific interactions with the curcumin molecules, thus reducing the film's structural integrity [74]. On the other hand, Parveen et al. [75] reported that as curcumin concentration was increased in whey protein isolate films TS values increased, while EB decreased. This behavior has been attributed to the formation of hydrogen bonds between hydroxyl groups of curcumin molecules and the whey protein isolate chains, which reduces molecular mobility and improves the rigidity of films [75]. Figure 8.2 shows a schematic representation of curcumin applications in food packaging.

As mentioned before, one of the main characteristics of curcumin is its high hydrophobicity, which greatly affects the WS and water vapor permeability (WVP) of the curcumin-based polymeric films. Reports have shown that adding curcumin to both polysaccharide and protein-based films leads to significant a reduction in WS and WVP values. As the interactions between the polyphenol and the polymeric matrix reduce the number of polar groups available to interact with water molecules [77].

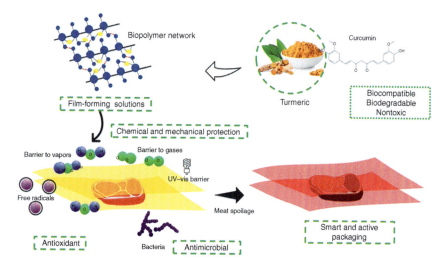

Figure 8.2 Schematic representation of curcumin applications in food packaging. Source: Oliveira Filho et al. [76] / with permission of Elsevier.

8.5 Curcumin-Based Nanocomposite Food Packaging

Nanocarriers have been used to increase the curcumin's bioavailability and WS. They have also been used to improve the incorporation of curcumin into different types of food packaging. Nieto-Suaza et al. [78] used native and acetylated banana SNPs as carriers of curcumin and incorporated them into composite films of *Aloe vera* gel and banana starch. They reported that WS and WVP decrease due to the hydrophobic nature of curcumin, while TS and EB increase as nanoencapsulation reduces the formation of curcumin agglomerates [78]. Similar behaviors have beenreported for composite films formed by zein, potato–starch, and curcumin-loaded ChNPs and pectin films incorporated with curcumin-loaded ChNPs [72, 79].

8.6 Curcumin-Based Active Food Packaging

Active packaging can be defined as the packaging in which subsidiary constituents have been included to enhance the performance of the packaging system [80]. Most of the active packaging systems developed include antioxidant, antimicrobial, or antifungal molecules that are slowly released, increasing the food products' shelf life. Due to the aforementioned biological properties of curcumin, this molecule has generated great interest in the development of active food packaging. The antimicrobial and antifungal activity of films made from chitosan and curcumin was measured against two common food pathogenic agents: *S. aureus* and *Rhizoctonia solani*, Results showed that curcumin slightly increased the antimicrobial activity of the chitosan films against both the microorganism [65]. Further studies have shown curcumin–chitosan films have antimicrobial activity against *Salmonella* and *E. coli* [81, 82]. Likewise, films made from curcumin-loaded chitosan/microcrystalline cellulose exhibited considerable antibacterial and antifungal activity toward *E. coli* and *Candida albicans* [83].

8.7 Curcumin-Based Intelligent Food Packaging

Intelligent food packaging materials are a type of food packaging that allows the monitoring of the food products' quality during storage [84]. They can be used to monitor changes in pH, temperature, humidity, and chemical composition of food products and transmit that information to consumers and distributors through color changes or electric signals. One of the most common types of intelligent food packaging is colorimetric sensors that change color as the food product suffers pH changes due to fermentation, protein degradation, or the presence of bacteria [85]. This type of sensor can be constructed using a polymeric matrix embedded with a synthetic or a natural colorant, mostly anthocyanins extracted from plants and fruits [86]. Over the last years, several authors have reported that curcumin can be used in the development of colorimetric sensors as this molecule is highly sensitive to the surrounding pH, changing from orange to red in alkaline media. One of the first reports of intelligent food packaging using curcumin as a colorimetric sensor used gelatin as the polymeric matrix, exhibiting a distinct color change at pH 6 and 11 [11]. Later, it was reported that curcumin/gelatin films can be used as pH sensors in fatty food products and even sensors of total volatile base nitrogen in meat and seafood products [87]. Similar behavior was observed for a curcumin–carrageenan composite sensor with a pH sensitivity in the 3–10 range, and a visible color change from yellow to red [69]. Other examples include sensors made from mucilage of *Lallemantia Iberica,* chitosan, cellulose, and pectin [88–90]. Figure 8.3 shows the application of curcumin as an intelligent food packaging.

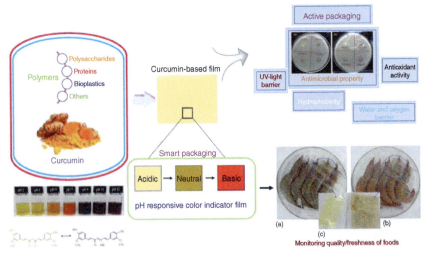

Figure 8.3 Application of curcumin as a pH sensor in intelligent food packaging. Source: Reproduced with permission from Roy et al. [91] / Reproduced with permission from Elsevier.

8.8 Perspectives

Curcumin has become one of the most interesting molecules in the development of both active and intelligent food packaging due to its well-known biological activities, especially its antimicrobial, antifungal, antioxidant, and pH sensitivity. Even though recent studies have shown great prospects for curcumin-based food packaging, several challenges need to be addressed, to continue their development. The first one is the low WS of curcumin; however, nanotechnology has been very helpful in overcoming this problem. The second one is the poor thermal and light stability of curcumin, which could limit the industrial development of this type of packaging. To overcome this problem, further studies are needed, creating an industrial process that could be used without any degradation of curcumin.

References

1 Alagawany, M., Farag, M.R., Abdelnour, S.A. et al. (2021). Curcumin and its different forms: a review on fish nutrition. *Aquaculture* 532: 736030. https://doi.org/10.1016/j.aquaculture.2020.736030.
2 Liu, X., Ma, B., Tan, H. et al. (2021). Utilization of turmeric residue for the preparation of ceramic foam. *Journal of Cleaner Production* 278: 123825. https://doi.org/10.1016/j.jclepro.2020.123825.
3 Siviero, A., Gallo, E., Maggini, V. et al. (2015). Curcumin, a golden spice with a low bioavailability. *Journal of Herbal Medicine* 5 (2): 57–70. https://doi.org/10.1016/j.hermed.2015.03.001.
4 Karthikeyan, A., Senthil, N., and Min, T. (2020). Nanocurcumin: a promising candidate for therapeutic applications. *Frontiers in Pharmacology* 11: 487. https://doi.org/10.3389/fphar.2020.00487.
5 Gupta, S.C., Patchva, S., and Aggarwal, B.B. (2013). Therapeutic roles of curcumin: lessons learned from clinical trials. *The AAPS Journal* 15 (1): 195–218. https://doi.org/10.1208/s12248-012-9432-8.
6 EFSA Panel on Food Additives and Nutrient Sources added to Food (ANS) (2010). Scientific opinion on the re-evaluation of curcumin (E 100) as a food additive. *EFSA Journal* 8 (9): 1679. https://doi.org/10.2903/j.efsa.2010.1679.
7 Rafiee, Z., Nejatian, M., Daeihamed, M., and Jafari, S.M. (2019). Application of curcumin-loaded nanocarriers for food, drug and cosmetic purposes. *Trends in Food Science & Technology* 88: 445–458. https://doi.org/10.1016/j.tifs.2019.04.017.
8 Nelson, K.M., Dahlin, J.L., Bisson, J. et al. (2017). The essential medicinal chemistry of curcumin. *Journal of Medicinal Chemistry* 60 (5): 1620–1637. https://doi.org/10.1021/acs.jmedchem.6b00975.
9 Cheng, Y.-J., Li, C.-W., Kuo, C.-L. et al. (2022). Improved synthesis of asymmetric curcuminoids and their assessment as antioxidants. *Molecules* 27 (8): https://doi.org/10.3390/molecules27082547.

10 Slika, L. and Patra, D. (2020). A short review on chemical properties, stability and nano-technological advances for curcumin delivery. *Expert Opinion on Drug Delivery* 17 (1): 61–75. https://doi.org/10.1080/17425247.2020.1702644.

11 Musso, Y.S., Salgado, P.R., and Mauri, A.N. (2017). Smart edible films based on gelatin and curcumin. *Food Hydrocolloids* 66: 8–15. https://doi.org/10.1016/j.foodhyd.2016.11.007.

12 Suresh, K. and Nangia, A. (2018). Curcumin: pharmaceutical solids as a platform to improve solubility and bioavailability. *CrystEngComm* 20 (24): 3277–3296. https://doi.org/10.1039/C8CE00469B.

13 Lestari, M.L.A.D. and Indrayanto, G. (2014). Chapter three – curcumin. In: *Profiles of Drug Substances, Excipients and Related Methodology*, vol. 39 (ed. H.G. Brittain), 113–204. Academic Press.

14 Jiang, T., Liao, W., and Charcosset, C. (2020). Recent advances in encapsulation of curcumin in nanoemulsions: a review of encapsulation technologies, bioaccessibility and applications. *Food Research International* 132: 109035. https://doi.org/10.1016/j.foodres.2020.109035.

15 Priyadarsini, K.I. (2009). Photophysics, photochemistry and photobiology of curcumin: studies from organic solutions, bio-mimetics and living cells. *Journal of Photochemistry and Photobiology C: Photochemistry Reviews* 10 (2): 81–95. https://doi.org/10.1016/j.jphotochemrev.2009.05.001.

16 Zheng, D., Huang, C., Huang, H. et al. (2020). Antibacterial mechanism of curcumin: a review. *Chemistry & Biodiversity* 17 (8): e2000171. https://doi.org/10.1002/cbdv.202000171.

17 Zorofchian Moghadamtousi, S., Abdul Kadir, H., Hassandarvish, P. et al. (2014). A review on antibacterial, antiviral, and antifungal activity of curcumin. *BioMed Research International* 2014: 186864. https://doi.org/10.1155/2014/186864.

18 da Silva, A.C., de Freitas Santos, P.D., do Prado Silva, J.T. et al. (2018). Impact of curcumin nanoformulation on its antimicrobial activity. *Trends in Food Science & Technology* 72: 74–82. https://doi.org/10.1016/j.tifs.2017.12.004.

19 Dias, L.D., Blanco, K.C., Mfouo-Tynga, I.S. et al. (2020). Curcumin as a photosensitizer: from molecular structure to recent advances in antimicrobial photodynamic therapy. *Journal of Photochemistry and Photobiology C: Photochemistry Reviews* 45: 100384. https://doi.org/10.1016/j.jphotochemrev.2020.100384.

20 Praditya, D., Kirchhoff, L., Brüning, J. et al. (2019). Anti-infective properties of the golden spice curcumin. *Frontiers in Microbiology* 10: 912.

21 Adamczak, A., Ożarowski, M., and Karpiński, T.M. (2020). Curcumin, a natural antimicrobial agent with strain-specific activity. *Pharmaceuticals (Basel, Switzerland)* 13 (7): 153. https://doi.org/10.3390/ph13070153.

22 Sharma, G., Raturi, K., Dang, S. et al. (2014). Combinatorial antimicrobial effect of curcumin with selected phytochemicals on *Staphylococcus epidermidis*. *Journal of Asian Natural Products Research* 16 (5): 535–541. https://doi.org/10.1080/10286020.2014.911289.

23 Teow, S.-Y., Liew, K., Ali, S.A. et al. (2016). Antibacterial action of curcumin against *Staphylococcus aureus*: a brief review. *Journal of Tropical Medicine* 2016: 2853045. https://doi.org/10.1155/2016/2853045.

24 Neyestani, Z., Ebrahimi, S.A., Ghazaghi, A. et al. (2019). Review of anti-bacterial activities of curcumin against *Pseudomonas aeruginosa*. 29 (5): 377–385. https://doi.org/10.1615/CritRevEukaryotGeneExpr.2019029088.

25 Yun, D.G. and Lee, D.G. (2016). Antibacterial activity of curcumin via apoptosis-like response in *Escherichia coli*. *Applied Microbiology and Biotechnology* 100 (12): 5505–5514. https://doi.org/10.1007/s00253-016-7415-x.

26 Chen, E., Benso, B., Seleem, D. et al. (2018). Fungal-host interaction: curcumin modulates proteolytic enzyme activity of *Candida albicans* and inflammatory host response *in vitro*. *International Journal of Dentistry* 2018: 2393146. https://doi.org/10.1155/2018/2393146.

27 Martins, C.V.B., da Silva, D.L., Neres, A.T.M. et al. (2009). Curcumin as a promising antifungal of clinical interest. *Journal of Antimicrobial Chemotherapy* 63 (2): 337–339. https://doi.org/10.1093/jac/dkn488.

28 Neelofar, K., Shreaz, S., Rimple, B. et al. (2011). Curcumin as a promising anticandidal of clinical interest. *Canadian Journal of Microbiology* 57 (3): 204–210. https://doi.org/10.1139/W10-117.

29 Dovigo, L.N., Carmello, J.C., de Souza Costa, C.A. et al. (2013). Curcumin-mediated photodynamic inactivation of *Candida albicans* in a murine model of oral candidiasis. *Medical Mycology* 51 (3): 243–251. https://doi.org/10.3109/13693786.2012.714081.

30 Ma, S., Moser, D., Han, F. et al. (2020). Preparation and antibiofilm studies of curcumin loaded chitosan nanoparticles against polymicrobial biofilms of *Candida albicans* and *Staphylococcus aureus*. *Carbohydrate Polymers* 241: 116254. https://doi.org/10.1016/j.carbpol.2020.116254.

31 Jahanshiri, Z., Shams-Ghahfarokhi, M., Allameh, A., and Razzaghi-Abyaneh, M. (2012). Effect of curcumin on *Aspergillus parasiticus* growth and expression of major genes involved in the early and late stages of aflatoxin biosynthesis. *Iranian Journal of Public Health* 41 (6): 72–79.

32 Schamberger, B. and Plaetzer, K. (2021). Photofungizides based on curcumin and derivates thereof against *Candida albicans* and *Aspergillus niger*. *Antibiotics* 10 (11): https://doi.org/10.3390/antibiotics10111315.

33 Temba, B.A., Fletcher, M.T., Fox, G.P. et al. (2019). Curcumin-based photosensitization inactivates *Aspergillus flavus* and reduces aflatoxin B1 in maize kernels. *Food Microbiology* 82: 82–88. https://doi.org/10.1016/j.fm.2018.12.013.

34 Ak, T. and Gülçin, İ. (2008). Antioxidant and radical scavenging properties of curcumin. *Chemico-Biological Interactions* 174 (1): 27–37. https://doi.org/10.1016/j.cbi.2008.05.003.

35 Jayaprakasha, G.K., Jaganmohan Rao, L., and Sakariah, K.K. (2006). Antioxidant activities of curcumin, demethoxycurcumin and bisdemethoxycurcumin. *Food Chemistry* 98 (4): 720–724. https://doi.org/10.1016/j.foodchem.2005.06.037.

36 Wright, J.S. (2002). Predicting the antioxidant activity of curcumin and curcuminoids. *Journal of Molecular Structure: THEOCHEM* 591 (1): 207–217. https://doi.org/10.1016/S0166-1280(02)00242-7.

37 Del Prado-Audelo, M.L., Caballero-Florán, I.H., Meza-Toledo, J.A. et al. (2019). Formulations of curcumin nanoparticles for brain diseases. *Biomolecules* 9 (2): 56. https://doi.org/10.3390/biom9020056.

38 Liczbiński, P., Michałowicz, J., and Bukowska, B. (2020). Molecular mechanism of curcumin action in signaling pathways: review of the latest research. *Phytotherapy Research* 34 (8): 1992–2005. https://doi.org/10.1002/ptr.6663.

39 Sanchez, L.T., Pinzon, M.I., and Villa, C.C. (2022). Development of active edible films made from banana starch and curcumin-loaded nanoemulsions. *Food Chemistry* 371: 131121. https://doi.org/10.1016/j.foodchem.2021.131121.

40 Singh, Y., Meher, J.G., Raval, K. et al. (2017). Nanoemulsion: concepts, development and applications in drug delivery. *Journal of Controlled Release* 252: 28–49. https://doi.org/10.1016/j.jconrel.2017.03.008.

41 Safaya, M. and Rotliwala, Y.C. (2020). Nanoemulsions: a review on low energy formulation methods, characterization, applications and optimization technique. *Materials Today: Proceedings* 27: 454–459. https://doi.org/10.1016/j.matpr.2019.11.267.

42 Sanidad, K.Z., Sukamtoh, E., Xiao, H. et al. (2019). Curcumin: recent advances in the development of strategies to improve oral bioavailability. *Annual Review of Food Science and Technology* 10 (1): 597–617. https://doi.org/10.1146/annurev-food-032818-121738.

43 Ban, C., Jo, M., Park, Y.H. et al. (2020). Enhancing the oral bioavailability of curcumin using solid lipid nanoparticles. *Food Chemistry* 302: 125328. https://doi.org/10.1016/j.foodchem.2019.125328.

44 Sadegh Malvajerd, S., Azadi, A., Izadi, Z. et al. (2019). Brain delivery of curcumin using solid lipid nanoparticles and nanostructured lipid carriers: preparation, optimization, and pharmacokinetic evaluation. *ACS Chemical Neuroscience* 10 (1): 728–739. https://doi.org/10.1021/acschemneuro.8b00510.

45 Wang, W., Chen, T., Xu, H. et al. (2018). Curcumin-loaded solid lipid nanoparticles enhanced anticancer efficiency in breast cancer. *Molecules* 23 (7): https://doi.org/10.3390/molecules23071578.

46 Müller, R.H., Mäder, K., and Gohla, S. (2000). Solid lipid nanoparticles (SLN) for controlled drug delivery – a review of the state of the art. *European Journal of Pharmaceutics and Biopharmaceutics* 50 (1): 161–177. https://doi.org/10.1016/S0939-6411(00)00087-4.

47 Paliwal, R., Paliwal, S.R., Kenwat, R. et al. (2020). Solid lipid nanoparticles: a review on recent perspectives and patents. *Expert Opinion on Therapeutic Patents* 30 (3): 179–194. https://doi.org/10.1080/13543776.2020.1720649.

48 Li, Z.-I., Peng, S.-f., Chen, X. et al. (2018). Pluronics modified liposomes for curcumin encapsulation: sustained release, stability and bioaccessibility. *Food Research International* 108: 246–253. https://doi.org/10.1016/j.foodres.2018.03.048.

49 Cheng, C., Wu, Z., McClements, D.J. et al. (2019). Improvement in stability, loading capacity, and sustained release of rhamnolipids-modified curcumin liposomes.

Colloids and Surfaces B: Biointerfaces 183: 110460. https://doi.org/10.1016/j.colsurfb.2019.110460.

50 Ng, Z.Y., Wong, J.-Y., Panneerselvam, J. et al. (2018). Assessing the potential of liposomes loaded with curcumin as a therapeutic intervention in asthma. *Colloids and Surfaces B: Biointerfaces* 172: 51–59. https://doi.org/10.1016/j.colsurfb.2018.08.027.

51 Basnet, P., Hussain, H., Tho, I., and Skalko-Basnet, N. (2012). Liposomal delivery system enhances anti-inflammatory properties of curcumin. *Journal of Pharmaceutical Sciences* 101 (2): 598–609. https://doi.org/10.1002/jps.22785.

52 Cheng, C., Peng, S., Li, Z. et al. (2017). Improved bioavailability of curcumin in liposomes prepared using a pH-driven, organic solvent-free, easily scalable process. *RSC Advances* 7 (42): 25978–25986. https://doi.org/10.1039/C7RA02861J.

53 Li, L., Braiteh, F.S., and Kurzrock, R. (2005). Liposome-encapsulated curcumin. *Cancer* 104 (6): 1322–1331. https://doi.org/10.1002/cncr.21300.

54 Ahmad, M., Mudgil, P., Gani, A. et al. (2019). Nano-encapsulation of catechin in starch nanoparticles: characterization, release behavior and bioactivity retention during simulated in-vitro digestion. *Food Chemistry* 270: 95–104. https://doi.org/10.1016/j.foodchem.2018.07.024.

55 Farrag, Y., Ide, W., Montero, B. et al. (2018). Preparation of starch nanoparticles loaded with quercetin using nanoprecipitation technique. *International Journal of Biological Macromolecules* 114: 426–433. https://doi.org/10.1016/j.ijbiomac.2018.03.134.

56 Gutiérrez, G., Morán, D., Marefati, A. et al. (2020). Synthesis of controlled size starch nanoparticles (SNPs). *Carbohydrate Polymers* 250: 116938. https://doi.org/10.1016/j.carbpol.2020.116938.

57 Acevedo-Guevara, L., Nieto-Suaza, L., Sanchez, L.T. et al. (2018). Development of native and modified banana starch nanoparticles as vehicles for curcumin. *International Journal of Biological Macromolecules* 111: 498–504. https://doi.org/10.1016/j.ijbiomac.2018.01.063.

58 Miskeen, S., An, Y.S., and Kim, J.-Y. (2021). Application of starch nanoparticles as host materials for encapsulation of curcumin: effect of citric acid modification. *International Journal of Biological Macromolecules* 183: 1–11. https://doi.org/10.1016/j.ijbiomac.2021.04.133.

59 Qin, Y., Wang, J., Qiu, C. et al. (2019). Effects of degree of polymerization on size, crystal structure, and digestibility of debranched starch nanoparticles and their enhanced antioxidant and antibacterial activities of curcumin. *ACS Sustainable Chemistry & Engineering* 7 (9): 8499–8511. https://doi.org/10.1021/acssuschemeng.9b00290.

60 Naskar, S., Kuotsu, K., and Sharma, S. (2019). Chitosan-based nanoparticles as drug delivery systems: a review on two decades of research. *Journal of Drug Targeting* 27 (4): 379–393. https://doi.org/10.1080/1061186X.2018.1512112.

61 Qu, B. and Luo, Y. (2020). Chitosan-based hydrogel beads: preparations, modifications and applications in food and agriculture sectors – a review. *International Journal of Biological Macromolecules* 152: 437–448. https://doi.org/10.1016/j.ijbiomac.2020.02.240.

62 Chuah, L.H., Billa, N., Roberts, C.J. et al. (2013). Curcumin-containing chitosan nanoparticles as a potential mucoadhesive delivery system to the colon. *Pharmaceutical Development and Technology* 18 (3): 591–599. https://doi.org/10.3109/10837450.2011.640688.

63 Khan, M.A., Zafaryab, M., Mehdi, S.H. et al. (2016). Characterization and antiproliferative activity of curcumin loaded chitosan nanoparticles in cervical cancer. *International Journal of Biological Macromolecules* 93: 242–253. https://doi.org/10.1016/j.ijbiomac.2016.08.050.

64 Aliabbasi, N., Fathi, M., and Emam-Djomeh, Z. (2021). Curcumin: a promising bioactive agent for application in food packaging systems. *Journal of Environmental Chemical Engineering* 9 (4): 105520. https://doi.org/10.1016/j.jece.2021.105520.

65 Liu, Y., Cai, Y., Jiang, X. et al. (2016). Molecular interactions, characterization and antimicrobial activity of curcumin–chitosan blend films. *Food Hydrocolloids* 52: 564–572. https://doi.org/10.1016/j.foodhyd.2015.08.005.

66 Almeida, C.M.R., Magalhães, J.M.C.S., Souza, H.K.S., and Gonçalves, M.P. (2018). The role of choline chloride-based deep eutectic solvent and curcumin on chitosan films properties. *Food Hydrocolloids* 81: 456–466. https://doi.org/10.1016/j.foodhyd.2018.03.025.

67 Roy, S. and Rhim, J.-W. (2020). Preparation of carbohydrate-based functional composite films incorporated with curcumin. *Food Hydrocolloids* 98: 105302. https://doi.org/10.1016/j.foodhyd.2019.105302.

68 Baek, S.-K. and Song, K.B. (2019). Characterization of active biodegradable films based on proso millet starch and curcumin. *Starch – Stärke* 71 (3–4): 1800174. https://doi.org/10.1002/star.201800174.

69 Liu, J., Wang, H., Wang, P. et al. (2018). Films based on κ-carrageenan incorporated with curcumin for freshness monitoring. *Food Hydrocolloids* 83: 134–142. https://doi.org/10.1016/j.foodhyd.2018.05.012.

70 Luo, N., Varaprasad, K., Reddy, G.V.S. et al. (2012). Preparation and characterization of cellulose/curcumin composite films. *RSC Advances* 2 (22): 8483–8488. https://doi.org/10.1039/C2RA21465B.

71 Xie, Y., Niu, X., Yang, J. et al. (2020). Active biodegradable films based on the whole potato peel incorporated with bacterial cellulose and curcumin. *International Journal of Biological Macromolecules* 150: 480–491. https://doi.org/10.1016/j.ijbiomac.2020.01.291.

72 Xin, S., Xiao, L., Dong, X. et al. (2020). Preparation of chitosan/curcumin nanoparticles based zein and potato starch composite films for *Schizothorax prenati* fillet preservation. *International Journal of Biological Macromolecules* 164: 211–221. https://doi.org/10.1016/j.ijbiomac.2020.07.082.

73 Said, N.S. and Sarbon, N.M. (2020). Response surface methodology (RSM) of chicken skin gelatin based composite films with rice starch and curcumin incorporation. *Polymer Testing* 81: 106161. https://doi.org/10.1016/j.polymertesting.2019.106161.

74 Tongnuanchan, P., Benjakul, S., and Prodpran, T. (2013). Physico-chemical properties, morphology and antioxidant activity of film from fish skin gelatin incorporated with root essential oils. *Journal of Food Engineering* 117 (3): 350–360. https://doi.org/10.1016/j.jfoodeng.2013.03.005.

75 Parveen, S., Ghosh, P., Mitra, A. et al. (2019). Preparation, characterization, and in vitro release study of curcumin-loaded cataractous eye protein isolate films. *Emergent Materials* 2 (4): 475–486. https://doi.org/10.1007/s42247-019-00036-6.

76 de Oliveira Filho, J.G., Bertolo, M.R.V., Rodrigues, M.Á.V. et al. (2021). Curcumin: a multifunctional molecule for the development of smart and active biodegradable polymer-based films. *Trends in Food Science & Technology* 118: 840–849. https://doi.org/10.1016/j.tifs.2021.11.005.

77 Salarbashi, D., Noghabi, M.S., Bazzaz, B.S.F. et al. (2017). Eco-friendly soluble soybean polysaccharide/nanoclay Na^+ bionanocomposite: properties and characterization. *Carbohydrate Polymers* 169: 524–532. https://doi.org/10.1016/j.carbpol.2017.04.011.

78 Nieto-Suaza, L., Acevedo-Guevara, L., Sánchez, L.T. et al. (2019). Characterization of Aloe vera-banana starch composite films reinforced with curcumin-loaded starch nanoparticles. *Food Structure* 22: 100131. https://doi.org/10.1016/j.foostr.2019.100131.

79 Basit, H.M., Ali, M., Shah, M.M. et al. (2021). Microwave enabled physically cross linked sodium alginate and pectin film and their application in combination with modified chitosan-curcumin nanoparticles. A novel strategy for 2nd degree burns wound healing in animals. *Polymers* 13 (16): https://doi.org/10.3390/polym13162716.

80 Soltani Firouz, M., Mohi-Alden, K., and Omid, M. (2021). A critical review on intelligent and active packaging in the food industry: research and development. *Food Research International* 141: 110113. https://doi.org/10.1016/j.foodres.2021.110113.

81 Kalaycıoğlu, Z., Torlak, E., Akın-Evingür, G. et al. (2017). Antimicrobial and physical properties of chitosan films incorporated with turmeric extract. *International Journal of Biological Macromolecules* 101: 882–888. https://doi.org/10.1016/j.ijbiomac.2017.03.174.

82 Wu, C., Zhu, Y., Wu, T. et al. (2019). Enhanced functional properties of biopolymer film incorporated with curcumin-loaded mesoporous silica nanoparticles for food packaging. *Food Chemistry* 288: 139–145. https://doi.org/10.1016/j.foodchem.2019.03.010.

83 Bajpai, S.K., Chand, N., and Ahuja, S. (2015). Investigation of curcumin release from chitosan/cellulose microcrystals (CMC) antimicrobial films. *International Journal of Biological Macromolecules* 79: 440–448. https://doi.org/10.1016/j.ijbiomac.2015.05.012.

84 Ghaani, M., Cozzolino, C.A., Castelli, G., and Farris, S. (2016). An overview of the intelligent packaging technologies in the food sector. *Trends in Food Science & Technology* 51: 1–11. https://doi.org/10.1016/j.tifs.2016.02.008.

85 Xiao-wei, H., Xiao-bo, Z., Ji-yong, S. et al. (2018). Colorimetric sensor arrays based on chemo-responsive dyes for food odor visualization. *Trends in Food Science & Technology* 81: 90–107. https://doi.org/10.1016/j.tifs.2018.09.001.

86 Weston, M., Phan, M.A.T., Arcot, J., and Chandrawati, R. (2020). Anthocyanin-based sensors derived from food waste as an active use-by-date indicator for milk. *Food Chemistry* 326: 127017. https://doi.org/10.1016/j.foodchem.2020.127017.

87 Etxabide, A., Kilmartin, P.A., and Maté, J.I. (2021). Color stability and pH-indicator ability of curcumin, anthocyanin, and betanin-containing colorants under different storage conditions for intelligent packaging development. *Food Control* 121: 107645. https://doi.org/10.1016/j.foodcont.2020.107645.

88 Pereira, P.F. and Andrade, C.T. (2017). An optimized pH-responsive film based on eutectic mixture-plasticized chitosan. *Carbohydrate Polymers* 165: 238–246. https://doi.org/10.1016/j.carbpol.2017.02.047.

89 Taghinia, P., Abdolshahi, A., Sedaghati, S., and Shokrollahi, B. (2021). Smart edible films based on mucilage of *Lallemantia Iberica* seed incorporated with curcumin for freshness monitoring. *Food Science & Nutrition* 9 (2): 1222–1231. https://doi.org/10.1002/fsn3.2114.

90 Xiao, Y., Liu, Y., Kang, S. et al. (2021). Development of pH-responsive antioxidant soy protein isolate films incorporated with cellulose nanocrystals and curcumin nanocapsules to monitor shrimp freshness. *Food Hydrocolloids* 120: 106893. https://doi.org/10.1016/j.foodhyd.2021.106893.

91 Roy, S., Priyadarshi, R., Ezati, P., and Rhim, J.-W. (2022). Curcumin and its uses in active and smart food packaging applications – a comprehensive review. *Food Chemistry* 375: 131885. https://doi.org/10.1016/j.foodchem.2021.131885.

9

Sustainable Materials from Starch-Based Plastics

Asanda Mtibe[1] and Maya J. John[1,2]

[1] *CSIR, Centre for Nanostructures and Advanced Materials, PO Box 395, Meiring Naudé Road, Brummeria, Pretoria, 0001, South Africa*
[2] *Nelson Mandela University, Department of Chemistry, PO Box 77000, University Way, Gqeberha, 6031, South Africa*

9.1 Introduction

9.1.1 Starch

Starch is a polysaccharide comprising linked glucose molecules. The common sources of starch include rice, maize, wheat, potatoes, and cassava. Chemically, it is a mixture of two polymers, amylase, which is a linear polysaccharide, and amylopectin, a highly branched polysaccharide. The amount of amylopectin and amylose varies between starch types. The common availability of starch crops (staple in many countries in Asia, Africa, and South America) has generated considerable interest in the potential for starch-based products to replace conventional plastics.

Starch-based products are continuing to garner considerable interest among research community and consumers in recent years. This is due to their excellent properties, such as nontoxicity, biodegradability, and sustainability, which are advantages in some applications, as well as low density with low production cost and abundantly available [1]. Additionally, strategies such as United Nations Sustainable Development Goals of responsible material consumption and production (Goal 12) [2] and the new plastics circular economy vision [3], which stress innovation and circulation to ensure plastics, are recycled and/or biodegradable and are kept in the economy and not ending up in the environment support the transition to bio-based materials. Starch also consists of hydroxyl groups making it easy to modify by converting hydroxyl group to other functional groups [4]. These unique properties make starch a suitable candidate to replace petroleum-based polymers in several applications.

Natural Materials for Food Packaging Application, First Edition.
Edited by Jyotishkumar Parameswaranpillai, Aswathy Jayakumar,
E. K. Radhakrishnan, Suchart Siengchin, and Sabarish Radoor.
© 2023 WILEY-VCH GmbH. Published 2023 by WILEY-VCH GmbH.

Despite the usefulness of starch, it has displayed some drawbacks. These drawbacks include (i) its characteristics and properties vary with the source, geographic area, and time of harvest, (ii) it has high moisture absorption, which limits its applications, especially in advanced industrial sectors, (iii) the polymer chains of starch are rigid, making it difficult to process in traditional processing techniques of thermoplastics, such as extrusion, injection molding, and compression molding, and (iv) it has numerous hydroxyl groups leading to strong hydrogen bonds that create crystalline polymers and thus make it difficult to process [5].

To achieve good processability of starch and mitigate their challenges, debonding of starch by derivatization as well as plasticization is required. The process of modifying starch using plasticizers under high temperature, shear, and pressure is called thermoplasticization and the end product is called thermoplastic starch (TPS) [6]. The properties of resultant TPS, such as superior mechanical, thermal, and flow properties, with the added advantage of biodegradability, make it an environmental-friendly alternative to petroleum-based polymers [7].

9.1.2 Preparation of Thermoplastic Starch (TPS)

The preparation of TPS can be achieved in two steps, namely (i) modification of starch using plasticizers and other materials such as urea and amines and (ii) thermoprocessing of plasticized starch into TPS. Modifying starch using the aforementioned materials penetrates starch and disrupts its crystalline structure, thus forming an amorphous structure. The mixing of starch with modifiers is performed at ambient temperatures. The amount of the modifiers added into starch can determine the behavior of the ensued TPS, whether the material will be brittle or flexible. For instance, the addition of high plasticizer content results in flexible TPS. Depending on the source of starch, the addition of adequate plasticizers under gelatinization can form a gel-like or dough-like mixture. Mościcki et al. [6] reported that addition of plasticizers in starch increases the activation energy for the melting of starch granules, which improves the glass transition temperature and interaction between the materials.

9.1.3 Plasticization of Starch

Starch in its natural form is not regarded as thermoplastic polymer due to its rigid polymer chains and strong hydrogen bonds in the amylose and amylopectin. Various plasticizers, such as water, glycerol, ethylene glycol, and sorbitol, have been reported in the literature to modify starch to form TPS. The role of plasticizers is to improve the flexibility and the applicability of resultant TPS [8, 9]. Plasticizers should be compatible with starch to achieve plasticization. The widely used plasticizers are water and glycerol due to their properties, such as low molecular weight, hydrophilic, polar, ease of adding, and compatibility with the three-dimensional network of starch. Glycerol is a by-product of diesel production and it has been recently used in high-value-added products, such as development of biopolymers [10].

Many researchers have investigated the effect of water as a plasticizer on the properties of TPS [11, 12]. For instance, Meng et al. [11] prepared TPS using water as both blowing agent and plasticizer using twin-screw extruder and single-screw extruder.

The authors reported that low water contents (<10%) negatively affected both glass transition and melting temperature. Also, low water contents affected the processability of the material in a twin-screw extruder resulting in high extrusion torque. Conversely, high water contents (>10%) resulted in good processability and the resultant material possesses high melt strength and low melting flow which make the material blowable. In addition, it was noticed that the introduction of low water contents resulted in high compressive strength. However, the compressive strength decreases with an increase in water content.

Although water has been used as a plasticizer and blowing agent, the disadvantage of using water as a plasticizer is that it evaporates fast and thereby affects processability, thus limiting the applications of TPS. Also, when water is used as a plasticizer, TPS becomes brittle at room temperature, which affects the mechanical and barrier properties [6]. To overcome these challenges, water must be mixed with other plasticizers or nonvolatile plasticizers should be used to achieve good processability.

Over the past decades, polyols, such as glycerol, have been employed for plasticization of starch to increase their flexibility and improve their processability. Numerous researchers [1, 6, 13, 14] have demonstrated that plasticization of starch with glycerol produces TPS with better mechanical properties because it is compatible with amylose. For instance, Tarique et al. [13] investigated the effect of glycerol on the properties of TPS. The authors reported that introduction of glycerol in arrowroot starch improves the flexibility of material. It is worth noting that the increase in glycerol content from 15% to 45% resulted in decrease in both tensile modulus and strength, whereas elongation at break was increasing with the increase in glycerol content. In addition, the increase in glycerol content resulted in increased water content and solubility, whereas water absorption exhibited adverse effects. Similar results were also reported by Sahari et al. [14] who also observed that the increase in glycerol content decreased the glass transition temperature of the material. The addition of higher glycerol content normally results in excellent properties of TPS; however, minimal concentrations of glycerol are required to plasticize starch [9]. The amount of glycerol can be reduced by mixing it with water while maintaining the properties of starch.

The mixture of glycerol and water has been extensively investigated as a plasticizer for the gelatinization of various types of starch [4, 15–20]. Jaramillo et al. [15] prepared TPS from cassava starch using glycerol and water as plasticizers with the concentration of 1.5 and 93.5 g, respectively. TPS films were transparent and demonstrated a surface roughness value of 3.7 ± 0.5. Contact angle results suggested that TPS films were hydrophilic. The films presented high value of water vapor permeability (WVP) (8.8 g/smPa ($\times 10^{-10}$), which indicates that films have high water vapor transmission. On the other hand, films demonstrated higher moisture content and solubility of about 38% and 35%, respectively. In terms of mechanical properties, TPS did not show plastic behavior indicating that the films were elastic with the applied load. The values of tensile modulus, tensile strength, elongation at break, and tensile toughness were 1.9 ± 0.2 MPa, 0.51 ± 0.04 MPa, 107 ± 3, and 3.8 ± 0.4 J/m^3 ($\times 10^6$), respectively. The increase in glycerol content results in improvement of mechanical properties of TPS. Bergel et al. [4] prepared TPS made from cassava starch and plasticized with glycerol and water concentration of 5% and 33%, respectively, and investigated their mechanical properties. The researchers discovered that TPS films containing 5% glycerol were brittle. The values of tensile modulus, tensile strength, elongation at break, and tensile toughness were 39.88 ± 6.3 MPa, 1.01 ± 0.17 MPa, 1.03 ± 0.16%, and 7.06 ± 0.98 J/m, respectively. The increase in glycerol content to a range of 20–25% resulted in an increase in tensile strength and elongation at break [19, 20]. It was reported in another study that the increase in glycerol content above 30% resulted in decrease in tensile modulus and tensile strength while elongation at break increased. This indicates that the addition of higher concentration of glycerol resulted in flexible films that can be used without breaking [17]. However, the concentration of glycerol can vary depending on the intended applications. For instance, rigid applications require low content of glycerol, whereas flexible applications require high content of glycerol.

Table 9.1 shows the source of starch, type of plasticizer, and properties of the resultant TPS used in different studies [21–35].

Table 9.1 Source of starch, plasticizers, and properties of TPS.

Source	Plasticizer(s)	Effect of plasticizer(s)
Tapioca starch	1-Ethyl-3-methylimidazolium acetate ([Emim] Ac) and 1-ethyl-3-methylimidazolium chloride ([Emim] Cl)	• Promotes thermal degradation
Corn starch	Sorbitol and glycerol	• Improves mechanical properties, moisture content, and water absorption
Banana starch	Sorbitol and glycerol	• Increase water barrier, elongation at break, tensile toughness • Films became transparent
Sweet potato starch	Sorbitol and glycerol	• Increase in plasticizer content led to increasing elongation at break, water solubility, and water vapor permeability
Maize starch	Poly(trimellitic glyceride) (PTG)	• Enhance mechanical and thermal properties and decrease moisture absorption
Sugar palm starch	Sorbitol and glycerol	• Improve flexibility • Increase in plasticizer content led to increasing water solubility and water vapor permeability • Decrease tensile strength
Cassava starch	Fructose, urea, tri-ethylene glycol, and triethanolamine	• Reducing glass transition temperature and crystallinity • Decreases tensile strength and increases elongation at break

Source	Plasticizer(s)	Effect of plasticizer(s)
Maize starch	Isosorbide	• Improve thermal stability of TPS • Good mechanical properties suitable for edible films
Mild-oxidized cornstarch	Urea	• Films were smooth and transparent • High urea content results in low tensile strength and high elongation at break
Wheat starch	Glycerol, sorbitol, and urea/ethanolamine	• Improves flexibility. However, flexibility of TPS decreases with increase in aging time
Anchote starch	1-Ethyl-3-methylimidazolium acetate, sorbitol, and triethylene glycol	• Enhance elongation at break and improves thermal stability
Cassava starch	Glycerol, glycerol/xylitol, and glycerol/sorbitol	• Glycerol alone resulted in low tensile strength and poor water vapor barrier properties. • However, mixing glycerol with other plasticizers improves tensile strength tensile modulus, water vapor, and oxygen barrier properties
Corn starch	Polymeric ionic liquid (PIL)	• Improves thermoplasticity and displays good compatibility with starch
Corn starch	Ethylene glycol, glycerol, sorbitol, formamide, and urea	• Increasing plasticizer content reduces tensile strength while improving flexibility and water absorption

9.1.4 Processing of TPS

There are several TPS processing methods reported in literature. These methods include solution casting, melt extrusion, and hot-pressing methods.

During the preparation of TPS by solution casting method, starch and plasticizers are manually mixed and stirred at ambient temperatures. For plasticization to occur, the original structure network of starch should be disrupted by applying heat in temperatures ranging from 70 to 100 °C and vigorous stirring until gel-like or dough-like material is formed [1, 13, 15, 22]. The gel-like solution is then dispensed into a petri dish or trays and allowed to dry either in an oven or at laboratory temperature. For instance, Tarique et al. [13] schematically demonstrated TPS preparation method in Figure 9.1 using solution casting method.

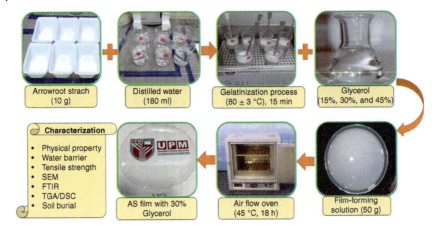

Figure 9.1 Schematic diagram of TPS preparation by solution casting method. Source: Tarique et al. [13] / Springer Nature / Licensed under CC BY 4.0.

Hot pressing method is similar to solution casting method. Unlike in solution casting method, the gel-like material is transferred into the mold and placed between two thick plates. The material is then compression molded at a certain temperature and time to produce films [4]. Bendaoud and Chalamet [36] prepared TPS using compression molding where the premixed plasticized starch gel was transferred into a mold and placed between 2-mm-thick plates. The material was compression molded at 120 °C for 4 minutes at 50 kN and 2 minutes at 150 kN. The plates were allowed to dry, and the film was peeled off and characterized using a range of characterization techniques. The major drawback associated with solution casting method and hot-pressing method is that they are not industrial scalable.

The most viable method for preparing TPS is melt extrusion because it can be industrial scalable. In this method, the original starch network is disrupted by applying heat at high temperatures, in high shear conditions, and with limited plasticizers [8]. Melt extrusion process produces TPS pellets as shown in Figure 9.2, which can be blown into films or sheets, molded into desired shape, and/or injection molded into test specimens as depicted in Figure 9.2 [6, 20, 29, 31, 33, 36].

In short, starch and plasticizers are premixed manually or in a mechanical mixer to attain homogeneity of the materials and allow plasticizers to penetrate chains of starch matrix. The premixed material will then feed in the main hooper, which is placed at the beginning of the screw through a barrel. During extrusion process, the temperature profile ranges from 80 to 130 °C, and the screw rotation speed ranges from 60 to 125 rpm [6, 29]. These parameters allow material to travel through the barrel to the die subject to friction and pressure. To achieve good processability, the processing parameters, such as screw speed, content of plasticizer, barrel temperature, die diameter, energy input, and pressure at the die [9], should be optimized to obtain optimized process parameters. TPS pellets obtained from optimized parameters are further processed into blown films, injection-molded specimens, and compression-molded sheets [6].

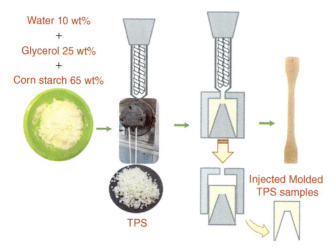

Figure 9.2 Schematic representation of the thermoplastic starch (TPS) preparation by melt extrusion. Source: Aldas et al. [20] / MDPI / CC BY 4.0.

9.1.5 Properties of TPS

The properties of TPS depend on various factors, such as source of starch, starch composition, chemical structure, type of modifiers/plasticizers, and processing method. The imperative properties of TPS studied in the literature are mechanical, thermal, and barrier.

9.1.5.1 Mechanical Properties

The mechanical properties measured gives values of tensile strength, tensile modulus, elongation at break, and tensile toughness. As aforementioned, mechanical properties of TPS depend on the source of starch, processing method, plasticizers, crystallinity of starch, and the ratio between amylose and amylopectin [9]. For example, Aldas et al. [20] and Mościcki et al. [6] prepared TPS from corn starch and potato starch using extrusion followed by injection molding and blown film, respectively. In the case of Aldas et al. [20], the developed TPS demonstrated tensile strength and elongation at break of 2 MPa and 63.46%, respectively. On the other hand, Mościcki et al. [6] reported that TPS exhibited tensile strength and elongation at break of 5.4 MPa and 79.28%, respectively. Furthermore, the decrease in glycerol content resulted in decreasing mechanical properties of the ensued TPS. It is worth noting that from these studies' that TPS produced from potato starch exhibited better mechanical properties when compared to that produced from corn starch. Basiak et al. [17] studied mechanical properties of starch containing 25% amylose and 33% glycerol. The films investigated exhibited tensile strength, tensile modulus, and elongation at break of 3.29 MPa, 0.12 MPa, and 15.21%, respectively. However, when glycerol content increased to 50%, the tensile strength and modulus of the material decreased, whereas elongation at break increased slightly. The reduction in mechanical properties when high amount of glycerol was added could be due to weak interaction between polymers. On the other hand, starch containing 18% amylose displayed tensile strength, tensile modulus, and elongation at break of 1.9 MPa, 0.51 MPa, and 107%, respectively [15].

9.1.5.2 Thermal Properties

Thermal behavior of TPS is determined by thermogravimetric analysis (TGA) and differential scanning calorimetry (DSC). TGA investigates thermal degradation and stability of the TPS, whereas DSC focuses on glass transition temperatures (T_g) and melting behavior. In the case of TGA, degradation of material is studied against temperature. Studies have indicated that TPS displayed three degradation steps [13]. The first step is at temperatures between 100 and 150 °C and is related to moisture evaporation [13, 17, 20]. The second step occurs between 260 and 450 °C, which represents degradation of starch. In this step, dehydration and breakdown of the glucose ring occurred, which resulted in the formation of C—C bonds and aldehyde [20]. The last degradation step takes place up to 600 °C related to the organic residues. It is worth mentioning that the stability of the material depends on the plasticizers and the content of plasticizer used. For example, the introduction of glycerol as plasticizer led to an increase in weight loss. However, increasing the content of glycerol reduces thermal stability of the material [13, 17].

Glass transition temperature plays a tremendous role in determining thermophysical transitions in polymer materials [9]. Glass transition temperature is regarded as a temperature where polymer chains are starting to be mobile. Below T_g, chains are in a glassy state, whereas above T_g, the chains become rubbery and can move freely [9]. T_g of TPS depends on the type of plasticizers and the content that is used. Tarique et al. [13] reported that T_g of unplasticized starch was recorded at 117 °C and the introduction of plasticizers into native starch reduces T_g of starch. This is in agreement with other studies which reported that incorporation of plasticizers decreases T_g of TPS. In addition, starch becomes more flexible and improves chain mobility [27]. However, increasing plasticizer content results in further reduction of T_g [9, 13, 27].

9.1.5.3 Barrier Properties

TPS is widely used in food packaging where barrier properties are very crucial in determining the suitability of WVP, which is regarded as a crucial property to understand if the material is suitable for food packaging. WVP regulates water vapor transportation between food and the surrounding environment [9]. Zhang et al. [9] reported that the permeation of gases such as water vapor occurs in starch amorphous domains. In general, for material to be qualified to be suitable for food packaging, the material must have low WVP [13, 27] and exhibit hydrophobic characteristics. Due to the hydrophilic nature of native starch and the plasticizer employed, TPS normally has high WVP [9]. Tarique et al. [13] and Sanyang et al. [27] reported that unplasticized native starch did not show the required values of WVP as the introduction of small content of plasticizers increases WVP. In addition, further increasing plasticizer content usually results in increased values of WVP. The researchers noticed that low plasticizer content (15%) results to lower values of WVP in comparison to high contents due to strong interaction between starch molecules. On the other hand, high plasticizer content (45%) improves the mobility and flexibility of starch chains and, therefore, increases the WVP values. Sanyang et al. [27] also mentioned that the values of WVP increase regardless of the type of plasticizer. The WVP values of TPS can be decreased by modifying starch either by blending with hydrophobic polymers or chemically modifying hydroxyl groups to impart hydrophobicity.

9.2 TPS-Biopolymer Blends

TPS has been demonstrated to be a suitable candidate to be employed as a replacement for petroleum-based polymers for food packaging subject to certain modifications. Notwithstanding TPS properties, they have shown some drawbacks, such as poor WVP, high water absorption, and poor mechanical properties, which limit their applications. Therefore, there is a necessity to improve the properties of TPS so that they can be suitable for food packaging applications. This can be achieved by either reinforcing or blending with other biodegradable polymers without compromising its biodegradation advantage. Blending TPS with biodegradable polymers, such as polylactic acid (PLA), poly-ϵ-caprolactone (PCL), polybutylene succinate (PBS), and poly(butylene adipate-co-terephthalate) (PBAT), has become a research hotspot because they are abundantly available, odorless, and nontoxic [37]. The main challenge in blending polymers is immiscibility, which can be mitigated by grafting compatibilizers or by adding reinforcing agent. Studies have shown that TPS and PBS are immiscible, and they display poor compatibility [38, 39], which affects their properties. To improve the compatibility between TPS and PBS, TPS was modified using N,N'-dicyclohexylcarbodiimide (DCC) and then blended with PBS [39]. The results demonstrated that modification led to good dispersion of TPS into a PBS matrix and enhanced tensile strength and tear resistance. These findings suggest that this blend has displayed potential for food packaging. Similarly, PBAT and TPS blends displayed a similar behavior as PBS and TPS blends. In the case of PBAT and TPS blends, the compatibility, mechanical, rheological, and thermal properties were improved by grafting maleic anhydride and clay [40, 41]. Müller et al. [42] and Ferri et al. [43] discovered that blending PLA and TPS resulted in weak interaction between them, which hinder stress transfer through the interface leading to poor properties. However, the blend of PLA and modified TPS with maleinized linseed oil exhibited the improvement in ductility but the compatibility was still poor [43]. Blending up to 50% of PCL and TPS resulted in good dispersion of TPS in PCL matrix. The blending of these polymers demonstrated improved mechanical properties and thermal stabilities. However, the best results were reported when low content of PCL was added [44, 45].

One of the most widely used commercially available TPS is Mater-Bi®, which mainly consists of corn starch blended with biopolymers and other compounds including natural plasticizers. Although the exact composition of Mater-Bi is not known, the following formulations have been elucidated: (i) Mater-Bi Y, composed of starch and cellulose acetate blends, whose properties resemble those of polystyrene (PS); (ii) noncompostable Mater-Bi A, constituted by a strong complex between

TPS and copolymers of polyvinyl alcohol (PVA); (iii) Mater-Bi V, having a TPS content greater than 85% and high solubility in water; (iv) Mater-Bi Z, having a PCL matrix; and (v) Mater-Bi N whose base polymeric matrix is PBAT [46].

Other developments include development of compostable resins from starch that is extracted from algae and compounded with biopolymers by Eranova in France. Recently, the U.S. company BioLogiQ, Inc., announced the launch of three new grades of its plant-based plastic, NuPlastiQ, developed from waste potato starch recovered from potato chips production. The new grades are blends with polyethylene (PE), polypropylene (PP), and PS and exhibit better durability than pure NuPlastiQ GP.

9.3 TPS-Biopolymer Composites

Cellulosic reinforcements – The high strength, renewability factor, and presence of hydroxyl groups on micro- and nanocellulose making it amenable to chemical modifications have made it one of the most popular bio-based reinforcement. The incorporation of nanocellulose in TPS results in completely biodegradable and renewable materials and hence several studies have focused on the use of nanocellulose in TPS matrices [47]. Researchers recently developed cassava nanofibrillated cellulose (CBN) incorporated in thermoplastic cassava starch by melt mixing and hot pressing with a combination of plasticizers, namely glycerol and sorbitol. The use of both plasticizers (1 : 1) resulted in better tensile strength and elastic modulus along with increased water uptake. The increasing cellulose loading did not essentially increase the tensile strength and elastic modulus but was favorable for water uptake because of the hydrophobicity properties of CBN [48].

The reinforcement of nanocellulose into TPS films from different vegetal sources for potential short shelf-life packaging applications was investigated by Montero et al. [49] The different starches explored were thermoplastic pea starch (TPeS), thermoplastic potato starch (TPoS), and thermoplastic corn starch (TCS). TPoS and TCS demonstrated good water resistance and thermal stability but poor stiffness. The incorporation of NCC decreased the water absorption rate in all the systems due to the good interfacial bonding between nano-sized filler and starch matrix. The authors concluded that TPoS can be a potential alternative for short-shelf-life packaging applications.

Studies have also focused on using microcrystalline cellulose as a filler in TPS resulting in a double-fold increase in tensile strength at medium-level loadings of MCC and consequent decrease in elongation at break values (Figure 9.3) [50]. This was mainly attributed to the formation of a more hydrogen-bonded cellulose network which leads to stronger interactions between MCC and TPS. At higher levels of MCC, tensile properties are seen to decrease due to poor dispersion and agglomeration of MCC. Interestingly, the WVP values of MCC-reinforced TPS composite films were found to be lower than the neat TPS films with the addition of the MCC. This was attributed to the fact that MCC is less hydrophilic than the TPS matrix.

The influence of lignin on TPS has been studied by several researchers [51–53]. The addition of Kraft lignin in TPS was investigated by De Freitas et al. [54]. Lignin is one

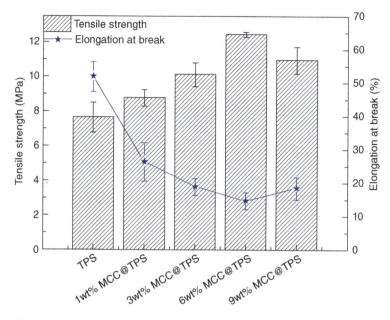

Figure 9.3 Mechanical properties of MCC-reinforced TPS composite films.

of the by-products of the pulp and paper industry and is considered to be an excellent filler since it has excellent mechanical and thermal properties, and its addition leads to a reduction in cost for the final product. The inclusion of lignin in TPS matrix resulted in a tensile strength increment for composites containing 4% Kraft lignin. The dynamic mechanical analysis showed that storage modulus increased, and that loss modulus decreased with higher loading of lignin. Biodegradation studies as shown in Figure 9.4 showed a continuous and progressive biodegradation process with

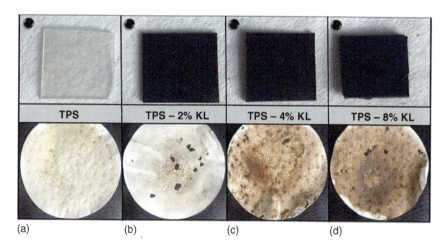

Figure 9.4 Samples for four types of films produced and the results of biodegradation test, (a) TPS, (b) TPS − 2% KL, (c) TPS − 4% KL, and (d) TPS − 8% KL.

complete disintegration of materials occurring on the tenth day. The control film exhibited 41.6% reduction in percentage of biodegraded material, which decreased progressively to 25.4%, 17.5%, and 9.2%, with an increased amount of lignin addition of 2%, 4%, and 8%, respectively. This was attributed to the fact that lignin contains phenolic groups that have bactericidal action, and one of the functions of lignin in plants is to act as a protective agent against attack by microorganisms. Hence, the presence of lignin at high content retards the biodegradation process.

Ternary composites of PBAT, TPS, and lignin containing varying loadings of TPS/lignin fillers were the subject of study by Li et al. [55]. Lignin reinforcement improved the miscibility, thermal stability, comprehensive mechanical properties, and hydrophobicity of the system. This was attributed to the amphiphilic structure of lignin molecules which is rich in polar functionalities (hydroxyl and carboxyl groups) which are able to interact with TPS by forming hydrogen bonds, and also contains aromatic moieties with affinity to PBAT segments via π-electron interactions as shown in Figure 9.5.

Another example of a ternary composite comprising biochar as a filler in polyhydroxy butyrate (PHB)–TPS composites was investigated by Haeldermans et al. [56]. Biochar is formed by the thermal decomposition of biomass and is now being explored for use as a filler in biocomposites. It was observed that TPS could act as an excellent intermediator between biochar and PHB at biochar concentrations up to 20 wt%, by limiting the degradation and reduction in molecular weight as compared to that of the PHB/char sample. These results also suggested biochar-reinforced PHB–TPS systems had the potential to act as slow-release fertilizers.

The additive manufacturing of biopolymers is gaining traction as customized products can be manufactured at a rapid pace and with minimum wastage. The utilization of starch or starch-based polymers, e.g. TPS and PLA, as feedstocks in 3D printing technologies for a variety of applications, has been explored by Li et al. [57]. Recently PLA–TPS filaments were developed by Haryńska et al. [58] who observed that the developed filaments were more hydrophilic and also demonstrated excellent printability (Figure 9.6) when compared to commercial PLA grades. Additionally, the composting study revealed that the commercial PLA printouts remained intact, whereas the PLA/TPS samples showed a mass loss of 19%.

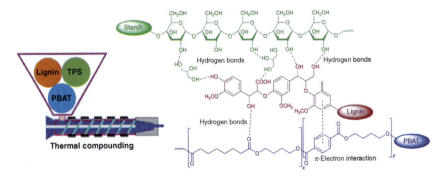

Figure 9.5 Schematic illustration of lignin-induced compatibilization upon thermal compounding of PBAT/TPS/lignin.

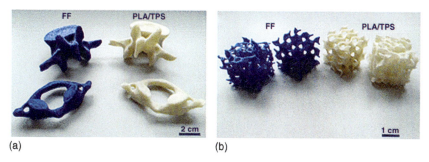

Figure 9.6 3D-printed FF (commercial PLA grade) and PLA/TPS complex structures: (a) personalized anatomical models of vertebrae C1 (atlas) and L3, (b) porous gyroid, and cancellous bone-like structure printouts.

9.4 Global Producers, Market Volumes, and Applications of Starch-Based Plastics

The global producers of starch-based plastics are presented in Table 9.2. Novamont is the largest producer, and its product line includes grades for food packaging as well as materials for injection molding and transparent mulching films.

The global market volume for starch-based plastics by application in 2018 and the predicted values in 2023 are presented in Figure 9.7. The packaging industry remains the single largest field of application for starch plastics and almost 65% of

Table 9.2 Global producers of starch-based plastics.

Company	Country	Brand name	Capacity
Novamont	Italy	Mater-Bi	Operates a biorefinery with 150 000 metric tpa
Wuhan Huali Environmental Technology Co.	China	Huali Eco-Tech	Capacity of 60 000 metric tpa
China Green Material Technologies	China	Eco-friendly	Capacity of 9 000 metric tpa
Biotec	Germany	Bioplast	Supplies Sphere and Biome Bioplastics; capacity of 20 000 metric tpa
Nihon Shokuhin Kako	Japan	Placorn	Mitsubishi subsidiary
Japan Corn Starch	Japan	Evercorn	–
Rodenburg	Netherlands	Solanyl	Produces Solanyl BP based potato starch. Annual capacity of 60 000 metric tpa

Source: Adapted from BCC Research (2019).

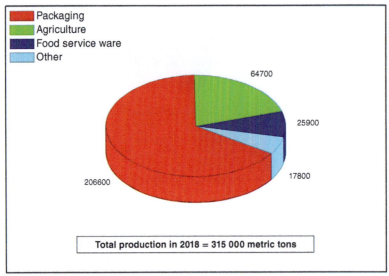

(a)

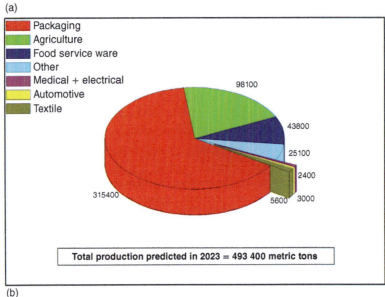

(b)

Figure 9.7 Global market volume for starch-based plastics by application (a) in 2018 (b) predicted in 2023. Source: Adapted from BCC Research (2019).

starch plastics are used in packaging sector (Figure 9.4). The total production of starch-based plastics is predicted to increase from 315 000 to 493 000 mt in 2023 with the packaging and food service ware sector reflecting a compounded annual growth rate (CAGR) of 8 and 11%, respectively. Asian countries like China and Japan are leading the consumption of starch-based plastics due to government supporting the use of bio-based products. For example, both in China (Wuhan region) and Italy, the banning of nonbiodegradable carrier bags has resulted in the conversion of more than 1 million shopping bags from oil-based to starch-based plastics [59].

Starch-based plastics also find applications in the agricultural sector. The major products include agricultural mulch films, drip irrigation, silage wrap for hay, heavy bags for animal feed and fertilizers, bail wrap for laying turf sod, and ground covers for roadside tree plantings. The global consumption of agricultural mulch films is around 4 million metric tpa of plastics with China alone consuming around 1.5 million metric tpa of agricultural mulch films. Regulatory measures are driving the opportunity of starch plastics in the agriculture sector in Europe, the United States, and China as there are bans on the use of fossil fuel plastics on farms and government provides 40–50% subsidy for farmland use.

9.5 Conclusions

Starch polymers are inherently hydrophilic due to presence of hydroxyl groups and cannot be processed on conventional processing equipment and therefore needs to be physically or chemically modified to form TPS. The properties of TPS can be further improved by blending with biopolymers or using it as a filler in biopolymer matrices. Several studies as elaborated in this chapter have shown that this leads to materials with superior properties, especially improved water resistance and barrier properties. However, more research is required on the use of novel plasticizers and nanoparticles that focus on decreasing retrogradation of the material to avoid decreasing mechanical strength or stiffness during storage in the development of TPSs. Additionally, processing techniques, such as 3D printing, need to be explored so that commercialization of new starch-based products occurs at a rapid pace.

References

1 Santana, R.F., Bonomo, R.C.F., Gandolfi, O.R.R. et al. (2018). Characterization of starch-based bioplastics from jackfruit seed plasticized with glycerol. *Journal of Food Science and Technology* 55 (1): 278–286. https://doi.org/10.1007/s13197-017-2936-6.
2 https://www.un.org/sustainabledevelopment/sustainable-consumption-production/ (accessed 01 February 2023).
3 https://www.newplasticseconomy.org/ (accessed 01 February 2023).
4 Bergel, B.F., da Luz, L.M., and Santana, R.M.C. (2017). Comparative study of the influence of chitosan as coating of thermoplastic starch foam from potato, cassava and corn starch. *Progress in Organic Coatings* 106: 27–32. https://doi.org/10.1016/j.porgcoat.2017.02.010.
5 Vidéki, B., Klébert, S., and Pukánszky, B. (2007). External and internal plasticization of cellulose acetate with caprolactone: structure and properties. *Journal of Polymer Science, Part B: Polymer Physics* 45 (8): 873–883. https://doi.org/10.1002/polb.21121.

6 Mościcki, L., Mitrus, M., Wójtowicz, A. et al. (2012). Application of extrusion-cooking for processing of thermoplastic starch (TPS). *Food Research International* 47 (2): 291–299. https://doi.org/10.1016/j.foodres.2011.07.017.

7 Bangar, S.P., Whiteside, W.S., Ashogbon, A.O., and Kumar, M. (2021). Recent advances in thermoplastic starches for food packaging: a review. *Food Packaging and Shelf Life* 30: 100743.

8 Nafchi, A.M., Moradpour, M., Saeidi, M., and Alias, A.K. (2013). Thermoplastic starches: properties, challenges, and prospects. *Starch/Staerke* 65 (1–2): 61–72. https://doi.org/10.1002/star.201200201.

9 Zhang, Y., Rempel, C., and Liu, Q. (2014). Thermoplastic starch processing and characteristics - a review. *Critical Reviews in Food Science and Nutrition* 54 (10): 1353–1370. https://doi.org/10.1080/10408398.2011.636156.

10 Bilck, A.P., Müller, C.M.O., Olivato, J.B. et al. (2015). Using glycerol produced from biodiesel as a plasticiser in extruded biodegradable films. *Polimeros* 25 (4): 331–335. https://doi.org/10.1590/0104-1428.1803.

11 Meng, L., Liu, H., Yua, L. et al. (2019). How water acting as both blowing agent and plasticizer affect on starch-based foam. *Industrial Crops and Products* 134: 43–49. https://doi.org/10.1016/j.indcrop.2019.03.056.

12 Ismail, S., Mansor, N., Majeed, Z., and Man, Z. (2016). Effect of water and [Emim][OAc] as plasticizer on gelatinization of starch. *Procedia Engineering* 148: 524–529. https://doi.org/10.1016/j.proeng.2016.06.542.

13 Tarique, J., Sapuan, S.M., and Khalina, A. (2021). Effect of glycerol plasticizer loading on the physical, mechanical, thermal, and barrier properties of arrowroot (*Maranta arundinacea*) starch biopolymers. *Scientific Reports* 11 (1): https://doi.org/10.1038/s41598-021-93094-y.

14 Sahari, J., Sapuan, S.M., Zainudin, E.S., and Maleque, M.A. (2013). Thermo-mechanical behaviors of thermoplastic starch derived from sugar palm tree (*Arenga pinnata*). *Carbohydrate Polymers* 92 (2): 1711–1716. https://doi.org/10.1016/j.carbpol.2012.11.031.

15 Jaramillo, C.M., Seligra, P.G., Goyanes, S. et al. (2015). Biofilms based on cassava starch containing extract of yerba mate as antioxidant and plasticizer. *Starch/Staerke* 67 (9–10): 780–789. https://doi.org/10.1002/star.201500033.

16 Lara, S.C. and Salcedo, F. (2016). Gelatinization and retrogradation phenomena in starch/montmorillonite nanocomposites plasticized with different glycerol/water ratios. *Carbohydrate Polymers* 151: 206–212. https://doi.org/10.1016/j.carbpol.2016.05.065.

17 Basiak, E., Lenart, A., and Debeaufort, F. (2018). How glycerol and water contents affect the structural and functional properties of starch-based edible films. *Polymers* 10 (4): https://doi.org/10.3390/polym10040412.

18 Sessini, V., Arrieta, M.P., Fernández-Torres, A., and Peponi, L. (2018). Humidity-activated shape memory effect on plasticized starch-based biomaterials. *Carbohydrate Polymers* 179: 93–99. https://doi.org/10.1016/j.carbpol.2017.09.070.

19 Peng, X.L., Kuo, M.C., Huang, C.Y. et al. (2018). Melt-processing, moisture-resistance and strength retention properties of supercritical CO_2-processed thermoplastic starch resins. *Express Polymer Letters* 12 (5): 462–478. https://doi.org/10.3144/expresspolymlett.2018.39.

20 Aldas, M., Pavon, C., López-Martínez, J., and Arrieta, M.P. (2020). Pine resin derivatives as sustainable additives to improve the mechanical and thermal properties of injected moulded thermoplastic starch. *Applied Sciences (Switzerland)* 10 (7): https://doi.org/10.3390/app10072561.

21 Ismail, S., Mansor, N., and Man, Z. (2017). A study on thermal behaviour of thermoplastic starch plasticized by [Emim] Ac and by [Emim] Cl. *Procedia Engineering* 184: 567–572. https://doi.org/10.1016/j.proeng.2017.04.138.

22 Hazrol, M.D., Sapuan, S.M., Zainudin, E.S. et al. (2021). Corn starch (*Zea mays*) biopolymer plastic reaction in combination with sorbitol and glycerol. *Polymers* 13 (2): 1–22. https://doi.org/10.3390/polym13020242.

23 Orsuwan, A. and Sothornvit, R. (2018). Effect of banana and plasticizer types on mechanical, water barrier, and heat sealability of plasticized banana-based films. *Journal of Food Processing and Preservation* 42 (1): https://doi.org/10.1111/jfpp.13380.

24 Ballesteros-Mártinez, L., Pérez-Cervera, C., and Andrade-Pizarro, R. (2020). Effect of glycerol and sorbitol concentrations on mechanical, optical, and barrier properties of sweet potato starch film. *NFS Journal* 20: 1–9. https://doi.org/10.1016/j.nfs.2020.06.002.

25 Zhang, K., Cheng, F., Lin, Y. et al. (2018). Effect of hyperbranched poly(trimellitic glyceride) with different molecular weight on starch plasticization and compatibility with polyester. *Carbohydrate Polymers* 195: 107–113. https://doi.org/10.1016/j.carbpol.2018.04.080.

26 Sanyang, M.L., Sapuan, S.M., Jawaid, M. et al. (2016). Effect of plasticizer type and concentration on physical properties of biodegradable films based on sugar palm (*Arenga pinnata*) starch for food packaging. *Journal of Food Science and Technology* 53 (1): 326–336. https://doi.org/10.1007/s13197-015-2009-7.

27 Sanyang, M.L., Sapuan, S.M., Jawaid, M. et al. (2015). Effect of plasticizer type and concentration on tensile, thermal and barrier properties of biodegradable films based on sugar palm (*Arenga pinnata*) starch. *Polymers* 7 (6): 1106–1124. https://doi.org/10.3390/polym7061106.

28 Edhirej, A., Sapuan, S.M., Jawaid, M., and Zahari, N.I. (2017). Effect of various plasticizers and concentration on the physical, thermal, mechanical, and structural properties of cassava-starch-based films. *Starch/Staerke* 69 (1–2): https://doi.org/10.1002/star.201500366.

29 Battegazzore, D., Bocchini, S., Nicola, G. et al. (2015). Isosorbide, a green plasticizer for thermoplastic starch that does not retrograde. *Carbohydrate Polymers* 119: 78–84. https://doi.org/10.1016/j.carbpol.2014.11.030.

30 Wang, J.L., Cheng, F., and Zhu, P.X. (2014). Structure and properties of urea-plasticized starch films with different urea contents. *Carbohydrate Polymers* 101 (1): 1109–1115. https://doi.org/10.1016/j.carbpol.2013.10.050.

31 Schmitt, H., Guidez, A., Prashantha, K. et al. (2015). Studies on the effect of storage time and plasticizers on the structural variations in thermoplastic starch. *Carbohydrate Polymers* 115: 364–372. https://doi.org/10.1016/j.carbpol.2014.09.004.

32 Abera, G., Woldeyes, B., Demash, H.D., and Miyake, G. (2020). The effect of plasticizers on thermoplastic starch films developed from the indigenous Ethiopian tuber crop Anchote (*Coccinia abyssinica*) starch. *International Journal of Biological Macromolecules* 155: 581–587. https://doi.org/10.1016/j.ijbiomac.2020.03.218.

33 Dang, K.M. and Yoksan, R. (2021). Thermoplastic starch blown films with improved mechanical and barrier properties. *International Journal of Biological Macromolecules* 188: 290–299. https://doi.org/10.1016/j.ijbiomac.2021.08.027.

34 Wang, J., Liang, Y., Zhang, Z. et al. (2021). Thermoplastic starch plasticized by polymeric ionic liquid. *European Polymer Journal* 148: https://doi.org/10.1016/j.eurpolymj.2021.110367.

35 Zuo, Y., Gu, J., Tan, H., and Zhang, Y. (2015). Thermoplastic starch prepared with different plasticizers: Relation between degree of plasticization and properties. *Journal Wuhan University of Technology, Materials Science Edition* 30 (2): 423–428. https://doi.org/10.1007/s11595-015-1164-z.

36 Bendaoud, A. and Chalamet, Y. (2013). Effects of relative humidity and ionic liquids on the water content and glass transition of plasticized starch. *Carbohydrate Polymers* 97 (2): 665–675. https://doi.org/10.1016/j.carbpol.2013.05.060.

37 Diyana, Z.N., Jumaidin, R., Selamat, M.Z. et al. (2021). Physical properties of thermoplastic starch derived from natural resources and its blends: a review. *Polymers* 13: 1396. https://doi.org/10.3390/polym13091396.

38 Fahrngruber, B., Fortea-Verdejo, M., Wimmer, R., and Mundigler, N. (2020). Starch/poly(butylene succinate) compatibilizers: effect of different reaction-approaches on the properties of thermoplastic starch-based compostable films. *Journal of Polymers and the Environment* 28 (1): 257–270. https://doi.org/10.1007/s10924-019-01601-0.

39 Yun, I.S., Hwang, S.W., Shim, J.K., and Seo, K.H. (2016). A study on the thermal and mechanical properties of poly (butylene succinate)/thermoplastic starch binary blends. *International Journal of Precision Engineering and Manufacturing – Green Technology* 3 (3): 289–296. https://doi.org/10.1007/s40684-016-0037-z.

40 Lendvai, L., Apostolov, A., and Karger-Kocsis, J. (2017). Characterization of layered silicate-reinforced blends of thermoplastic starch (TPS) and poly(butylene adipate-co-terephthalate). *Carbohydrate Polymers* 173: 566–572. https://doi.org/10.1016/j.carbpol.2017.05.100.

41 Fourati, Y., Tarrés, Q., Mutjé, P., and Boufi, S. (2018). PBAT/thermoplastic starch blends: effect of compatibilizers on the rheological, mechanical and morphological properties. *Carbohydrate Polymers* 199: 51–57. https://doi.org/10.1016/j.carbpol.2018.07.008.

42 Müller, P., Bere, J., Fekete, E. et al. (2016). Interactions, structure and properties in PLA/plasticized starch blends. *Polymer* 103: 9–18. https://doi.org/10.1016/j.polymer.2016.09.031.

43 Ferri, J.M., Garcia-Garcia, D., Sánchez-Nacher, L. et al. (2016). The effect of maleinized linseed oil (MLO) on mechanical performance of poly(lactic acid)-thermoplastic starch (PLA-TPS) blends. *Carbohydrate Polymers* 147: 60–68. https://doi.org/10.1016/j.carbpol.2016.03.082.

44 Ninago, M.D., López, O.V., Lencina, M.M.S. et al. (2015). Enhancement of thermoplastic starch final properties by blending with poly(ε-caprolactone). *Carbohydrate Polymers* 134: 205–212. https://doi.org/10.1016/j.carbpol.2015.08.007.

45 Correa, A.C., Carmona, V.B., Simão, J.A. et al. (2017). Biodegradable blends of urea plasticized thermoplastic starch (UTPS) and poly(ε-caprolactone) (PCL): morphological, rheological, thermal and mechanical properties. *Carbohydrate Polymers* 167: 177–184. https://doi.org/10.1016/j.carbpol.2017.03.051.

46 Aldas, M., Rayón, E., López-Martínez, J., and Arrieta, M.P. (2020). A deeper microscopic study of the interaction between gum rosin derivatives and a Mater-Bi type bioplastic. *Polymers* 12: 226. https://doi.org/10.3390/polym12010226.

47 Rivadeneira-Velasco, K.E., Utreras-Silva, C.A., Díaz-Barrios, A. et al. (2021). Green nanocomposites based on thermoplastic starch: a review. *Polymers* 13: 3227. https://doi.org/10.3390/polym13193227.

48 de Teixeira, E.M., Pasquini, D., Curvelo, A.A.S. et al. (2009). Cassava bagasse cellulose nanofibrils reinforced thermoplastic cassava starch. *Carbohydrate Polymers* 78: 422–431. https://doi.org/10.1016/j.carbpol.2009.04.034.

49 Montero, B., Rico, M., Rodríguez-Llamazares, S. et al. (2017). Effect of nanocellulose as a filler on biodegradable thermoplastic starch films from tuber, cereal and legume. *Carbohydrate Polymers* 157: 1094–1104. https://doi.org/10.1016/j.carbpol.2016.10.073.

50 Chen, J., Wang, X., Long, Z. et al. (2020). Preparation and performance of thermoplastic starch and microcrystalline cellulose for packaging composites: extrusion and hot pressing. *International Journal of Biological Macromolecules* 165: 2295–2302.

51 Kaewtatip, K. and Thongmee, J. (2013). Effect of Kraft lignin and esterified lignin on the properties of thermoplastic starch. *Materials & Design* 49 (701–704): https://doi.org/10.1016/j.matdes.2013.02.010.

52 Baumberger, S., Lapierre, C., Monties, B., and Valle, G.D. (1998). Use of Kraft lignin as filler for starch films. *Polymer Degradation and Stability* 59: 273–277. https://doi.org/10.1016/s0141-3910(97)00193-6.

53 Majeed, Z., Mansor, N., Ajab, Z. et al. (2018). Kraft lignin ameliorates degradation resistance of starch in urea delivery biocomposites. *Polymer Testing* 65: 398–406.

54 de Freitas, A.d.S.M., Rodrigues, J.S., Maciel, C.C. et al. (2021). Improvements in thermal and mechanical properties of composites based on thermoplastic starch and Kraft lignin. *International Journal of Biological Macromolecules* 184: 863–873.

55 Li, M., Jia, Y., Shen, X. et al. (2021). Investigation into lignin modified PBAT/thermoplastic starch composites: thermal, mechanical, rheological and water absorption properties. *Industrial Crops and Products* 171: 113916.

56 Haeldermans, T., Samyn, P., Cardinaels, R. et al. (2021). Bio-based poly (3-hydroxybutyrate)/thermoplastic starch composites as a host matrix for biochar fillers. *Journal of Polymers and the Environment* 29: 2478–2491. https://doi.org/10.1007/s10924-021-02049-x.

57 Lam, C.X.F., Mo, X.M., Teoh, S.H., and D.W. (2002). Hutmacher scaffold development using 3D printing with a starch-based polymer. *Materials Science and Engineering: C* 20: 49–56.

58 Haryńska, A., Janik, H., Sienkiewicz, M. et al. (2021). PLA–potato thermoplastic starch filament as a sustainable alternative to the conventional PLA filament: processing, characterization, and FFF 3D printing. *ACS Sustainable Chemistry & Engineering* 9: 6923–6938. https://doi.org/10.1021/acssuschemeng.0c09413.

59 BCC Report: Global Markets and Technologies for Bioplastics (2019). https://www.bccresearch.com/market-research/plastics/biopolymers-market-report.html (accessed 01 February 2023).

10

Main Marine Biopolymers for Food Packaging Film Applications

Jesús Rubén Rodríguez-Núñez[1], Diana Gabriela Montoya-Anaya[2], Judith Fortiz-Hernández[3], Yolanda Freile-Pelegrín[4], and Tomás Jesús Madera-Santana[3]

[1] *Universidad de Guanajuato, Programa de Biotecnología, Mutualismo #303, Colonia La Suiza, Celaya, Guanajuato, 38060, México*
[2] *División de Estudios de Posgrado e Investigación, Tecnológico Nacional de México/IT de Roque, Carr. Juventino Rosas-Celaya km. 8, Celaya, Guanajuato, 38110, México*
[3] *Centro de Investigación en Alimentación y Desarrollo, A.C, Coordinación de Tecnología de Alimentos de Origen Vegetal, Carr. a La Victoria No. 46,, Hermosillo, Sonora 83304, México*
[4] *Centro de Investigación y de Estudios Avanzados del IPN. CINVESTAV-Unidad Mérida, Departamento de Recursos del Mar, Carr. Mérida-Progreso km. 6, Mérida, Yucatán, 97310, México*

10.1 Introduction

More than 70% of the earth's surface is covered by the oceans, which contain enormous biodiversity. Marine organisms live in complex habitats exposed to extreme conditions. Ever-changing conditions (temperature, salinity, tides, pressures, radiation, light, predators, etc.) have forced species to develop extraordinary physical strategies for exploiting diverse ecological niches. Many organisms have consequently developed chemical defense systems that facilitate their survival in extremely competitive environments.

Therefore, the broad variety of marine species provides an abundant source of natural resources and functional materials, such as antioxidants, minerals, vitamins, bioactive peptides, enzymes, polyunsaturated fatty acids (PUFA), and polysaccharides [1]. Polysaccharides are biopolymers found in a variety of animals, plants, algae, and microorganisms. In particular, marine macroalgae (or seaweeds), which have lower strength or rigidity than terrestrial plants, contain large amounts of special polysaccharides (phycocolloids) needed to support the seaweed thallus in the water. They comprise high-molecular-weight polysaccharides and are the main structural components of seaweed cell walls. Because of their complexity

Natural Materials for Food Packaging Application, First Edition.
Edited by Jyotishkumar Parameswaranpillai, Aswathy Jayakumar,
E. K. Radhakrishnan, Suchart Siengchin, and Sabarish Radoor.
© 2023 WILEY-VCH GmbH. Published 2023 by WILEY-VCH GmbH.

"unconventional" and heterogeneous sugar content, sulfation, and other changes, seaweed polysaccharides provide a difficulty as industrial feedstocks.

Seaweeds are grouped into three classes based on their pigmentation: brown (Phaeophyceae), green (Chlorophytae), and red (Rhodophyta) algae. Some species are edible and are frequently consumed as food in coastal regions of the world. Polysaccharides, proteins, lipids, pigments, vitamins, minerals, and bioactive compounds are abundant in seaweeds.

Special polysaccharides – such as agar and carrageenan from Rhodophyta, fucoidan and alginate from Phaeophyceae, and ulvan from Chlorophyta – are characterized by their gelling, viscosifying, and emulsifying properties. They are commonly used in agricultural fertilizers, medicine, energy, and industrial applications. Aside from that, there has been growing interest in their potential use in biological and medical applications. Polysaccharides have a range of biological functional qualities and are critical to sustain life. They are well-known for their nutritional benefits and undeniable effects on human immunity, digestion, and detoxification. Additionally, polysaccharides are inexpensive biomaterials with minimal toxicity, excellent biocompatibility and biodegradability; they are proving to be a promising alternative to synthetic plastics used in food films and coatings.

Chitosan is a polysaccharide conformed by D-glucosamine units; it has been demonstrated to be biodegradable, nontoxic, and biocompatible. Also, due to its functional groups ($-NH_2$ and $-OH$) it has bioactivity (i.e. antimicrobial and antifungal), and it is possible to improve its properties with chemical modifications. Moreover, the chemical reactions used to enhance chitosan's properties should not have an effect on its sustainability, because chitosan should preserve its biodegradability, nontoxicity, antimicrobial and antifungal properties. In recent times, modified chitosan has been studied in a variety of fields, including the cosmetics industry, the biomedical field, water engineering, drug delivery systems, and food technology [2].

This review presents the current state of research of marine biopolymers used as food packaging in edible films and coatings – including agar, alginate, carrageenan, ulvan, fucoidan, and modified chitosan – and their commercial applications in foods such as fruits, vegetables, and meat. Moreover, the benefits of using seaweed polysaccharides in food packaging are shown by summarizing their mechanical, physical, thermal, antioxidant, antibacterial, and chemical properties, as well as their ability to release bioactive compounds.

10.2 Polysaccharides from Seaweeds

There are about 10 000 distinct species of seaweed across the globe, and it can grow at depths of up to 180 m on solid substrates (benthic species), or floating (pelagic species). It is one of the most significant organisms in the ocean for marine biodiversity. There is a growing interest in marine seaweeds, including macroalgae (seaweeds) and microalgae, that has spiraled in the last decade; this is due to the fact that marine seaweed provides a diversity of protein foods and polysaccharides with a variety of applications in several industries. It is well known that marine environments are enormously complex and diverse in terms of available nutrients, pressure, temperature, marine currents, organisms living in the environment, and adaptation strategies developed by those organisms, among many other factors. From this point of view, seaweeds are a source of valuable compounds such as pigments, lipids, fatty acids, sterols, polysaccharides, proteins, and peptides, along with many secondary metabolites (amino acids, phenolic compounds, and terpenes) [3]. Additionally, they are a notable source of bioactive compounds with characteristics such as anti-inflammatory, antioxidant, anticancer, antiviral, antibacterial, antifungal, antiobesity, and antidiabetic properties [4].

10.2.1 Main Seaweed Polysaccharides

Seaweeds are sometimes referred to as edible marine algae; this is because they are good sources of nutrients with one distinguishing feature, namely the abundance of sulfated polysaccharides (SPs). SPs are a complex polysaccharide-containing sulfate groups and are found in marine invertebrates, microorganisms, and seaweeds [5]. The major source of nonanimal SPs is seaweeds (Figure 10.1), and the chemical

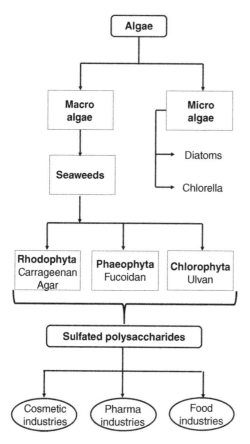

Figure 10.1 General representation of the derivation of sulfated polysaccharides from seaweeds and their uses in different industries.

structure of these polymers differs depending on the algal species [6, 7]. SPs have numerous applications for food, packaging, cosmetics, fertilizers, medicine, energy, and biotechnology, among others [1, 8, 9]. Nevertheless, isolation of SPs from seaweeds is still difficult, but recent advances in screening, isolation, and analytical procedures have made it possible to boost overall production while also ensuring batch quality and uniformity.

Seaweeds have several advantages over other biomass in that they do not require pesticides or large amount of land, and they grow quickly, easily, and inexpensively. As a result, employing seaweeds as biomass material to replace traditional plastic is a potential method that is not only cost-effective but also environmentally beneficial. Films from seaweed polysaccharides can be produced and mixed with other hydrosoluble polysaccharides by dissolution to improve physicochemical, antioxidant, and antimicrobial properties. These films and coatings are edible and offer a potential alternative to traditional plastic because, unlike conventional plastics that pollute water bodies, they will not degrade into microplastics. A summary of the main applications of seaweed polysaccharides is described below.

10.2.2 Alginate

Alginates are natural polymers extracted from brown seaweed, which belong to the *Phaeophyceae* class [6]; they are found in the cell walls and intercellular matrix. Alginates provide the plant with the springiness and mechanical strength it needs to live in the water. They are naturally linked to all the salts found in saltwater (Ca^{2+}, Na^+, Mg^{2+}, Sr^{2+}, and Ba^{2+} ions). Alginates molecules are composed of 1-4 β-D-mannuronic acid (M block) and α-L-guluronic acid (G block) (Figure 10.2) [11, 12], where M and G blocks are homogenously or heterogeneously linked via 1–4 glycosidic bonds to form linear dimers, which make up the large polymer [13]. The ratios of G and M blocks in the alginate molecule define its molecular structure and

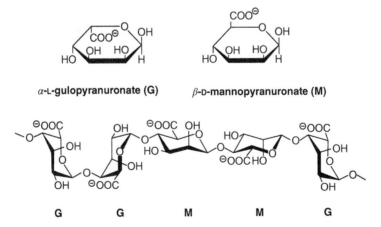

Figure 10.2 Alginate monomers G and M, and macromolecular conformation of alginate polymer. Source: Adapted from Smidsrød and Draget [10].

physical properties. The content of G and M can affect the performance of alginate biopolymers, those containing a high content of GG blocks showing higher solubility than those with MM blocks. Alginates with MG or GM blocks are soluble at low pH levels; however, alginates rich in MM or GG blocks are insoluble [14]. The GG content can impact gelation by interacting with divalent ions such as calcium (Ca^{2+}), which can act as a crosslinker between the functional groups of alginate molecules. Moreover, G content is considered one of the key physical and chemical characteristics of sodium alginate. The term "high G" refers to an alginate hydrogel with a G content greater than 70%, an along G chain, and high stiffness [13], whereas an alginate hydrogel with a low G content is more elastic. As a result, by varying the M/G ratio in the product, a gel of varying strength may be generated, which makes possible a wide variety of applications.

In polymer chains, monomers are arranged alternately in GG and MM blocks, together with MG blocks. Alginates used in commerce are produced from the biomass of marine macroalgae such as *Laminaria hyperborea*, *Laminaria japónica*, *Laminaria digitata*, *Macrocystis pyrifera*, *Ascophyllum nodosum*, *Eclonia máxima*, *Lessonia nigrescens*, *Durvillea antartica*, and *Sargassum* sp. [15]. The tendency of alginates to react with polyvalent metal cations, particularly calcium ions is the most interesting property. The ions form a cooperative interaction between M and G blocks, resulting in a tridimensional network, in which they may be packed together and be coordinated (Figure 10.3). This can be visualized as an "egg-box" arrangement [17]; it is a three-dimensional network structure, in the center of which are found Ca^{2+} ions [13]. Calcium alginate gels are heat-irreversible; in this respect, they have an advantage over sodium alginate gels and other colloids [18].

The industrial production is estimated to be around 30 000 mTon/year; this is likely to be less than 10% of annually biosynthesized material in the standing stock of macroalgae crops [19]. The extraction process of alginate from algae material is shown in Figure 10.4. The ion exchange with protons is performed as the initial stage in the production of alginate by extracting the milled algal tissue with 0.1–0.2 M mineral acid. Then, the alginic acid is neutralized with an alkali such as sodium carbonate or sodium hydroxide to produce the water-soluble sodium alginate. After

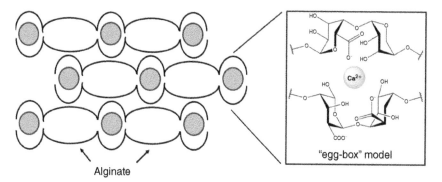

Figure 10.3 Alginate crosslinking with Ca^2 and "egg-box" formation. Source: Adapted from [16].

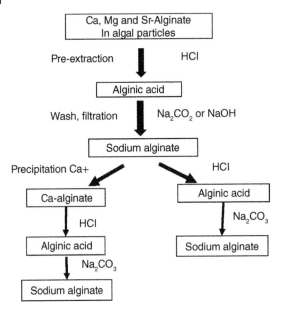

Figure 10.4 Extraction of alginate.

removing algal particles via sifting, flotation, centrifugation, and filtering, the soluble sodium alginate is precipitated straight by alcohol, calcium chloride, or mineral acid, and then dried and ground [19].

10.2.2.1 Properties and Limitations of Alginate

Polysaccharides have a polydisperse molecular weight, and alginate is no exception. Therefore, the biochemical and physical properties of alginate are strongly dependent on the molecular weight and the G/M ratios. Owing to their capacity to bind Ca^{2+} and H^+, G blocks are thought to be a crucial in the structure of alginate. Alginates have molecular weights ranging from 32 to 200 kDa [20, 21]. The methods for determining the average molecular weight are the number-average, Mn (the molecular weight of polymer molecules is calculated based on the number of molecules in a population with a specific molecular weight), and the weight average, Mw (the weight of polymer molecules in a population is based on the weight of molecules with a certain molecular weight). The molecular weight is related to the viscosity developed by alginate in solution; this may have consequences for the usage of alginates, since low molecular weight fragments comprising only short G blocks may not participate in the development of gel networks and hence do not contribute to gel strength.

The molecules of alginate in an aqueous solution are in random conformation; this solution is colorless, transparent, and has a variable range of viscosity. The viscosity of the solution increases with alginate concentration, molecular weight, G/M ratio, and interactions between alginate chains. Alginates have unique ion-binding characteristics, namely that their affinity for multivalent cations depends on their composition. Alginate's attraction to alkaline elements increases in the sequence Mg ≪ Ca < Sr < Ba; this is a characteristic that distinguishes alginates from other polyanions [19]. However, the solubility of alginates is affected by the pH of the solvent, the ionic strength of the solution, and the gelling ions in the medium; all of these can reduce the solubilization of alginates.

In dry powder form, sodium alginate has a long shelf life when it is stored in a dry, cool place without exposure to sunlight. It can also be kept in the deep freezer for a number of years without significantly losing molecular weight. Dried alginic acid has extremely low stability at room temperature, owing to an intramolecular acid-catalyzed degradation [22]. It is crucial to understand the parameters that influence and restrict the stability of aqueous alginate solutions, as well as the chemical process that causes the deterioration. The reduction in relative viscosity is indicative of the degradation of alginate in solution. Both acid and alkaline degradation, as well as oxidation by free radicals, can affect alginate chains, and this biopolymer can be digested by microorganisms (biodegradation).

10.2.2.2 Applications of Alginate in Edible Films and Coatings

Most often, alginate is used as a food ingredient (colloid, thickener, gelling agent, etc.), a medium for retaining moisture, and a means for providing a good appearance and consistency (by way of emulsifying effects) in baked food and canned products. Because of its high viscosity and gelling properties, alginate can modify food texture [15]. Alginates are used as binding agents, texturants, and thickeners in the industrial textile industry; this is because alginate as a substrate is safer, easy to break down, and enables easier wastewater disposal. Nowadays, the most common applications for alginate are in medicine (e.g. 3D-hydrogels and 2D-membranes for soft tissue regeneration), the environment (e.g. sponges for water purification), and food (rheological additives). Sodium alginate has been extensively used in drug delivery, wound dressing, and tissue engineering, due to its biocompatibility, low cost, and functional gelation by addition of polyvalent cations, such as Ca^{2+}, Ba^{2+}, Sr^{2+}, Zn^{2+}, and Cu^{2+}.

Sodium alginate films have relatively low tensile strength and antimicrobial activity, which limits their application in food packaging. Multivalent ions are used as crosslinkers to establish ionic interactions between alginate chains; this improves the mechanical and transport properties, among other things. Alginate films can be formulated with antimicrobials, which are natural extracts from plants (essential oils and compounds) that successfully limit bacterial deterioration and reduce the risk of food-borne infections [23]. Table 10.1 shows the uses of alginate in films, coatings, and composites in food packaging applications.

10.2.3 Agar

Red algae (agarophytes) are a morphologically diverse group of macroalgae, consisting of more than 700 genera and 6000 species, found at depths of up to 200 m. A variety of agarophytes are used around the world for agar production, the main genera being *Gelidium*, *Gelidiella*, *Pterocladia*, *Gracilaria*, and *Ahnfeltia*. However,

Table 10.1 Films, coatings, and composites formulated with alginates for food packaging applications.

Formulation	Properties	Application	Reference
Alginate film + essential oils of oregano, cinnamon and savory in beef muscle	The addition of essential oils into alginate-based edible films allowed to reduce E. coli O157 : H7 and Salmonella typhimurium growth on beef muscle during 5 days of storage. The films treated with 2% $CaCl_2$ preserved the active compounds better	Edible films for preservation of beef	[24]
Alginate films + sago starch + lemongrass oil	The films containing lemongrass oil are effective in inhibiting the growth of E. coli O157 : H7. The water vapor permeability (WVP) and percentage of elongation at break (%Eb) increased, and the tensile strength (TS) decreased, with the addition of different concentrations of lemongrass oil to the polymer matrix	Antimicrobial edible films	[25]
Alginate films + white, red, and extruded white ginseng extracts	Ginseng extract incorporated into alginate film decreased TS and elastic modulus (EM) but augmented the %Eb. The extruded white ginseng extract has good potential for use in antioxidant biodegradable films	Antioxidant biodegradable films or coatings for various food applications	[26]
Alginate films + $CaCO_3$ + oregano essential oil (OEO)	Improvement in mechanical properties (tensile strength, TS). Thickness and %Eb of the films was increased with the addition of OEO, while TS and WVP decreased. Films incorporated with OEO were more effective against Gram-positive bacteria than Gram-negative bacteria	Biodegradable edible films are developed for their alternative antibacterial properties	[11]
Alginate films + $CaCl_2$ + H_2O/C_2H_5OH solution	The addition of ethanol caused a moderate increase in the TS (119–135 MPa) and a slight decrease in the %Eb of the dry films (8.76–7.32%). Therefore, the TS of the films increased and the %Eb of the films decreased with addition of ethanol	Biodegradable food packaging	[27]

Formulation	Properties	Application	Reference
Alginate films + CaCl$_2$	Crosslinking had a substantial impact on the structure and characteristics of alginate, causing the alginate-based film's film thickness, moisture content, solubility, and water vapor permeability to decrease. An improvement in mechanical properties (TS) and lower values of %Eb	Biodegradable food packaging	[28]
Alginate films + cinnamon essential oil nanoemulsion (CEO-NE)	Improvement in the Young modulus, while %Eb was significantly decreased. The biocomposite films containing 20% CEO-NE also showed strong antibacterial effects against *Salmonella typhimurium*, *Bacillus aureus*, *E. coli*, and *Staphylococcus aureus*, achieving inhibition zones from 29.7 to 53 mm	Biocomposite films as antibacterial packaging for extending the shelf life of fresh food	[12]
Thyme/sodium alginate films in packaging for fresh-cut apple slices	Improvement in mechanical properties (high TS, %Eb, and UV-VIS light blocking capability), but low WVP, and low swelling ratio in comparison to sodium alginate film without thymol. This film exhibited inhibition of the growth of *Staphylococcus aureus* and *Escherichia coli*	Biocomposite films, application in fresh-cut apple packaging	[23]

the first four are the most used agarophytes in industry [29]. Based on their particular properties, there are several types of agars that are used for different applications. They have been developed to satisfy a range of needs, and they come from a variety of agarophyte algae produced using diverse methods. We must distinguish between "natural agars or native agars" and "industrial agars." In general, commercial agar contains 50–90% agarose. Araki [30] identified the structure of agarose, which was later confirmed by others.

Knutsen *et al.* [31] reported the structure of alternating 3-linked b-D-galactopyranose and 4-linked a-L-galactopyranose, naming it agaran (Figure 10.5A), while the repeating backbone is called agarose when all the 4-linked residues are in the 3,6-anhydro form (Figure 10.5B). Agarans may also include varying quantities of O-linked groups, such as methyl ether, sulfate ester, pyruvate acetal, or b-D-xylopyranosyl residues.

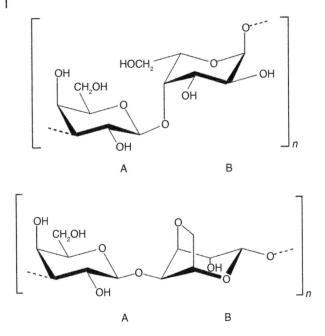

Figure 10.5 Repeating unit of an agaran and agarose. 3-linked and 4-linked units are indicated by A and B, respectively.

The most common polysaccharides identified from marine red macroalgae are sulfated galactans; these are composed of alternating 3-β-D-galactopyranosyl and 4-α-L-galactopyranosyl units in a linear chain. The physicochemical characteristics of the polysaccharide are affected by the amount and distribution of the abovementioned substituents in the skeletal chain of the galactan [32, 33]. The viscosity and gelling capability of these hydrocolloids – which enable various commercial uses, such as thickeners, stabilizers, and gelling agents – are of special interest [34].

Despite the fact that the finest quality agar comes from the genus *Gelidium*, the high cost of this polysaccharide and the gradual exhaustion of natural grasslands have made it necessary to look alternative natural sources of this polysaccharide. Consequently, *Gracilaria* (Gracilariales, Rhodophyta), a global genus, proved an excellent substitute for *Gelidium* agar in the food industry after it was discovered that alkaline treatment with sodium hydroxide, which converts L-galactose 6-sulfate into 3,6-anhydro-L-galactose, significantly improved low gel strength [35–37]. The species type, physiological variables, life cycle stage ambient circumstances, seasonal and geographical features, extraction techniques, and postharvest storage all influence the composition and qualities of *Gracilaria* agars [38–40].

Agars from various origins are used as food additives; they must be characterized according to international standards such as those set out by the FAO or the World Health Organization [35]. However, contemporary food business cannot function without packaging. It is required for preserving food and ensuring the food's safety and integrity. Packaging is an important element of our supply chain because of these benefits, but it also certain drawbacks, such as migration of residual contaminants, and issues around cost, energy efficiency, and sustainability [41]. Glass, paper, different metals, and plastic are all extensively used materials for food packaging; nevertheless, polymers, after paper, are the most utilized.

Several research projects have been conducted in recent years to find viable, environmentally acceptable alternatives such as biodegradable and/or biobased polymers, and substantial progress has been achieved in this area [42, 43]. Biopolymers, like polysaccharides, are nontoxic and biodegradable. Moreover, they are biocompatible and have the capacity to perform with an appropriate host response in a specific application [44]. These properties provide substantial benefits for food packaging, notably in edible coatings and films. Polysaccharides are a sustainable choice for active packaging because of their inherent antibacterial and antioxidant qualities, along with several other biological characteristics. As we have mentioned, seaweeds are an important source of polysaccharides, and these are a viable raw material for active packaging. When they are mixed with biodegradable polymers, they provide a more sustainable option for food packaging than conventional materials.

Agar's capacity to produce continuous films is based on its gelling capability. When agar powder is dissolved in hot water, it creates a viscous fluid. A thermoreversible gel (Figure 10.6) can be achieved by lowering the temperature below the gelling temperature (90–103 °C) [45, 46]. The physical gelation of agar can be established by hydrogen bonds between agarose molecules [29]. An agar gel consists of networks of agarose double helices, which are stabilized by water molecules; these are clumped together because of their hydroxyl groups [47]. This network of agar gels allows large amount of water to be held inside, which provides it with its unique gelling ability [29]. To prevent the gelation of agar solution when agar films are fabricated, the temperature of the casting surface and the film-forming solution must be higher than the temperature at which agar begins to gel. During the drying process of agar films, the linear structure of nonionic agarose can form hydrogen interactions and the film is structured in a strong network that can be peeled off from the plastic mold [8, 38].

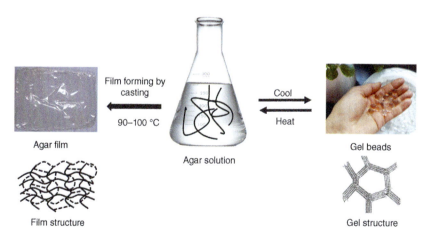

Figure 10.6 Gelling and film-forming process of agar gum. Gel beads image source: PortaUardin / Pixabay.

10.2.3.1 Applications of Agar in Edible Films and Coatings

The mechanical, thermal, optical, and chemical properties of edible films and coating are modified when seaweed polysaccharides are incorporated into them. The results are reliant on the amount and type of polysaccharide added. An extensive study is required to develop a safe and effective package. Table 10.2 provides a summary of the developments reported in edible films and coatings over the last ten years.

Table 10.2 Films, coatings, and composites formulated with agar for food packaging applications.

Formulation	Properties	Application	Reference
Apple sauce + agar as a plasticizer	Edible films based on apple sauce with the addition of agar showed a trend to increase in tensile strength (TS) as the amount of plasticizer was increased (from 1.32 to 1.70 MPa for agar), and the structure of the edible film became more homogeneous	Biodegradable edible films are developed as an alternative packaging material	[48]
Agar + nanoclay	Improvement in mechanical properties (TS). Decrease in surface properties (reduce contact angle of water), water vapor permeability (WVP), water vapor absorption and water solubility (WS). Increase in swelling ratio	Biodegradable food packaging	[49]
Agar + gelatin + cornstarch + citrus pectin + CO_2-extracts of parsley and ginger	An effective coating – composed of agar (0.25%), citrus pectin (0.5%), citric acid (0.5%), and CO_2 extract of parsley or ginger (0.1%) – was developed. The coating prolonged the shelf life of pork meat at a storage temperature ($\pm 1\,°C$) for 2 days. The physical and chemical properties of coated meat showed the peroxide number of its fat content is lower by 0.007% of iodine, and the acid number of its fat content is lower by 0.39 mg KOH/kg of fat	Edible coating for fresh meat	[50]
Starch + agar	The starch/agar film-forming solution at the ratio of 70/30 wt% had a low consistency index in comparison to pure starch, and the components are compatible. The mechanical parameters and water contact angle are highly dependent on their composition	Active packaging by applying holographic marker on the coating surface	[51]

Formulation	Properties	Application	Reference
Agar + sodium alginate (AS) and AS + ginger essential oil (AS + GEO)	Edible coatings of AS and AS + GEO were applied to fresh beef during refrigerated storage. Coating treatments significantly slowed in the oxidation of beef slices and a reduction in the microbial population was observed. AS and AS + GEO coatings extended the shelf life of fresh beef by 6 and 9 days, respectively	Antibacterial coating material for chilled beef storage	[52]
Agar + grapefruit seed extract (GSE)	GSE produced an increase in UV-barrier, color, moisture content, WS, and WVP, but a decrease in surface hydrophobicity, TS and modulus of the films. Agar/GSE films showed antimicrobial activity against *L. monocytogenes*, *B. cereus*, and *E. coli*	Active food packaging material for food safety and for extending the shelf-life of the packaged food	[53]
Agar + nanocrystalline cellulose + savory essential oil	An improvement in the elongation at break, but a reduction in mechanical properties (TS and modulus). Increase in water contact angle, swelling ratio, and viscosity of solutions. The WS and WVP both showed a decrease	Active packaging for improving the safety and shelf life of foodstuffs	[54]
Agar + green tea extract (GTE) + probiotic strains (*Lactobacillus paracasei* L26 and *Bifidobacterium lactis* B94)	Bioactive film composed was applied to hake fillets and evaluated over 15 days of storage. The films produced a decrease in the indexes of fish quality (total volatile basic nitrogen (TVB-N), trimethylamine nitrogen (TMA-N) and pH). The total viable counts, H_2S-producing microorganisms, and TVB-N were maintained within the limits of acceptability (15 days)	Bioactive films and coatings for extending shelf life of fish	[55]
Agar + banana powder + silver nanoparticles (A/B/AgNPs)	Binary blend films. Agar and banana powder reinforced with silver nanoparticles modified the film's properties such as color, transmittance, mechanical properties, moisture content, WS, water contact angle, WVP, thermal stability, and antioxidant levels. A/B/AgNPs films exhibited higher UV-screening function with strong bactericidal activity against Gram- (*E. coli*) food-borne pathogenic bacteria and bacteriostatic activity against Gram+ bacteria (*L. monocytogenes*)	Active food packaging material for maintaining the safety and extending the shelf life of packaged food	[56]

(Continued)

Table 10.2 (Continued)

Formulation	Properties	Application	Reference
Rice starch (RS) + hydroxypropyl cassava starch (HCS) + agar blends	Peelable shells of starch blend suspensions were produced using layer-by-layer mold dipping. HCS adsorbed water, which acted as plasticizer giving lower relaxation temperature. RS + agar increased crystallinity and non-homogeneity of films and reduced network flexibility and light transmission	Peelable coatings for edible films	[57]
Anti-*L. monocytogenes* enterocins synthezised by *Enterococcus avium* DSMZ17511 was supported on agar edible films	The consistency of the cheese plays an important role: soft cheese facilitated rapid diffusion of the antimicrobials, while semihard cheese produced a gradual release with prolonged inhibition of the pathogen. The application of these enterocin agar coatings is an effective, low cost, natural and safe alternative for controlling *L. monocytogenes* in cheeses	Antimicrobial coatings on different cheese matrices	[58]
PLA + agar + κ-carrageenan + clay	This bio-nano-composite showed high tensile strength, water vapor permeability, water solubility, and uptake ratio. The composite showed an improvement in thermal stability	Food packaging	[59]
Starch + agar + sorbitol + Tween-20	The increase in Tween-20 and sorbitol content produced a reduction in the water-vapor permeability of the film	Edible films and coatings	[60]
Agar + sugar palm starch	The composites showed an improvement in tensile strength and modulus, but a decrease in elongation at break. The thermal properties showed an increase in glass transition and melting temperature and an increase in moisture absorption and swelling. The composites exhibited a smooth surface, no clustering, and good miscibility	Material for food packaging and other applications	[61]
Agar + low density polyethylene (LDPE)	TS and elongation at break of composites decreased with increased agar content. In contrast, the tensile modulus (TM) showed a continuous increase. The crystallinity of LDPE increased as a result of the addition of agar particles	Composite material for food packaging	[62]
Agar + poly(butylene adipate-*co*-terephatalate) (PBAT)	Composites of agar particles and PBAT enhanced the TS and TM, but reduced the elongation at break	Composite material for food packaging	[63]

10.2.4 Carrageenan

Carrageenan is a complex family of anionic polysaccharides derived from red seaweeds (*Rhodophyta*) [64]. A wide variety of species of Rhodophycae are used in the commercial production of carrageenan, and the most relevant are iota (*i*)-, kappa (*k*)-, lamba (*λ*)-, Mu(*μ*)-, Nu()-, and theta (*θ*)- carrageenan; these are the six fundamental types of polysaccharides of this biopolymer. Carrageenan can be extracted from species such as *Chondus crispus* (Irish moss), the specie *Gigartina skottsbergi* (which contain *λ*-carrageenan), *Kappaphycus alvarezii* (a source of *κ*-carrageenan), *Eucheuma denticulatum*, *E. cottonii*, and *E. spinosum*. *E. Cottonii* yields *κ*- and *ι*-carrageenan, and *E. spinosum* contains *ι*-carrageenan [65–67].

Carrageenans are sulfated linear polysaccharides of D-galactose and 3,6-anhydro-D-galactose obtained from red seaweeds. They are extensively utilized as thickening, gelling, and protein-suspending agents in the food industry, as well as in nonfood industries such as cosmetics, printing, and textiles. Recently, carrageenans have been used in the pharmaceutical industry as excipients in pills and tablets [64, 68].

They are made up of alternating copolymers of α-(1-3)-D-galactose and β-(1-4)-3,6-anhydro-D or L-galactose that are divided into three subgroups (*κ*, *λ*, and *i*) depending on the number and distribution of sulfated ester patterns on 3,6-anhydro-D or L-galactose residues (Figure 10.7).

ι-carrageenan has approximately 30% of ester sulfate groups and 25–30% of anhydro-galactose units, is soluble in hot water and its solution shows thixotropic behavior. *κ*-carrageenan has 25–30% of ester sulfate groups and 28% of anhydro-galactose units, is partially soluble in cold water and fully soluble in hot water [70]. Carrageenans are water-soluble polymers whose solubility is determined by the amount of ester sulfate present as well as by the presence of any possible related cations. Moreover, 3,6-anhydro-galactose bridges are found in *ι*- and *κ*-carrageenans, but not in *k*-carrageenan; the presence or absence of anhydro-galactose bridges determines the gelation and rheological behavior. *ι*- and *κ*-carrageenans con form gels, while *κ*-carrageenan also acts as a thickener.

The molecular weight of carrageenan depends on extraction temperature; commercial carrageenans have a molecular weight from 100 to 1000 kDa. Other carbohydrate residues – such as xylose, glucose, and uronic acids, as well as certain substituents, such as methyl ethers and pyruvate groups – can be found in carrageenans.

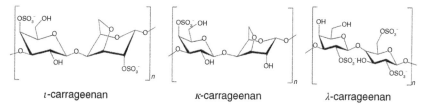

ι-carrageenan   *κ*-carrageenan   *λ*-carrageenan

Figure 10.7 Structures of the repeating dimeric units of *ι*-, *κ*-, and *λ*-carrageenans. Source: Adaptation from Van de Velde et al. [69].

In theory, the extraction and processing of the three carrageenans are quite similar. Impurities are firstly removed, the carrageenan is then separated from the aqueous extraction medium and other components, and finally, the carrageenan is dried. For the production of κ-carrageenan (from seaweed species such as *Eucheuma, Chondrus, Hypnea, Iridaea,* and *Furcellaria* [71], the following process is applied. The surface contaminants and other particles adhered to the thalli of dried seaweed are thoroughly cleaned with a hot alkali treatment (NaOH or KOH solutions), which is then used to eliminate any impurities that may still be present. Then, it is filtered through cold alcohol (isopropanol at 7 °C), which precipitates the polysaccharides. The polysaccharide is dried at 60 °C for two days and milled for later usage. Freile-Pelegrin et al. [72] reported the yield extraction and physicochemical properties of carrageenan from *Eucheuma isiforme* (Solieriaceae, Rhodophyta) in Yucatan, Mexico. Carrageenan was extracted using various alkali concentrations (0, 1, 3, 5, and 7% KOH) and treatment times (3, 4, and 5 hours). The extractions, which took place in 1% KOH for 3 hours, yielded carrageenan with suitable properties that make it as a viable substitute for traditional ι-carrageenan sources. *Eucheuma isiforme* from Yucatan may be a good source for pure ι-carrageenan if it is extracted under certain conditions. The procedure (1% KOH and 3 hours of extraction) produces carrageenan with sufficient quality for serving as a substitute for traditional carrageenan sources since sulfate levels fit the US Food and Drug Administration purity standard of 20–40% (dry weight).

Carrageenans are widely used in the food industry due to their excellent physical and functional properties, such as thickening, gelling, and stabilizing. In dairy products, they improve the texture of cottage cheese, they act as binders and stabilizers in the meat-processing sector to create patties, sausages, and other meat products, and they maintain the viscosity and texture of puddings and dairy desserts.

Carrageenan-based coatings can extend the shelf life of fresh-cut and fresh whole fruits because the carrageenan films show lower oxygen permeability than starch films. Furthermore, when compared to starch-coated strawberries, carrageenan-coated strawberries losses less weight and firmness, indicating that the carrageenan coating acts as a more effective moisture barrier. Carrageenan-derived materials have been extensively studied in recent decades for their possible uses – ranging from pharmacological and biomedical applications to edible films and coatings.

In flexible films, carrageenan shows limitations due to its inherent hydrophilicity; in this regard, blending with other biopolymers, the use of plasticizers, and reinforcements with nanomaterials has been reported as measures for improving the properties of carrageenan films. A comprehensive summary of the current status of carrageenan-based films along with the various strategies for obtaining desirable film properties with characteristics is presented in Table 10.3.

Table 10.3 Films, coatings, and composites formulated with carrageenans for food packaging applications.

Formulation	Properties	Application	Reference
Carrageenan film + grapefruit seed extract GSE	The tensile strength (TS) and tensile modulus (TM) decrease with increasing content of GSE; however, the elongation at break (%Eb) increased significantly up to 6.6 µg/ml of GSE. The carrageenan/GSE composite films showed higher antibacterial activity against food-borne pathogens	Antimicrobial or active food packaging applications	[73]
Blend film (κ-carrageenan, λ-carrageenan, and alginate)	λ-carrageenan impaired the mechanical, barrier, and optical properties of the films; this was attributed to the formation of aggregates instead of a continuous homogeneous film. κ-carrageenan and alginate produced interesting blend films, the addition of λ-carrageenan ability to improve moisture barrier and overall tensile to improve elongation and transparency	Biodegradable edible films	[74]
κ-carrageenan (k) film + cassava starch (c)	The film presented high transparency. The sample (κ-carrageenan) supported a film with high water solubility (39.22%) and low swelling degree (391%). The lowest WVP was observed for 50 κ-carrageenan-cassava starch (3.01×10^{-8} g Pa m/s)	Edible films as food packs	[75]
κ-carrageenan films + pomegranate flesh PFE and peel extracts PPE.	To increase the tensile strength, water vapor and UV light barrier abilities of κ-carrageenan films were improved by the incorporation of PFE or PPE. Due to different polyphenolic compositions, κ-carrageenan PPE film presented a higher UV light barrier	Active and intelligent packaging	[76]
Carrageenan edible films + Aloe vera	The films containing optimum levels of carrageenan (1.5%) and glycerol (15%) showed the highest antioxidant and antimicrobial potential	Film improved the microbial and lipid oxidative stability of kulfi	[77]

(Continued)

Table 10.3 (Continued)

Formulation	Properties	Application	Reference
ι-carrageenan (IC) film + arrow root starch (AS)	XRD analysis indicated fraction of IC contributed to increase in degree of crystallinity (28.35%), and FTIR signaled strong hydrogen bond interactions between polymers. AS/IC samples successfully inhibited weight loss of cherry tomatoes at room temperature and extended their shelf-life to 10 days	Biodegradable edible films	[78]
κ-carrageenan + sodium carboxymethyl starch (CMS) with carboxylated-cellulosa nanocrystal C-CNC	The addition of C-CNC in κ-carrageenan–CMS films showed a significant improvement in the TS and %Eb. The films maintained good thermal stability and barrier performance	Biodegradable edible films	[79]

10.2.5 Fucoidan

Fucoidan is a naturally occurring SP found in the extracellular matrix of brown seaweeds; Kylin performed the first isolation of this biopolymer in 1913, and it was named "Fucoidin." [80] The International Union of Pure and Applied Chemistry (IUPAC) renamed it as "Fucoidan," although it is also known as sulfated fucan, fucan, and fucosan [81, 82]. According to Pal et al. [83], brown seaweeds vary greatly in size, species, and general shape, but they are always brown due to their photosynthetic pigment. Common brown seaweeds include kelp species (such as *Laminaria pallida*) and species of Fucus and Zonaria; alginate, laminarians, and fucoidans are polysaccharides found in brown seaweed. Other components found in seaweed include lipids, fatty acids, sterols, phenolic compounds, pigments, alkaloids, terpenes, and halogenated chemicals [84].

Fucoidan has a complex structure owing to the fact that it is mostly made up of the sugar L-fucose (monomeric unit) (Figure 10.8); this homopolysaccharide can connect to rhamnose (rhamnofucan), galactose (galactofucan), or a combination of rhamnose and galactose (rhamnogalactofucan). However, other monosaccharides

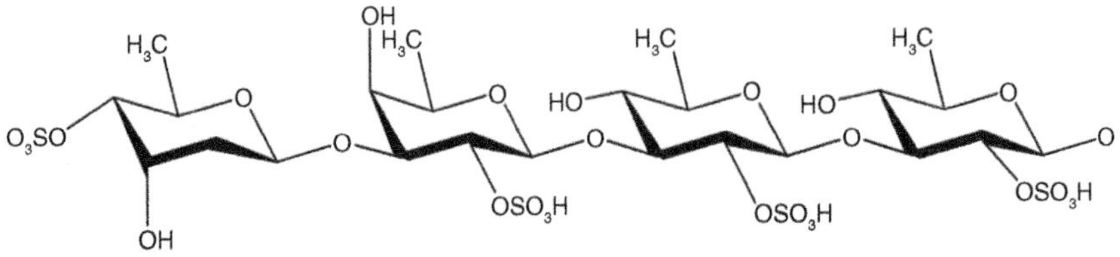

Figure 10.8 Structure of fucoidan.

(glucose, galactose, mannose, xylose, arabinose, uronic acids, acetyl groups, etc.) may be included in the structure of fucoidans, although the proportions can vary among species [85, 86]. According to several authors, the fucoidan has two major backbones $(1 \rightarrow 3)$-linked-α-L-fucose (fucopyranose) residues or alternating $(1 \rightarrow 3)$- and $(1 \rightarrow 4)$-linked-α-L-fucose (fucopyranose) residues [87, 88].

The content and characteristics of polysaccharides are mainly dependent on species, season, geographical origin, population age, and extraction procedure. Fucoidan in solution does not produce viscous solutions. As a result, unlike many other polysaccharides, fucoidan is not employed as a gelling agent or thickener in the food industry. Partially neat fucoidan from species of *Undaria pinnatifida*, *Sargassum fulvellum*, *Hizikia fusiforme*, and *Laminaria religiosa* has reported low-viscosity aqueous solutions with a pseudoplastic behavior [89, 90]. Fucoidan cannot form gels or films on its own; however, combining fucoidan with other biopolymers can provide additional benefits [91]. On the other hand, the molecular weight of this polysaccharide, which is around 100 000 Da [92], could vary to a significant degree. Some authors have reported fucoidans with a low-molecular weight of 6800 Da [93]; meanwhile, Rupérez et al. [94] have reported fucoidans with a molecular weight as high as 1 600 000 Da, with fractions of 43,000 Da. The molecular weight, sulfate content, and structure of fucoidan all have a notable influence on the biological properties of this polysaccharide. Several reports in the literature have documented the various properties of fucoidan (antitumor, antiviral, anticoagulant, anti-inflammatory, antiobesity, etc.), which make it particularly suitable for use in biomedical disciplines [95–99].

Natural active edible films with promising antioxidant properties were produced by the blending of alginate and fucoidan (extracted from the brown macroalga *Sargassum latifolium*) with chitosan derived from the filamentous fungus *Aspergillus niger* [91]. The authors found that adding fucoidan and/or Ca^{2+} into the alginate–chitosan films lowered the water solubility, but enhanced film thickness, water vapor permeability, and oxygen permeability. The films presented good UV light barrier properties. The moisture level of films at equilibrium was increased with the addition of fucoidan. Furthermore, an effective diffusion coefficient was used to represent how water vapor and polyphenol release are diffused, following Fick's second law and Peleg's model, respectively [100]. The films often had strong antioxidant qualities as determined by the total antioxidant assay, ferric-reducing antioxidant power, and hydroxyl radical scavenging activity.

Fruits, such as strawberries are perishable and usually lose antioxidants during storage. To prevent this loss, Duan et al. [101] and Luo et al. [102] used carboxymethylated fucoidan (CMFL) and fucoidan from *Laminaria japonica*, respectively; the coating was applied to the strawberries, and they were subjected to cold storage (4 °C for 5 days). Strawberries coated with CMFL showed reduced water loss and decay in comparison to uncoated fruit. In terms of antioxidant capacity and ascorbic acid content, fruits coated with CMFL and fucoidan showed better results than the control fruit [101]. Lou et al. (2020) reported that the fucoidan coating on strawberries slowed the degradation of antioxidant activity, as well as inhibiting the endogenous antioxidant capacity and extending the shelf life of this fruit.

Recently, Xu and Wu [103] reported the use of fucoidan as a coating for mango fruit to extend its shelf life. At the highest concentration of fucoidan (3% wt), the optimal preservation properties of fucoidan coatings on mango fruits were found. During storage at 20 °C and 80% humidity, fucoidan coatings successfully prevented mango fruit respiration, nutritional loss, and weight loss, as well as protecting mango fruits from physical and biological damage.

10.2.6 Ulvan

As previously described, alginates from Phaeophyceae, and agar and carrageenan from red seaweeds, constitute a large fraction of the world's commercially exploited seaweed polysaccharides. However, within this context, green algae (Cholorophyta) are still relatively unexploited. Despite the fact that *Ulva* species of green seaweed have been consumed for many centuries ("sea lettuce"), it was the discovery of an important polysaccharide from their cell wall, namely ulvan, which dramatically increased interest in these green algae. More recently, there has been a sudden surge in interest around developing and utilizing this underexploited genus for novel materials. From an environmental sustainability point of view, it is noteworthy that the *Ulva* genus represents a valid source for obtaining materials due to its abundance and widespread availability worldwide. Because of its very high growth rates and productivity, it can result in the extensive accumulation of biomass related with the marine eutrophication, and the formation of problematic "green tides." [104] Although these *Ulva* blooms have detrimental effects on ecosystems, environments, and the economies of coastal areas, this waste biomass is nevertheless proposed as an excellent raw material for ulvan production, mitigating the excessive growth of green algae.

Ulvan is a cell wall polysaccharide that makes up 9–36% of the dry weight of the biomass of the *Ulva* species. It possesses attractive physicochemical properties, peculiar self-aggregation molecular behavior, and numerous biological activities, allowing for a variety of innovative applications. Glucuronic and iduronic acids, as well as xylose, constitute the majority of its chemical structure. The effects of rhamnose on dermal biosynthesis pathways and plant immunity are of importance, whereas uronic acids (glucuronic and iduronic acids) and their sulfate esters are crucial components of mammalian glycosaminoglycans (e.g. heparin, heparan sulfate, and dermatan sulfate). Therefore, ulvan is an ideal option for the procedures and roles performed by mammalian polysaccharides. In this context, ulvan shows multifaceted applications, as has been reviewed by Chiellini and Morelli [105] and Tziveleka et al. [106], including biomaterial science (tissue engineering, biofilm prevention, and excipients), pharmaceutical, and nutraceutical uses (antiviral, antioxidant, antihyperlipidemic, anticancer, and immunostimulatory). In addition to their biocompatibility and biodegradability, ulvan have a high number of functional groups that can be easily modified or tailored to provide desirable gelling and functional properties. In this regard, some interesting examples have been mentioned in the recent review by Lakshmi et al. [107] involving the preparation of a hydrogel from functionalized ulvan using photopolymerization under UV irradiation.

Glycidyl methacrylate was used to functionalize ulvan and add unsaturated groups to the ulvan structure. As a result, a material with antioxidant activity that is suited for cell encapsulation was produced [108]. Also, Morelli et al. [109] grafted poly (N-isopropylacrylamide) chains onto ulvan as a thermosensitive component which modified the sol–gel transition of ulvan, producing a material suitable for biomedical applications.

Regarding food packaging applications, ulvan has also gained considerable attention due to both its environmental benefits and its functional aspects, the latter being based on its many inherent biological properties – especially antioxidant activity, which makes it suitable for active packaging. As defined by Commission Regulation (EC) No 450/2009, active packaging comprises packaging systems that interact with the food in such a way as to "deliberately incorporate components that would release or absorb substances into or from the packaged food or the environment surrounding the food." [110] The use of active packaging instead of the direct addition to the food of active ingredients such as antimicrobials and antioxidants may bring about a reduction for substances needed.

In this context, very recent studies have demonstrated the feasibility of producing active food packaging using ulvan. According to Guidara et al. [111], films made with ulvan from the species *Ulva lactuca* and 2% sorbitol as a plasticizer generated robust antioxidant packaging with appealing optical and structural qualities as well as functional thermal properties. Further studies by the same authors [112] revealed that smart films were also produced by varying the proportions of glycerol or sorbitol in specific plasticizer concentrations. Each of the produced films; all of which are necessary for packaging food and biomedical products displayed excellent solubility, barrier, optical, and mechanical qualities. The effective method used in this research revealed that the addition of acidic extract (HCl at pH 2) or enzymatic–chemical extract (with Celluclast and protease) into the films has the biggest effect on their mechanical, moisture resistance, solubility, transparency, water vapor barrier, and other properties, as well as their ability to block UV and visible light. It has been observed that after the addition of sorbitol, the films showed more favorable water-related properties: low-moisture content, high levels of solubility, and water vapor permeability; as well as visible and UV light transmission. Moreover, the addition of glycerol to the films not only contributed to an improvement in their transparency but also highly enhanced their mechanical properties. Amin [113] extracted ulvan from *Ulva lactuca*, producing cheap, novel, nontoxic, and eco-friendly bio-nanocomposite films using green synthesized silver nanoparticles. In addition to showing good levels of antioxidant activity, the formed films demonstrated good antimicrobial properties against a number of gram-positive and gram-negative bacteria that are responsible for the decomposition of packaged fruits, fish, vegetables, meat, poultry, fish, baked goods, and dairy products.

According to the promising results describe above, ulvan-based films have found use in the food packaging and in the medical industry. A number of contemporary and safe bio-nanocomposite active films for food packaging are anticipated to be created.

10.3 Modified Chitosan for Food Film Applications

10.3.1 Chemical Modifications of Chitosan for Food Packaging

Chitin is one of the most abundant biopolymers in nature after cellulose and starch, and its deacetylated form is called chitosan (poly-D-glucosamine). This product is widely used in the cosmetics, water treatment, and food industries, as well as in the development of sustainable packaging [114] in the form of either edible coatings or films [115]. Chitosan has excellent properties, such as nontoxicity, biocompatibility, biodegradability, and high adsorption capacity [116]. However, it also its disadvantages, such as low thermal and mechanical stability, and high sensitivity to water, additionally, it has a shorter shelf life than conventional food packaging materials, thus limiting its applications for food packaging [117].

The chemical modification of chitosan is a frequently used method of improving its chemical and physical properties. Table 10.4 shows the chemical modifications

Table 10.4 Chemical modifications of chitosan for use in food packaging.

Method	Modification	Results	References
Quaternization	Carboxymethyl cellulose/quaternized chitosan (2-N-Hydroxypropyl-3-trimethylammonium chloride chitosan	Presented low water permeability, with better tensile strength, thermal stability, and water resistance	[118]
Acetylation	N-acetylated chitosan	Greater water retention capacity and elongation at break, lower crystallinity and thermal stability	[119]
Sulfonation	Sulfonated chitosan	Inhibits the formation of biofilms of E. coli and S. aureus and improves the efficiency in the elimination of bacteria embedded in biofilms	[120]
	Zwitterionic Sulfonated Chitosan	Inhibits biofilm formation, decreases metabolic activity and exopolysaccharide secretion from biofilm	[121]
Carboxymethyl chitosan	Carboxymethyl chitosan/polyvinyl alcohol with citric acid	Films were antifog, antibacterial, biodegradable, and improved mechanical properties	[122]
	Rice starch mixed with carboxymethyl chitosan	Improves tensile strength, flexibility and thermal stability	[123]

10.3 Modified Chitosan for Food Film Applications

Method	Modification	Results	References
Graft	Hypromellosa chitosan (HC) graft by polyelectrolyte complexation	They allow real-time monitoring of the status of frozen foods, have good barrier properties, high transparency, intrinsic antibacterial effects against gram negative and positive bacteria	[124]
	Through derivatives of Schiff's base using different aldehydes	Improves better antimicrobial activities	[125]
Reticulation	Chitosan/silica crosslinked by genipin and glutaraldehyde	Genipin improved mechanical properties and crystallinity compared to glutaraldehyde	[126]
Graft copolymerization	Chitosan with polyvinyl alcohol	Good mechanical and barrier properties, antibacterial activity, and biofilm inhibition against foodborne pathogens	[127]
Graft copolymerization	Chitosan copolymerized with poly (acryloyloxy) ethyltrimethylammonium chloride	Improved antimicrobial properties against bacteria and fungi. Very low tensile strength and thermal stability	[128]

of chitosan for use in films as a food packaging system [129]. These modifications improve the oxidation resistance, thermal stability, rheological stability, and antibacterial properties, and can be carried out without affecting the degree of polymerization due to the presence of the following functional groups: the primary amine group ($-NH_2$); and the primary and secondary hydroxyl groups ($-OH$) in the C-6 and C-3 positions, respectively (Figure 10.9) [130, 131].

Regarding the reactivity of chitosan, the amino groups are more reactive than the hydroxyl groups due to the fact that nitrogen is an excellent base; at the same time, the primary hydroxyl group (C6-OH) is more reactive than the secondary hydroxyl

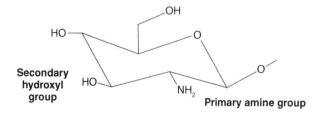

Figure 10.9 Functional groups in chitosan monomer (*N*-glucosamine).

group (C3–OH) owing to a lower degree of steric impediment, then the chemical modification can occur in the amino, hydroxyl or both groups [131, 132]. In this sense, the most widely used method for the chemical modification of chitosan is the N-substitution (–NH$_2$) or O-substitution (–OH) in which the amino groups or hydroxyl groups react; the O-substitution commonly requires the protection and deprotection of primary amino groups because these are more reactive [2].

Modified chitosan is used as a coating for fruits to extend their shelf life, in this sense, Vu et al. [133] used chemically modified chitosan by acylation with palmitoyl chloride and incorporated limonene to create coatings that they tested on strawberries to extend shelf life during storage. The results showed that the decomposition was less in the fruits with coating, also, they are not phytotoxic, do not affect the appearance of the strawberries, and helped to maintain a good appearance and coloration. Da Mata et al. [134] modified the chitosan with 2-chloro-N,N-diethylaminoethyl hydrochloride. The derivative (diethylaminoethyl chitosan) was then transformed into high molecular weight amphiphilic derivative using an alkylation reaction procedure, which was applied as a coating on strawberries, bananas, and apples, the results showed good efficacy in inhibiting the growth of *P. expansum*. In addition, *A. alternata* and *A. solani* reached an inhibition rate of 100%, which indicates that derivatization with this hydrophobic groups is important to increase the antifungal activity of chitosan.

Taştan et al. [135] studied a method of surface decontamination of cucumber slices inoculated with *E. coli*, by combining pulsed light treatments with a modified chitosan coating containing carvacrol nanoemulsions, the authors reported an increase in antimicrobial activity and a reduction of up to 4 logarithmic cycles of *E. coli*. Elbarbary and Mostafa [136] synthesized carboxymethyl chitosan of different molecular weights by alkylating chitosan with monochloroacetic acid, to use it as a coating on peaches; the lower molecular weight had a good effect in delaying deterioration and reducing the content of malondialdehyde, suggesting its possible use as an antioxidant and preservative coating. Also, Benhabiles et al. [137] obtained N,O-carboxymethyl chitosan (NOCC), reported that a 2% (w/v) solution was the most effective in prolonging the useful life of the tomato, in addition, the coating helped to keep the fruit firm and provide higher levels of titratable acidity than uncoated control fruits at the end of storage. Table 10.4 shows different methods of chemical modification of chitosan used in food packaging.

10.3.2 Chitosan Blends/Composites for Films and Coating for Food Applications

The functionality of chitosan for food applications has been improved by blends/composites with biomaterials as collagen, alginate, gelatin, starch, pectin, carrageenan, and xanthan guar that permit to obtain films (>30 μm) or coatings (<30 μm). Also, the use of additives as essential oils, phenolics compounds, and organic acids, enhancing the functional properties such as antioxidant, antimicrobial and antifungal activities of the chitosan-blends based films. These films/coatings increase the water retention, generate a favorable microenvironment improving the delays ripening, optimizing the gas concentration, avoid the oxidation of phenols, and increasing the shelf life of fruits and vegetables [138–140].

The use of these combined technologies to improve the films/coatings for food applications has allowed to increase the shelf life and quality attributes of fruits and vegetables. In this sense, Zhang et al. [141] reported the use of carboxymethyl chitosan–gelatin–$CaCl_2$ for coating sweet cherry decreasing the weight loss, decay ratio and pedicle browning incidence, and increasing the antioxidant activity. Indumathi et al. [142] designed a chitosan–cellulose-acetate phthalate film incorporated with ZnO and extending the shelf life of that black grape for 9 days due that work as protection for the fruit. Nguyen et al. [143] decreased the weight loss and reduced the disease infection of the dragon fruit until 30 days, using a coating of chitosan-k-carrageenan-based composite combined with gibberellic acid or methyl jasmonate. In apples, it is possible to increase 4 weeks of the shelf life with good organoleptic properties and also to decrease the weight loss of the fruit, using coatings of chitosan–starch from purple yam [144].

Moreover, the edible chitosan/blends coating applied in seeds and vegetables help to preserve the organoleptic properties and improve the shelf life. In this sense, the edible films of chitosan–thermoplastic starch-alpha tocopherol–bentonite in green coffee beans improved the antioxidant activity, the water vapor permeability, and the compressive loads of the seeds [145]. The use of coating of starch/chitosan/lemongrass in chilies significantly reduced the weight loss, due to the semipermeable barrier formed over the fruit, reducing the water vapor permeability and the bacterial growth after 4 days of storage [146]. Araújo et al. [147] reported the used of Cassava starch–chitosan edible coating enriched with essential oil of *Lippia sidoides* Cham and pomegranate peel extract for preservation of tomatoes, delaying the ripening, lowering the total soluble solids and weight loss, also, maintaining the firmness at 12 days of storage. Also, Kumar et al. [138] pointed out the formulation of edible coating with chitosan–pullulan enriched with pomegranate peel extract, the results showed that the green bell paper with this coating reduced the loss weight, color browning, maintained the pH, soluble solids, titratable acidity, firmness, sensorial attributes, flavonoid content, and antioxidant activity.

The biodegradable films to obtain active packaging (i.e. antioxidant and antimicrobial activity) enhanced the food preservation Regulation EC 1935, [148]. The chitosan has been used for this technology because of its film-forming properties, biocompatibility, biodegradability, good mechanical properties, and selective gas permeability [149]. The technology of active packaging is useful to increase the shelf life from food of animal origin. In this way, Fiore et al. [149] elaborated an active packaging of poly(lactic acid) (PLA) coating with chitosan–caseinate enriched with rosemary essential oil, and was used for packing fresh minced chicken, this packaging reduced the meat oxidation, maintained constant the malondialdehyde and color for 14 days, and decrease the concentration of heptanal and ethanol. Hu et al. [118] reported the use of quaternary ammonium chitosan–fish gelatin blends films enriched with betalains from amaranth and was used as active and intelligent packaging monitoring the ammonia concentration produced for fresh shrimp, also, reported antioxidant and antimicrobial activity. Furthermore, an active

Table 10.5 Trends of chitosan blends/composites for biodegradable food packaging.

Formulation	Results	References
Polylactic acid-chitosan-tributyl o-acetyl citrate	Antifungal activity against: *Aspergillus brasiliensis*, *Penicillium corylophilum*, and *Fusarium graminearum* Antibacterial activity against: *S. aureus* and *E. coli* on the contact surfaces	[151]
Poly(ε-caprolactone)-chitosan-grapefruit seed extract by extrusion method	Antibacterial activity against: *Escherichia coli* and *Pseudomonas aeruginosa* tested in salmon. Also, no mold growth was observed on the bread packaged with this film (after 7 days)	[152]
Polycaprolactone-quaternized chitosan by melt processing	Antimicrobial activity against: *Escherichia coli* and *Staphylococcus aureus* until 99.9% with 20% of quaternized chitosan	[153]
Polylactic acid-chitosan (60/40)	Excellent mechanical properties: tensile strength (17.8 MPa), elongation at break (300.11%), and good thermal stability	[154]
Chitosan-collagen-lemon essential oil	The films showed lower oxygen permeability, higher tensile strength, and higher elongation at break. Inhibit the lipid oxidation, prevent the microbial growth, increasing the shelf life of meat pork	[155]
Crosslinked chitosan-guar gum composite films	High water stability, high mechanical strength (39 MPa), high contact angle (98.2°), and lower water vapor permeability. Showing good properties to be applied as food packaging	[156]
Chitosan-gelatin with lemongrass essential oil	Extended the shelf life of stored raspberries from four to eight days. Showed antimicrobial activity in vitro against *E. coli*, *B. subtilis*, and *S. aureus*	[157]
Chitosan-corn starch-pluronic F127	The films were homogeneous and smoother. The water vapor permeability decreasing with the pluronic F127. These materials were reported as a good alternative for product packaging	[158]

packaging of cellulose–chitosan–citric acid showed good results for the preservation of fresh pork meat, decreasing the total viable counts and volatile basic nitrogen in the fresh meat [150]. The Table 10.5 show the trends of chitosan blends/composites for biodegradable food packaging.

10.3.3 Nanomaterials of Chitosan for Food Packaging

Nanotechnology refers to the design, production, and characterization of systems, structures, and devices by controlling the size and shape at the nanometric scale ($\leq 1 \times 10^{-9}$ m), then, nanomaterials are those that have one or more dimensions with

10.3 Modified Chitosan for Food Film Applications

a measure of ≤100 nm. Nanotechnology is considered an inter and multidisciplinary science in which chemistry, biotechnology, food, and industrial process technology coincide, providing solutions to various industrial sectors, including the development of containers of biodegradable material [159, 160].

The use of biopolymers in food packaging materials has drawbacks regarding thermal, mechanical and barrier properties compared to conventional plastics. In this sense, the literature suggests that the use of nanofillers such as metallic nanoparticles (Ag, Cu, Au, S, Pt, and Pd) or metal oxide (ZnO, MgO, Ag_2O, SiO_2, and TiO_2), chitosan nanoparticles, nanocellulose (cellulose nanofibers or cellulose nanocrystals), nanoclays (silicates), nanorods, and nanotubes (carbon nanotubes) improve the properties of the resulting nanocomposites [161, 162].

The use of chitosan nanoparticles for food packaging are an excellent choice because are biodegradable, biocompatible, and nontoxic. Table 10.6 shows the trends

Table 10.6 Trends of use chitosan nanocomposites in food packaging.

Material	Blends	Results	References
Film	Nanoparticles of carboxymethyl chitosan–ZnO and sodium alginate	Presented greater resistance to traction and water vapor, improved mechanical properties, and antibacterial activity	[163]
Film	Chitosan/boron nitride nanobiocompounds	It increased the thermal, mechanical, and barrier properties, in addition, the cytotoxicity of the films decreased	[164]
Film	Quinoa/chitosan protein film with chitosan thymol nanoparticles	Improved the barrier to water vapor, in addition to acting as an antimicrobial potential for the preservation of fresh fruits	[165]
Coating	5% nano-chitosan, 1.5% chitosan, 0.45% glycerol, and a microporous packaging	Improved the gas environment in the packing bag, and reduced the microbial count by more than 1.29 log CFU/g. Delayed darkening, maintained a high visual index, and significantly inhibited softening, mass loss, and respiration rate during cold storage	[166]
Coating	*Eryngium campestre* essential oil in chitosan nanoparticles	Decreased the microbial count during refrigeration storage. The pH, firmness, antioxidant capacity, and total phenolic of cherries enhanced moderately	[167]
Coating	Chitosan nanoparticles with α-pinene	Prevented the growth of *A. alternata* during the cold storage of the pepper and conserve the physicochemical properties	[168]
Coating	Chitosan-cellulose nanocrystals-oleic acid by Pickering emulsion	Delayed 2 month the ripening of pear, and significantly decreased the ethylene production	[169]
Coating	Chitosan-nanoemulsion of mandarin essential oil	The coating reduced *L. innocua* from 1 to 3 logarithmic cycles in green beans	[170]

of use of chitosan nanoparticles in food packaging. Among the methods for obtaining are deconstruction strategies, ionic gelling modified with radical polymerization, polyelectrolytic complexation, and crosslinking by emulsification, reverse micellization, spray drying, coalescence of emulsion droplets, desolvation, nanoprecipitation, emulsion–diffusion of solvent and ionotropic gelation that is the best-known process. Generally, the incorporation of these nanoparticles in the film matrices is carried out by the emptying method. Some biodegradable matrices used are tara gum, banana puree, carboxymethyl cellulose, fish gelatin, hydroxypropyl methyl cellulose, PLA, pectin, and starch [129, 171].

10.4 Conclusions and Future Trends

Traditionally, edible films and coatings based on polysaccharides have been used to protect certain foods (fruits, vegetables, fish, meat, etc.) and to extend their shelf life. Polysaccharides have become more important in food structure, in terms of both generating distinct sensory experiences based on textural and rheological properties, and increasing nutritional properties. With regard to their influence on food structure and their interactions with food components, polysaccharides offer a variety of advantageous applications in foods. Several such applications are innovative, including the reduction of glycemic response, offer satiety, enable controlled release of micronutrients and flavor compounds, and function as a "scaffold" for 3D food printing, which are innovative applications.

One of these innovative applications is the use of agar, starch, gelatin, and its combinations, which are functionalized to be used as holographic marker as coatings, to obtain active packaging sensitivities to storage conditions. This technology was made using nanoimprint method from film-forming solution, and the morphology depends of the miscibility, composition, and consistency. Polysaccharides from seaweeds are used improve properties (mechanical, thermal, and antibacterial) in biopolymer blends. They not only enhance hydrophilicity and elongation at break, but they also make possible to use these blends as active packaging. It is possible owing to the natural existing antioxidant properties, which can reduce lipid oxidation, and reduce free radicals' production that may have mutagenic, cytotoxic, and carcinogenic effects.

On the other hand, the use of nanomaterials for food packaging/films improves the physical barrier and antimicrobial properties of manufactured films and coatings, increasing the shelf life and improving food quality, while reducing the use of synthetic plastics and promotes a healthy environment. This approach appears to have an excellent future for innovative food packaging design, as it will reduce the use of synthetic plastics for packaging foods [117, 172].

The modified properties of chitosan by chemical modifications or blending with other compounds (i.e. polymers, additives) have shown significant progress in the last decade. However, it is still incomplete and requires the development of more chitosan formulations with natural and synthetic polymers to affect the food, pharmaceutical, bioprocessing, and cosmetic industries, which marks an important change and promotes the use of chitosan on an industrial scale.

References

1 Wijesekara, I., Pangestuti, R., and Kim, S.-K. (2011). Biological activities and potential health benefits of sulfated polysaccharides derived from marine algae. *Carbohydrate Polymers* 84: 14–21. https://doi.org/10.1016/j.carbpol.2010.10.062.

2 Madera-Santana, T.J., Herrera-Méndez, C.H., and Rodríguez-Núñez, J.R. (2018). An overview of the chemical modifications of chitosan and their advantages. *Green Materials* 6 (4): 131–142. https://doi.org/10.1680/jgrma.18.00053.

3 Schiener, P., Black, K.D., Stanley, M.S., and Green, D.H. (2015). The seasonal variation in the chemical composition of the kelp species *Laminaria digitata*, *Laminaria hyperborea*, *Saccharina latissimi* and *Alaria esculenta*. *Journal of Applied Phycology* 27 (1): 363–373. https://doi.org/10.1007/s10811-014-0327-1.

4 Wang, L., Wang, X., Wu, H., and Liu, R. (2014). Overview on biological activities and molecular characteristics of sulfated polysaccharides from marine green algae in recent years. *Marine Drugs* 12: 4984–5020. https://doi.org/10.3390/md12094984.

5 Nunes, C., Rocha, A., Quiterio, P. et al. (2019). Salt pan brine water as a sustainable source of sulphated polysaccharides with immunostimulatory activity. *International Journal of Biological Macromolecules* 133: 235–242. https://doi.org/10.1016/j.ijbiomac.2019.04.021.

6 Costa, L.S., Fidelis, G.P., Cordeiro, S.L. et al. (2010). Biological activities of sulfated polysaccharides from tropical seaweeds. *Biomedicine & Pharmacotherapy* 64 (1): 21–28. https://doi.org/10.1016/j.biopha.2009.03.005.

7 Mourao, P.A. (2007). A carbohydrate-based mechanism of species recognition in sea urchin fertilization. *Brazilian Journal of Medical and Biological Research* 40 (1): 5–17. https://doi.org/10.1590/S0100-879X2007000100002.

8 Madera-Santana, T.J., Freile-Pelegrín, Y., and Azamar-Barrios, J.A. (2014). Physicochemical and morphological properties of plasticized poly(vinyl alcohol)-agar biodegradable films. *International Journal of Biological Macromolecules* 69: 176–184. https://doi.org/10.1016/j.ijbiomac.2014.05.044.

9 Sellimi, S., Younes, I., Ayed, H.B. et al. (2015). Structural, physicochemical and antioxidant properties of sodium alginate isolated from a Tunisian brown seaweed. *International Journal of Biological Macromolecules* 72: 1358–1367. https://doi.org/10.1016/j.ijbiomac.2014.10.016.

10 Smidsrød, O. and Draget, K.I. (1996). Chemistry and physical properties of alginates. *Carbohydrates in Europe* 14: 6–13.

11 Benavides, S., Villalobos-Carvajal, R., and Reyes, J.E. (2012). Physical, mechanical and antibacterial properties of alginate film: effect of the crosslinking degree and oregano essential oil concentration. *Journal of Food Engineering* 110 (2): 232–239. https://doi.org/10.1016/j.jfoodeng.2011.05.023.

12 Frank, K., Garcia, C.V., Shin, G.H., and Kim, J.I. (2018). Alginate biocomposite films incorporated with cinnamon essential oil nanoemulsions; physical, mechanical and antibacterial properties. *International Journal of Polymer Science* 2018 1519407: 8. https://doi.org/10.1155/2018/1519407.

13 Guo, X., Wang, Y., Qin, Y. et al. (2020). Structures, properties and application of alginic acid: a review. *International Journal of Biological Macromolecules* 162: 618–628. https://doi.org/10.1016/j.ijbiomac.2020.06.180.

14 Shimokawa, T., Yoshida, S., Takeuchi, T. et al. (1996). Preparation of two series of oligo-guluronic acids from sodium alginate by acid hydrolysis and enzymatic degradation. *Bioscience Biotechnology & Biochemistry* 60: 1532–1534. https://doi.org/10.1271/bbb.60.1532.

15 Kim, H.S., Lee, C.G., and Lee, E.Y. (2011). Alginate lyase: structure, property, and application. *Biotechnology and Bioprocess Engineering* 16: 843–851. https://doi.org/10.1007/s12257-011-0352-8.

16 Kühbeck, D., Mayr, J., Häring, M. et al. (2015). Evaluation of the nitroaldol reaction in the presence of metal ion-crosslinked alginates. *New Journal of Chemistry* 39 (3): 2306–2315. https://doi.org/10.1039/C4NJ02178A.

17 Grant, G.T., Morris, E.R., Rees, D.A., and Smith, P.J.C. (1973). Biological interactions between polysaccharides and divalent cations: the egg-box model. *FEBS Letters* 32 (1): 195–198.

18 Clark, A.H. and Ross-Murphy, S.B. (1987). Structural and mechanical properties of biopolymer gels. *Advances in Polymer Science* 87: 157–192. https://doi.org/10.1007/BFb0023332.

19 Draget, K.I. (2009). Chapter 29-Alginates. In: *Handbook of Hydrocolloids*, Woodhead Publishing Series in Food Science, Technology and Nutrition, 2e (ed. G.O. Phillips and P.A. Williams), 807–828. New York, USA: Woodhead Publishing https://doi.org/10.1533/9781845695873.807.

20 Owusu-Apenten, R.K. (2004). *Introduction to Food Chemistry*, 55. Boca Raton, FL: CRC Press.

21 Venugopal, V. (2011). *Marine Polysaccharides, Food Applications*, 89–134. Boca Raton, FL, USA: CRC Press, Taylor & Francis Group.

22 Zhang, Z., Yu, G., Zhao, X. et al. (2006). Sequence analysis of alginate-derived oligosaccharides by negative-ion electrospray tandem mass spectrometry. *Journal of the American Society for Mass Spectrometry* 17 (4): 621–630. https://doi.org/10.1016/j.jasms.2006.01.002.

23 Chen, J., Wu, A., Yang, M. et al. (2021). Characterization of sodium alginate-based films incorporated with thymol for fresh-cut apple packaging. *Food Control* 126: 108063. https://doi.org/10.1016/j.foodcont.2021.108063.

24 Oussalah, M., Caillet, S., Salmieri, S. et al. (2006). Antimicrobial effects of alginate-based film containing essential oils for the preservation of whole beef muscle. *Journal of Food Protection* 69 (10): 2364–2369. https://doi.org/10.4315/0362-028x-69.10.2364.

25 Maizura, M., Fazilah, A., Norziah, M.H., and Karin, A.A. (2007). Antibacterial activity and mechanical properties of partially hydrolyzed sag starch-alginate edible film containing lemon-grass oil. *Journal of Food Science* 72 (8): 324–330. https://doi.org/10.1111/j.1750-3841.2007.00427.

26 Norajit, K., Kim, K.M., and Ryu, G.H. (2010). Comparative studies on the characterization and antioxidant properties of biodegradable alginate films containing ginseng extract. *Journal of Food Engineering* 98: 377–384. https://doi.org/10.1016/j.jfoodeng.2010.01.015.

27 Li, J., He, J., Li, D., and Chen, X. (2015). Improving surface and mechanical properties of alginate films by using ethanol as a co-solvent during external gelation. *Carbohydrate Polymers* 123: 208–216. https://doi.org/10.1016/j.carbpol.2015.01.040.

28 Costa, M.J., Marques, A.M., Pastrana, L.M. et al. (2018). Physicochemical properties of alginate-based films: effect of ionic crosslinking and mannuronic and guluronic acid ratio. *Food Hydrocolloids* 31: 442–448. https://doi.org/10.1016/j.foodhyd.2018.03.014.

29 Armisén, R. and Galatas, F. (2009). Agar. In: *Handbook of Hydrocolloids*, 2e (ed. G.O. Phillips and P.A. Williams), 82–107. Cambridge, UK: Woodhead Publishing Ltd.

30 Araki, C. (1956). Structure of agarose constituent of Agar–Agar. *Bulletin Chemical Society of Japan* 29: 43–44.

31 Knutsen, S.H., Myslabodski, D., Larsen, B., and Usov, A.I. (1994). A modified system of nomenclature for red algal galactans. *Botanica Marina* 37 (2): 163–170. https://doi.org/10.1515/botm.1994.37.2.163.

32 Lahaye, M. and Rochas, C. (1991). Chemical structure and physico-chemical properties of agar. In: *International Workshop on Gelidium*, Developments in Hydrobiology, vol. 68 (ed. J.A. Juanes, B. Santelices and J.L. McLachlan). Dordrecht: Springer https://doi.org/10.1007/978-94-011-3610-5_13.

33 Yaphe, W. and Duckworth, M. (1972). The relationship between the structure and biological properties of agars. *Proceeding of International Seaweed Symposium* 7: 15–22.

34 Usov, A.I. (1992). Sulfated polysaccharides of the red seaweeds. *Food Hydrocolloids* 6 (1): 9–23. https://doi.org/10.1016/S0268-005X(09)80055-6.

35 Armisén, R. (1995). Worldwide use and importance of *Gracilaria*, communication presented in the workshop 'Gracilaria and its Cultivation', organized in the university of trieste (Italy) 10–12 April 1994, under the auspices of COST 48 of the CCEE. *Journal of Applied Phycology* 7: 231–243.

36 Freile-Pelegrín, Y. and Murano, E. (2005). Agars from three species of *Gracilaria* (Rhodophyta) from Yucatán Peninsula. *Bioresource Technology* 96 (3): 295–302. https://doi.org/10.1016/j.biortech.2004.04.010. PMID: 15474929.

37 Marinho-Soriano, E. (2001). Agar polysaccharides from *Gracilaria* species (Rhodophyta, Gracilariaceae). *Journal of Biotechnology* 89 (1): 81–84. https://doi.org/10.1016/S0168-1656(01)00255-3.

38 Freile-Pelegrín, Y., Madera-Santana, T., Robledo, D. et al. (2007). Degradation of agar films in a humid tropical climate: thermal, mechanical, morphological and structural changes. *Polymer Degradation and Stability* 92 (2): 244–252. https://doi.org/10.1016/j.polymdegradstab.2006.11.005.

39 Marinho-Soriano, E., Fonseca, P.C., Carneiro, M.A.A., and Moreira, W.S.C. (2006). Seasonal variation in the chemical composition of two tropical seaweeds. *Bioresource Technology* 97: 2402–2406. https://doi.org/10.1016/j.biortech.2005.10.014.

40 Romero, J.B., Villanueva, R.D., Nemesio, M., and Montaño, E. (2008). Stability of agar in the seaweed *Gracilaria eucheumatoides* (Gracilariales, Rhodophyta) during postharvest storage. *Bioresource Technology* 99 (17): 8151–8155. https://doi.org/10.1016/j.biortech.2008.03.017.

41 Karmaus, A.L., Osborn, R., and Krishan, M. (2018). Scientific advances and challenges in safety evaluation of food packaging materials: workshop proceedings. *Regulatory Toxicology and Pharmacology* 98: 80–87. https://doi.org/10.1016/j.yrtph.2018.07.017.

42 Latos-Brozio, M. and Masek, A. (2020). The application of (+)-catechin and polydatin as functional additives for biodegradable polyesters. *International Journal of Molecular Sciences* 21 (2): 414. https://doi.org/10.3390/ijms21020414.

43 Sharma, R., Jafari, S.M., and Sharma, S. (2020a). Antimicrobial bio-nanocomposites and their potential applications in food packaging. *Food Control* 112: 107086. https://doi.org/10.1016/j.foodcont.2020.107086.

44 Williams, P.A. (2007). Gelling agents. In: *Handbook of Industrial Water Soluble Polymers* (ed. P.A. Williams), 73–77. Oxford: UK Blackwell Publishing.

45 Mohammed, Z.H., Hember, M.W.N., Richardson, R.K., and Morris, E.R. (1998). Kinetic and equilibrium processes in the formation and melting of agarose gels. *Carbohydrate Polymers* 36 (1): 15–26. https://doi.org/10.1016/S0144-8617(98)00011-3.

46 Mostafavi, F.S. and Zaeim, D. (2020). Agar-based edible films for food packaging applications - a review. *International Journal of Biological Macromolecules* 159: 1165–1176. https://doi.org/10.1016/j.ijbiomac.2020.05.123.

47 Nieto, M.B. (2009). Structure and function of polysaccharide gum-based edible films and coatings. In: *Edible Films and Coatings for Food Applications* (ed. K. Huber and M. Embuscado), 57–112. New York, NY: Springer https://doi.org/10.1007/978-0-387-92824-1_3.

48 Bykov, D.E., Eremeeva, N.B., Makarova, N.V. et al. (2017). Influence of plasticizer content on organoleptic, physico-chemical and strength characteristics of apple sauce-based edible film. *Foods & Raw Materials* 5 (2): 5–14. https://doi.org/10.21179/2308-4057-2017-2-5-14.

49 Rhim, J.W. (2011). Effect of clay contents on mechanical and water vapor barrier properties of agar-based nanocomposite films. *Carbohydrate Polymers* 86: 691–699. https://doi.org/10.1016/j.carbpol.2011.05.010.

50 Soletska, A., Nistor, K., and Hevryk, V. (2018). Edible film-forming coating with CO_2-extracts of plants for meat products. *Food Science and Technology* 12 (3): 50–56. https://doi.org/10.15673/fst.v12i3.1039.

51 Podshivalov, A., Toropova, A., Fokina, M., and Uspenskaya, M. (2020). Surface morphology formation of edible holographic marker on potato starch with gelatin or agar thin coatings. *Polymers* 12 (5): 1123. https://doi.org/10.3390/polym12051123.

52 Zhang, B., Liu, Y., Wang, H. et al. (2021a). Effect of sodium alginate-agar coating containing ginger essential oil on the shelf life and quality of beef. *Food Control* 130: 108216. https://doi.org/10.1016/j.foodcont.2021.108216.

53 Kanmani, P. and Rhim, J.W. (2014a). Antimicrobial and physical-mechanical properties of agar-based films incorporated with grapefruit seed extract. *Carbohydrate Polymers* 102: 708–716. https://doi.org/10.1016/j.carbpol.2013.10.099.

54 Atef, M., Rezaei, M., and Behrooz, R. (2015). Characterization of physical, mechanical, and antibacterial properties of agar-cellulose bio-nanocomposite films incorporated with savory essential oil. *Food Hydrocolloids* 45: 150–157. https://doi.org/10.1016/j.foodhyd.2014.09.037.

55 de Lacey, A.M.L., Lopez-Caballero, M.E., and Montero, P. (2014). Agar films containing green tea extract and probiotic bacteria for extending fish shelf-life. *LWT - Food Science and Technology* 55 (2): 559–564. https://doi.org/10.1016/j.lwt.2013.09.028.

56 Orsuwan, A., Shankar, S., Wang, L.-F. et al. (2016). Preparation of antimicrobial agar/banana powder blend films reinforced with silver nanoparticles. *Food Hydrocolloids* 60: 476–485. https://doi.org/10.1016/j.foodhyd.2016.04.017.

57 Phinainitisatra, T. and Harnkarnsujarit, N. (2021). Development of starch-based peelable coating for edible packaging. *International Journal of Food Science and Technology* 56 (1): 321–329. https://doi.org/10.1111/ijfs.14646.

58 Guitian, M.V., Ibarguren, C., Soria, M.C. et al. (2019). Anti-Listeria monocytogenes effect of bacteriocin-incorporated agar edible coatings applied on cheese. *International Dairy Journal* 97: 92–98. https://doi.org/10.1016/j.idairyj.2019.05.016.

59 Rhim, J.W. (2013). Effect of PLA lamination on performance characteristics of agar/κ-carrageenan/clay bio-nanocomposite film. *Food Research International* 51: 714–722. https://doi.org/10.1016/j.foodres.2013.01.050.

60 Ramesh, R., Palanivel, H., Prabhu, S.V. et al. (2021). Process development for edible film preparation using avocado seed starch: response surface modeling and analysis for water-vapor permeability. *Advances in Materials Science and Engineering* 2021: 7859658. https://doi.org/10.1155/2021/7859658.

61 Jumaidin, R., Sapuan, S.M., Jawaid, M. et al. (2016). Characteristics of thermoplastic sugar palm Starch/Agar blend: thermal, tensile, and physical properties. *International Journal of Biological Macromolecules* 89: 575–581. https://doi.org/10.1016/j.ijbiomac.2016.05.028.

62 Madera-Santana, T.J., Robledo, D., Azamar, J.A. et al. (2010). Preparation and characterization of low-density polyethylene-agar biocomposites: torque-rheological, mechanical, thermal and morphological properties. *Polymer Engineering and Science* 50 (3): 585–591. https://doi.org/10.1002/pen.21574.

63 Madera-Santana, T.J., Misra, M., Drzal, L.T. et al. (2009). Preparation and characterization of biodegradable agar/poly(butylene adipate-*co*-terephatalate) composites. *Polymer Engineering and Science* 49 (6): 1117–1126. https://doi.org/10.1002/pen.21389.

64 Leira Campo, V., Kawano, D.F., da Silva, D.B., and Carvalho, I. (2009). Carrageenans: biological properties, chemical modifications and structural analysis – a review. *Carbohydrate Polymers* 77: 167–180. https://doi.org/10.1016/j.carbpol.2009.01.020.

65 McHugh, D.J. (2003). A Guide to the Seaweed Industry. In FAO Fisheries Technical Paper 441, Rome, Italy.

66 Rudolph, B. (2000). Seaweed product: red algae of economic significance. In: *Marine and Fresh Water Products Handbook* (ed. R.E. Martin, E.P. Carter, L.M. Davis and G.J. Flich), 515–529. Lancaster, USA: Technomic Publishing Company Inc.

67 Tavassoli-Kafrani, E., Shekarchizadeh, H., and Masoudpour-Behabadi, M. (2016). Development of edible films and coatings from alginates and carrageenans. *Carbohydrate Polymers* 137: 306–374. https://doi.org/10.1016/j.carbpol.2015.10.074.

68 Bhat, K.M., Jyothsana, R., Sharma, A., and Rao, N.N. (2020). Carrageenan-based edible biodegradable food packaging: a review. *International Journal of Food Science and Nutrition* 5 (4): 69–75.

69 van de Velde, F., Knutsen, S.H., Usov, A.I. et al. (2002). ^{1}H and ^{13}C high resolution NMR spectroscopy of carrageenans: application in research and industry. *Trends in Food Science & Technology* 13 (3): 73–92. https://doi.org/10.1016/S0924-2244(02)00066-3.

70 Muthukumar, J., Chidambaram, R., and Sukumaran, S. (2020). Sulfated polysaccharides and its commercial applications in food industries—a review. *Journal of Food Science and Technology* 58 (7): 2453–2466. https://doi.org/10.1007/s13197-020-04837-0.

71 Prajapati, P.M., Maheriya, G.K., Jani, G.K., and Solanki, H. (2014). Carrageenan: a natural seaweed polysaccharide and its applications. *Carbohydrate Polymers* 105 (1): 97–112. https://doi.org/10.1016/j.carbpol.2014.01.067.

72 Freile-Pelegrin, Y., Robledo, D., and Azamar, J. (2006). Carrageenan of *Eucheuma isiforme* (Solieriaceae, Rhodophyta) from Yucatan, México I. effect of extraction conditions. *Botanica Marina* 49: 65–71. https://doi.org/10.1007/s10811-007-9270-8.

73 Kanmani, P. and Rhim, J.W. (2014b). Development and characterization of carrageenan/grapefruit seed extract composite films for active packaging. *International Journal of Biological Macromolecules* 68: 258–266. https://doi.org/10.1016/j.ijbiomac.2014.05.011.

74 Paula, G.A., Benevides, N.M.B., Cunha, A.P. et al. (2015). Development and characterization of edible films from mixtures of κ-carrageenan, ι-carrageenan, and alginate. *Food Hydrocolloids* 47: 140–145. https://doi.org/10.1016/j.foodhyd.2015.01.004.

75 Lima, B.C., Crepaldi, M.I., de Oliveira, O., and Bonafe, E.G. (2020). Biodegradable films base don comercial k-carrageenan and cassava starch to achieve low productions costs. *International Journal of Biological Macromolecules* 165: 582–590. https://doi.org/10.1016/j.ijbiomac.2020.09.150.

76 Liu, T., Zhang, X., Yi, C. et al. (2020). Comparison of the structural, physical and functional properties of κ-carrageenan films incorporated with pomegranate flesh and peel extracts. *International Journal of Biological Macromolecules* 147: 1076–1088. https://doi.org/10.1016/j.ijbiomac.2019.10.075.

77 Mahajan, K., Kumar, S., Bhat, Z.F. et al. (2021). Functionalization of carrageenan based edible film using *Aloe vera* for improved lipid oxidative and microbial stability of frozen dairy products. *Food Bioscience* 43: 101336. https://doi.org/10.1016/j.fbio.2021.101336.

78 Abdillah, A.A. and Charles, L.A. (2021). Characterization of a natural biodegradable edible film obtained from arrowroot starch and iota-carrageenan and application in food packaging. *International Journal of Biological Macromolecules* 191: 618–626. https://doi.org/10.1016/j.ijbiomac.2021.09.141.

79 Zhang, C., Chi, W., Meng, F., and Wang, L. (2021b). Fabricating an anti-shrinking κ-carrageenan/sodium carboxymethyl starch film by incorporating carboxylate cellulose nanofibrils for fruit preservation. *International Journal of Biological Macromolecules* 191: 706–713. https://doi.org/10.1016/j.ijbiomac.2021.09.134.

80 Kylin, H. (1913). Zur Biochemie der Meeresalgen. 83 (3): 171–197. https://doi.org/10.1515/bchm2.1913.83.3.171.

81 Etmana, S.M., Elnaggar, Y.S.R., and Abdallah, O.Y. (2020). Fucoidan, a natural biopolymer in cancer combating: from edible algae to nanocarrier tailoring. *International Journal of Biological Macromolecules* 147: 799–808. https://doi.org/10.1016/j.ijbiomac.2019.11.191.

82 Senthilkumar, K., Manivasagan, P., Venkatesan, J., and Kim, S.-K. (2013). Review: brown seaweed fucoidan: biological activity and apoptosis, growth signaling mechanism in cancer. *International Journal of Biological Macromolecules* 60: 366–374. https://doi.org/10.1016/j.ijbiomac.2013.06.030.

83 Pal, A., Kamthania, M.C., and Kumar, A. (2014). Bioactive compounds and properties of seaweeds—a review. *Open Access Library Journal* 1: 1–17. https://doi.org/10.4236/oalib.1100752.

84 Pérez, M.J., Falqué, E., and Domínguez, H. (2016). Antimicrobial action of compounds from marine seaweed. *Marine Drugs* 14: https://doi.org/10.3390/md14030052.

85 Bilan, M.I., Grachev, A.A., Ustuzhanina, N.E. et al. (2002). Structure of a fucoidan from the brown seaweed *Fucus evanescens* C.Ag. *Carbohydrate Research* 337 (8): 719–730. https://doi.org/10.1016/S0008-6215(02)00053-8.

86 Li, B., Xin, J.W., Sun, J.L., and Xu, S.Y. (2006). Structural investigation of a fucoidan containing a fucose-free core from the brown seaweed *Hizikia fusiforme*. *Carbohydrate Research* 341: 1135e46. https://doi.org/10.1016/j.carres.2006.03.035.

87 Bilan, M.I. and Usov, A.I. (2008). Structural analysis of fucoidans. *Natural Product Communications* 3 (10): 1639–1648. https://doi.org/10.1177/1934578X0800301011.

88 Cumashi, A., Ushakova, N.A., Preobrazhenskaya, M.E. et al. (2007). A comparative study of the anti-inflammatory, anticoagulant, antiangiogenic, and antiadhesive activities of nine different fucoidans from brown seaweeds. *Glycobiology* 17 (5): 541–552. https://doi.org/10.1093/glycob/cwm014.

89 Azeem, M., Batool, F., Iqbal, N., and Haq, I.-u. (2017). Algal-based biopolymers. In: *Seaweed Polysaccharides: Isolation, Biological and Biomedical Applications* (ed. J. Venkatesan, S. Aniul and A.-K. Kim), 1–9. New York, USA: Elsevier, Inc. https://doi.org/10.1016/B978-0-12-812360-7.00001-X.

90 Rioux, L.E., Turgeon, S.L., and Beaulieu, M. (2007). Characterization of polysaccharides extracted from brown seaweeds. *Carbohydrate Polymers* 69 (3): 530–537. https://doi.org/10.1016/j.carbpol.2007.01.009.

91 Gomaa, M., Fawzy, M.A., Hifney, A.F., and Abdel-Gawad, K.M. (2018a). Use of the brown seaweed *Sargassum latifolium* in the design of alginate-fucoidan based films with natural antioxidant properties and kinetic modeling of moisture sorption and polyphenolic release. *Food Hydrocolloids* 82: 64–72. https://doi.org/10.1016/j.foodhyd.2018.03.053.

92 Saravana, P.S. and Chun, B.S. (2017). Seaweed polysaccharide isolation using subcritical water hydrolysis. In: *Seaweed Polysaccharides* (ed. J. Venkatesan, S. Anil and S.-K. Kim), 47–73. New York, USA: Elsevier Inc https://doi.org/10.1016/B978-0-12-809816-5.00004-9.

93 Nishino, T., Takabe, Y., and Nagumo, T. (1994). Isolation and partial characterization of a novel β-d-galactan sulfate from the brown seaweed *Laminaria angustata* var. longissima. *Carbohydrate Polymers* 23 (3): 165–173. https://doi.org/10.1016/0144-8617(94)90099-X.

94 Rupérez, P., Ahrazem, O., and Leal, J.A. (2002). Potential antioxidant capacity of sulfated polysaccharides from the edible marine brown seaweed *Fucus vesiculosus*. *Journal of Agricultural and Food Chemistry* 50 (4): 840–845. https://doi.org/10.1021/jf010908o.

95 Bit, A. (2021). 8 - Application of polysaccharides in tissue engineering. In: *Food, Medical, and Environmental Applications of Polysaccharides* (ed. K. Kunal Pal, I. Banerjee, P. Sarkar, et al.), 233–261. New York, USA: Elsevier Inc. https://doi.org/10.1016/B978-0-12-819239-9.00020-8.

96 Cunha, L. and Grenha, A. (2016). Sulfated seaweed polysaccharides as multifunctional materials in drug delivery applications. *Marine Drugs* 14 (3): 42. https://doi.org/10.3390/md14030042.

97 Pierre, G., Sopena, V., Juin, C. et al. (2011). Antibacterial activity of a sulfated galactan extracted from the marine alga *Chaetomorpha aerea* against *Staphylococcus aureus*. *Biotechnology and Bioprocess Engineering* 16: 937–945. https://doi.org/10.1007/s12257-011-0224-2.

98 Venkatesan, J., Aniul, S., and Kim, A.-K. (2017). Introduction to seaweed polysaccharides. In: *Seaweed Polysaccharides: Isolation, Biological and Biomedical Applications* (ed. J. Venkatesan, S. Aniul and A.-K. Kim), 1–9. New York, USA: Elsevier, Inc. https://doi.org/10.1016/B978-0-12-809816-5.00001-3.

99 Wang, J. and Zhang, Q. (2017). Chemical modification of fucoidan and their application. In: *Seaweed Polysaccharides: Isolation, Biological and Biomedical Applications* (ed. J. Venkatesan, S. Aniul and A.-K. Kim), 1–9. New York, USA: Elsevier, Inc. https://doi.org/10.1016/B978-0-12-809816-5.00009-8.

100 Gomaa, M., Hifney, A.F., Fawzy, M.A., and Abdel-Gawad, K.M. (2018b). Use of seaweed and filamentous fungus derived polysaccharides in the development of alginate-chitosan edible films containing fucoidan: study of moisture sorption, polyphenol release and antioxidant properties. *Food Hydrocolloids* 82: 239–247. https://doi.org/10.1016/j.foodhyd.2018.03.056.

101 Duan, Z., Duan, W., Li, F. et al. (2019). Effect of carboxymethylation on properties of fucoidan from *Laminaria japonica*: antioxidant activity and preservative effect on strawberry during cold storage. *Postharvest Biology and Technology* 151: 127–133. https://doi.org/10.1016/j.postharvbio.2019.02.008.

102 Luo, P., Li, F.P., Liu, H.Z. et al. (2020). Effect of fucoidan-based edible coating on antioxidant degradation kinetics in strawberry fruit during cold storage. *Journal of Food Processing and Preservation* 44: e14381. https://doi.org/10.1111/jfpp.14381.

103 Xu, B. and Wu, S. (2021). Preservation of mango fruit quality using fucoidan coatings. *LWT - Food Science and Technology* 143: 111150. https://doi.org/10.1016/j.lwt.2021.111150.

104 Kong, F., Mao, Y., Cui, F. et al. (2011). Morphology and molecular identification of Ulva forming green tides in Qingdao, China. *Journal of Ocean University of China* 10: 73–79. https://doi.org/10.1007/s11802-011-1728-2.

105 Chiellini, F. and Morelli, A. (2011). Ulvan: a versatile platform of biomaterials from renewable resources. In: *Biomaterialsphysics and Chemistry* (ed. R. Pignatello), 75–98. Riejeka: IntechOpen.

106 Tziveleka, L.-A., Ioannou, E., and Roussis, V. (2019). Ulvan, a bioactive marine sulphated polysaccharide as a key constituent of hybrid biomaterials: a review. *Carbohydrate Polymers* 218: 355–370. https://doi.org/10.1016/j.carbpol.2019.04.074.

107 Lakshmi, D.S., Sankaranarayanan, S., Gajaria, T.K. et al. (2020). A short review on the valorization of green seaweeds and ulvan: feedstock for chemicals and biomaterials. *Biomolecules* 10 (7): 991. https://doi.org/10.3390/biom10070991.

108 Morelli, A. and Chiellini, F. (2010). Ulvan as a new type of biomaterial from renewable resources: functionalization and hydrogel preparation. *Macromolecular Chemistry and Physics* 211 (7): 821–832. https://doi.org/10.1002/macp.200900562.

109 Morelli, A., Betti, M., Puppi, D., and Chiellini, F. (2016). Design, preparation and characterization of ulvan based thermosensitive hydrogels. *Carbohydrate Polymers* 136: 1108–1117. https://doi.org/10.1016/j.carbpol.2015.09.068.

110 European Commission (2009). *European commission regulation (EC) No 450/2009 of 29 May 2009 on active and intelligent materials and articles intended to come into contact with food*. Luxembourg: European Commission.

111 Guidara, M., Yaich, H., Richel, A. et al. (2019). Effects of extraction procedures and plasticizer concentration on the optical, thermal, structural and antioxidant properties of novel ulvan films. *International Journal of Biological Macromolecules* 135: 647–658. https://doi.org/10.1016/j.ijbiomac.2019.05.196.

112 Guidara, M., Yaich, H., Benelhadj, S. et al. (2020). Smart ulvan films responsive to stimuli of plasticizer and extraction condition in physico-chemical, optical, barrier and mechanical properties. *International Journal of Biological Macromolecules* 150: 714–726. https://doi.org/10.1016/j.ijbiomac.2020.02.111.

113 Amin, H.H. (2021). Safe ulvan silver nanoparticles composite films for active food packaging. *American Journal of Biochemistry and Biotechnology* 17 (1): 28–39. https://doi.org/10.3844/ajbbsp.2021.28.39.

114 Mutreja, R., Thakur, A., and Goyal, A. (2020). Chitin and chitosan: current status and future opportunities. In: *Preparation and Properties*, Handbook of Chitin and Chitosan, vol. 1, 401–417. https://doi.org/10.1016/b978-0-12-817970-3.00013-4.

115 Cardona, C.E.O. and Fernández, X.A. (2020). Chitosan: a sustainable alternative for food packaging. *University Digital Magazine* 21 (5): https://doi.org/10.22201/cuaieed.16076079e.2020.21.5.4.

116 El Knidri, H., Belaabed, R., Addaou, A. et al. (2018). Extraction, chemical modification and characterization of chitin and chitosan. *International Journal of Biological Macromolecules* 120 (A): 1181–1189. https://doi.org/10.1016/j.ijbiomac.2018.08.139.

117 Haghighi, H., Licciardello, F., Fava, P. et al. (2020). Recent advances on chitosan-based films for sustainable food packaging applications. *Food Packaging and Shelf Life* 26: 100551. https://doi.org/10.1016/j.fpsl.2020.100551.

118 Hu, H., Yao, X., Qin, Y. et al. (2020). Development of multifunctional food packaging by incorporating betalains from vegetable amaranth (*Amaranthus tricolor* L.) into quaternary ammonium chitosan/fish gelatin blend films. *International Journal of Biological Macromolecules* 159: 675–684. https://doi.org/10.1016/j.ijbiomac.2020.05.103.

119 Feng, F., Liu, Y., Zhao, B., and Hu, K. (2012). Characterization of half *N*-acetylated chitosan powders and films. *Procedia Engineering* 27: 718–732. https://doi.org/10.1016/j.proeng.2011.12.511.

120 Huang, J., Liu, Y., Yang, L., and Zhou, F. (2019). Synthesis of sulfonated chitosan and its antibiofilm formation activity against *E. coli* and *S. aureus*. *International Journal of Biological Macromolecules* 129: 980–988. https://doi.org/10.1016/j.ijbiomac.2019.02.079.

121 Liu, Y., Jiang, Y., Zhu, J. et al. (2019a). Inhibition of bacterial adhesion and biofilm formation of sulfonated chitosan against *Pseudomonas aeruginosa*. *Carbohydrate Polymers* 206: 412–419. https://doi.org/10.1016/j.carbpol.2018.11.015.

122 Wen, L., Liang, Y., Lin, Z. et al. (2021). Design of multifunctional food packaging films based on carboxymethyl chitosan/polyvinyl alcohol crosslinked network by using citric acid as crosslinker. *Polymer* 230: 124048. https://doi.org/10.1016/j.polymer.2021.124048.

123 Suriyatem, R., Auras, R.A., and Rachtanapun, P. (2018). Improvement of mechanical properties and thermal stability of biodegradable rice starch–based films blended with carboxymethyl chitosan. *Industrial Crops and Products* 122: 37–48. https://doi.org/10.1016/j.indcrop.2018.05.047.

124 Lai, W.F., Zhao, S., and Chiou, J. (2021). Antibacterial and clusteroluminogenic hypromellose-graft-chitosan-based polyelectrolyte complex films with high functional flexibility for food packaging. *Carbohydrate Polymers* 271: 118447. https://doi.org/10.1016/j.carbpol.2021.118447.

125 Sabaa, M.W., Elzanaty, A.M., Abdel-Gawad, O.F., and Arafa, E.G. (2018). Synthesis, characterization and antimicrobial activity of Schiff bases modified chitosan-graft-poly(acrylonitrile). *International Journal of Biological Macromolecules* 109: 1280–1291. https://doi.org/10.1016/j.ijbiomac.2017.11.129.

126 Liu, Y., Cai, Z., Sheng, L. et al. (2019b). Structure-property of crosslinked chitosan/silica composite films modified by genipin and glutaraldehyde under alkaline conditions. *Carbohydrate Polymers* 215: 348–357. https://doi.org/10.1016/j.carbpol.2019.04.001.

127 Bi, J., Tian, C., Zhang, G.L. et al. (2021). Novel procyanidins-loaded chitosan-graft-polyvinyl alcohol film with sustained antibacterial activity for food packaging. *Food Chemistry* 365: 130534. https://doi.org/10.1016/j.foodchem.2021.130534.

128 Hassan, M.M. (2018). Enhanced antimicrobial activity and reduced water absorption of chitosan films graft copolymerized with poly (acryloyloxy) ethyltrimethylammonium chloride. *International Journal of Biological Macromolecules* 118 (B): 1685–1695. https://doi.org/10.1016/j.ijbiomac.2018.07.013.

129 Ferreira, D.C.M., De Souza, A.L., Da Silveira, J.V.W. et al. (2020). Chapter 17 - Chitosan nanocomposites for food packaging applications. In: *Multifunctional Hybrid Nanomaterials for Sustainable Agri-Dood and Ecosystems* (ed. K.A. Abd-Elsalam), 393–435. New York, USA: Elsevier https://doi.org/10.1016/B978-0-12-821354-4.00017-0.

130 Negm, N.A., Hefni, H.H., Abd-Elaal, A.A. et al. (2020). Advancement on modification of chitosan biopolymer and its potential applications. *International Journal of Biological Macromolecules* 152: 681–702. https://doi.org/10.1016/j.ijbiomac.2020.02.196.

131 Wang, W., Xue, C., and Mao, X. (2020). Chitosan: structural modification, biological activity and application. *International Journal of Biological Macromolecules* 164: 4532–4546. https://doi.org/10.1016/j.ijbiomac.2020.09.042.

132 Ahmad, M., Zhang, B., Manzoor, K. et al. (2020). Chitin and chitosan-based bionanocomposites. In: *Bionanocomposites Green Synthesis and Applications* (ed. K.M. Zia, F. Jabeen, M.N. Anjum and S. Ikram), 145–156. Cambridge MA, USA: Elsevier Inc. https://doi.org/10.1016/B978-0-12-816751-9.00006-4.

133 Vu, K.D., Hollingsworth, R.G., Leroux, E. et al. (2011). Development of edible bioactive coating based on modified chitosan for increasing the shelf life of strawberries. *Food Research International* 44 (1): 198–203. https://doi.org/10.1016/j.foodres.2010.10.037.

134 Da Mata, C.O., Lima, A.M.F., Assis, O.B.G. et al. (2020). Amphiphilic diethylaminoethyl chitosan of high molecular weight as an edible film. *International Journal of Biological Macromolecules* 164: 3411–3420. https://doi.org/10.1016/j.ijbiomac.2020.08.145.

135 Taştan, Ö., Pataro, G., Donsì, F. et al. (2017). Decontamination of fresh-cut cucumber slices by a combination of a modified chitosan coating containing carvacrol nanoemulsions and pulsed light. *International Journal of Food Microbiology* 260: 75–80. https://doi.org/10.1016/j.ijfoodmicro.2017.08.011.

136 Elbarbary, A.M. and Mostafa, T.B. (2014). Effect of γ-rays on carboxymethyl chitosan for use as antioxidant and preservative coating for peach fruit. *Carbohydrate Polymers* 104: 109–117. https://doi.org/10.1016/j.carbpol.2014.01.021.

137 Benhabiles, M.S., Tazdait, D., Abdi, N. et al. (2013). Assessment of coating tomato fruit with shrimp shell chitosan and *N,O*-carboxymethyl chitosan on postharvest preservation. *Journal of Food Measurement and Characterization* 7 (2): 66–74. https://doi.org/10.1007/s11694-013-9140-9.

138 Kumar, N., Neeraj, P., Ojha, A. et al. (2021). Effect of active chitosan-pullulan composite edible coating enrich with pomegranate peel extract on the storage quality of green Bell pepper. *LWT- Food Science and Technology* 138: 110435. https://doi.org/10.1016/j.lwt.2020.110435.

139 Madera-Santana, T.J., De Dios-Aguilar, M.A., Colín-Chávez, C. et al. (2019). Recubrimiento a base de quitosano y extracto acuoso de hoja de *Moringa oleifera* obtenido por UMAE y su efecto en las propiedades fisicoquímicas de fresa (*Fragaria x ananassa*). *Biotecnia* XXI (2): 155–163.

140 Nair, M.S., Tomar, M., Punia, S. et al. (2020). Enhancing the functionality of chitosan- and alginate-based active edible coatings/films for the preservation of fruits and vegetables: a review. *International Journal of Biological Macromolecules* 164: 304–320. https://doi.org/10.1016/j.ijbiomac.2020.07.083.

141 Zhang, Y.L., Cui, Q.L., Wang, Y. et al. (2021c). Effect of carboxymethyl chitosan-gelatin-based edible coatings on the quality and antioxidant properties of sweet cherry during postharvest storage. *Scientia Horticulturae* 289: 110462. https://doi.org/10.1016/j.scienta.2021.110462.

142 Indumathi, M.P., Sarojini, K.S., and Rajarejeswari, G.R. (2019). Antimicrobial and biodegradable chitosan/cellulose acetate phthalate/ZnO nano composite films with optimal oxygen permeability and hydrophobicity for extending the shelf life of black grapefruits. *International Journal of Biological Macromolecules* 132: 1112–1120. https://doi.org/10.1016/j.ijbiomac.2019.03.171.

143 Nguyen, H.T., Boonyaritthongchai, P., Buanong, M. et al. (2020). Postharvest hot water treatment followed by chitosan- and κ-carrageenan-based composite coating induces the disease resistance and preserves the quality in Dragon fruit (*Hylocereus undatus*). *International Journal of Fruit Science* 20 (sup3): S2030–S2044. https://doi.org/10.1080/15538362.2020.1851342.

144 Da Costa, J.C.M., Miki, K.S.L., da Silva Ramos, A., and Teixeira-Costa, B.E. (2020). Development of biodegradable films based on purple yam starch/chitosan for food application. *Heliyon* 6 (4): e03718. https://doi.org/10.1016/j.heliyon.2020.e03718.

145 Ferreira, L.F., Figueiredo, L.P., Martins, M.A. et al. (2021). Active coatings of thermoplastic starch and chitosan with alpha-tocopherol/bentonite for special green coffee beans. *International Journal of Biological Macromolecules* 170: 810–819. https://doi.org/10.1016/j.ijbiomac.2020.12.199.

146 Perdana, M.I., Ruamcharoen, J., Panphon, S., and Leelakriangsak, M. (2021). Antimicrobial activity and physical properties of starch/chitosan film incorporated with lemongrass essential oil and its application. *LWT - Food Science and Technology* 141: 110934. https://doi.org/10.1016/j.lwt.2021.110934.

147 Araújo, J.M.S., de Siqueira, A.C.P., Blank, A.F. et al. (2018). A cassava starch–chitosan edible coating enriched with *Lippia sidoides* cham. Essential oil and pomegranate peel extract for preservation of Italian tomatoes (*Lycopersicon esculentum* mill.) stored at room temperature. *Food and Bioprocess Technology* 11 (9): 1750–1760. https://doi.org/10.1007/s11947-018-2139-9.

148 Regulation (EC) No 1935/2004 of the European Parliament and of the Council (2004). Regulation (EC) No 1935/2004 of the European Parliament and of the Council of 27 October 2004 on materials and articles intended to come into contact with food and repealing Directives 80/590/EEC and 89/109/EEC.

149 Fiore, A., Park, S., Volpe, S. et al. (2021). Active packaging based on PLA and chitosan-caseinate enriched rosemary essential oil coating for fresh minced chicken breast application. *Food Packaging and Shelf Life* 29: 100708. https://doi.org/10.1016/j.fpsl.2021.100708.

150 Song, Z., Ma, T., Zhi, X., and Du, B. (2021). Cellulosic films reinforced by chitosan-citric complex for meat preservation: influence of nonenzymatic browning. *Carbohydrate Polymers* 272: 118476. https://doi.org/10.1016/j.carbpol.2021.118476.

151 Râpă, M., Miteluţ, A.C., Tănase, E.E. et al. (2016). Influence of chitosan on mechanical, thermal, barrier and antimicrobial properties of PLA-biocomposites for food packaging. *Composites Part B: Engineering* 102: 112–121. https://doi.org/10.1016/j.compositesb.2016.07.016.

152 Wang, K., Lim, P.N., Tong, S.Y., and Thian, E.S. (2019a). Development of grapefruit seed extract-loaded poly(ε-caprolactone)/chitosan films for antimicrobial food packaging. *Food Packaging and Shelf Life* 22: 100396. https://doi.org/10.1016/j.fpsl.2019.100396.

153 Zeng, A., Wang, Y., Li, D. et al. (2021). Preparation and antibacterial properties of polycaprolactone/quaternized chitosan blends. *Chinese Journal of Chemical Engineering* 32: 462–471. https://doi.org/10.1016/j.cjche.2020.10.001.

154 Kasirajan, S., Umapathy, D., Chandrasekar, C. et al. (2019). Preparation of poly(lactic acid) from *Prosopis juliflora* and incorporation of chitosan for packaging applications. *Journal of Bioscience and Bioengineering* 128: 323–331. https://doi.org/10.1016/j.jbiosc.2019.02.013.

155 Jiang, Y., Lan, W., Sameen, D.E. et al. (2020). Preparation and characterization of grass carp collagen-chitosan-lemon essential oil composite films for application as food packaging. *International Journal of Biological Macromolecules* 160: 340–351. https://doi.org/10.1016/j.ijbiomac.2020.05.202.

156 Rahman, S., Konwar, A., Majumdar, G., and Chowdhury, D. (2021). Guar gum-chitosan composite film as excellent material for packaging application. *Carbohydrate Polymer Technologies & Applications* 2: 100158. https://doi.org/10.1016/j.carpta.2021.100158.

157 Jovanović, J., Ćirkovića, J., Radojković, A. et al. (2021). Chitosan and pectin-based films and coatings with active components for application in antimicrobial food packaging. *Progress in Organic Coatings* 158: 106349. https://doi.org/10.1016/j.porgcoat.2021.106349.

158 Fonseca-García, A., Jiménez-Regalado, E., and Aguirre-Loredo, R. (2021). Preparation of a novel biodegradable packaging film based on corn starch-chitosan and poloxamers. *Carbohydrate Polymers* 251: 117009. https://doi.org/10.1016/j.carbpol.2020.117009.

159 Cartelle, G.M. and Zurita, J. (2015). La nanotecnología en la producción y conservación de alimentos. *Revista Cubana de Alimentación y Nutrición* 25 (1): 184–207.

160 Mendoza, U.G. and Rodríguez-López, J.L. (2007). La nanociencia y la nanotecnología: una revolución en curso. *Perfiles Latinoamericanos* 14 (29): 161–186.

161 Landa-Salgado, P., Cruz-Monterrosa, R.G., Hernández-Guzmán, F.J., and Reséndiz-Cruz, V. (2017). Nanotecnología en la industria alimentaria: bionanocompuestos en empaques de alimenticios. *Agroproductividad* 10 (10): 34–40.

162 Othman, S.H. (2014). Bio-nanocomposite materials for food packaging applications: types of biopolymer and nano-sized filler. *Agriculture and Agricultural Science Procedia* 2: 296–303. https://doi.org/10.1016/j.aaspro.2014.11.042.

163 Wang, H., Gong, X., Miao, Y. et al. (2019b). Preparation and characterization of multilayer films composed of chitosan, sodium alginate and carboxymethyl chitosan-ZnO nanoparticles. *Food Chemistry* 283: 397–403. https://doi.org/10.1016/j.foodchem.2019.01.022.

164 Behera, K., Kumari, M., Chang, Y.H., and Chiu, F.C. (2021). Chitosan/boron nitride nanobiocomposite films with improved properties for active food packaging applications. *International Journal of Biological Macromolecules* 186: 135–144. https://doi.org/10.1016/j.ijbiomac.2021.07.022.

165 Medina, E., Caro, N., Abugoch, L. et al. (2019). Chitosan thymol nanoparticles improve the antimicrobial effect and the water vapour barrier of chitosan-quinoa protein films. *Journal of Food Engineering* 240: 191–198. https://doi.org/10.1016/j.jfoodeng.2018.07.023.

166 Wu, D., Zhang, M., Bhandari, B., and Guo, Z. (2021). Combined effects of microporous packaging and nano-chitosan coating on quality and shelf-life of fresh-cut eggplant. *Food Bioscience* 43: 101302. https://doi.org/10.1016/j.fbio.2021.101302.

167 Arabpoor, B., Yousefi, S., Weisany, W., and Ghasemlou, M. (2021). Multifunctional coating composed of *Eryngium campestre* L. essential oil encapsulated in nano-chitosan to prolong the shelf-life of fresh cherry fruits. *Food Hydrocolloids* 111: 106394. https://doi.org/10.1016/j.foodhyd.2020.106394.

168 Hernández-López, G., Ventura-Aguilar, R.I., Correa-Pacheco, Z.N. et al. (2020). Nanostructured chitosan edible coating loaded with α-pinene for the preservation of the postharvest quality of *Capsicum annuum* L. and *Alternaria alternata* control. *International Journal of Biological Macromolecules* 165 (B): 1881–1888. https://doi.org/10.1016/j.ijbiomac.2020.10.094.

169 Rosenbloom, R.A., Wang, W., and Zhao, Y. (2020). Delaying ripening of 'Bartlett' pears (*Pyrus communis*) during long-term simulated industrial cold storage: mechanisms and validation of chitosan coatings with cellulose nanocrystals pickering emulsions. *LWT - Food Science and Technology* 122 (3): 109053. https://doi.org/10.1016/j.lwt.2020.109053.

170 Donsì, F., Marchese, E., Maresca, P. et al. (2015). Green beans preservation by combination of a modified chitosan based-coating containing nanoemulsion of mandarin essential oil with high pressure or pulsed light processing. *Postharvest Biology and Technology* 106: 21–32. https://doi.org/10.1016/j.postharvbio.2015.02.006.

171 Caro-León, F.J., López-Martínez, L.M., Lizardi-Mendoza, J. et al. (2019). Métodos de preparación de nanopartículas de quitosano: una revisión. *Biotecnia* 21 (3): 13–25.

172 Kumar, S., Mukherjee, A., and Dutta, J. (2020). Chitosan based nanocomposite films and coatings: emerging antimicrobial food packaging alternatives. *Trends in Food Science & Technology* 97: 196–209. https://doi.org/10.1016/j.tifs.2020.01.002.

11

Chitosan-Based Food Packaging Films

Kunal Singha[1] and Kumar Rohit[2]

[1] *National Institute of Fashion Technology, Department of Textile Design, Salt Lake, Kolkata, 700098, West Bengal, India*
[2] *National Institute of Fashion Technology, Department of Textile Design, Mithapur farm, Patna, 800001, Bihar, India*

11.1 Introduction

Packaging is a science, art, and technology of securely delivering commodities to end consumers at low prices. When it comes to the preservation of fresh or processed foods, the packing procedure is concerned. It is critical in making food manufactured in one location available to consumers in another location after days to weeks, or sometimes even months have passed after the initial production or harvest. People used to collect food only when it was time to eat it. It became vital to preserving food for upcoming use as the demand for dwellings and shelters grew, as did agricultural development. That's when the need for storing and packaging became apparent [1]. Leaves, shells, woven grasses, hollowedwood, animal organs, and other items were used for this purpose, etc. [2]. However, these preservation methods were very primitive and the need for improvement/innovation continued to rise resulted in the start of food packaging research. These improvements/developments of packaging materials that are still in use today.

11.1.1 A Brief History of Food Packaging Materials Used

From ancient days, paper has been regarded as the most cost-effective packing material. The first commercial paper bags were created in Bristol, England in 1844. This was the beginning of the industrial usage of papers in packaging [3]. Glass containers ruled the food packaging marketplace from the early 1900s through the late 1960s, particularly for liquid products. For items that require significant flavor and fragrance protection, glass packaging remains the most effective and preferred

---

Natural Materials for Food Packaging Application, First Edition.
Edited by Jyotishkumar Parameswaranpillai, Aswathy Jayakumar,
E. K. Radhakrishnan, Suchart Siengchin, and Sabarish Radoor.
© 2023 WILEY-VCH GmbH. Published 2023 by WILEY-VCH GmbH.

option [2, 3]. Glass was the ideal choice of packaging materials based on material properties and food packaging needs due to its good barrier qualities, inertness, and transparency. However, it falls behind due to its brittleness, heavyweight, and limited availability. Metal and metal sheets compete well with glass, although they suffer from low inertness, nontransparency, and lower portability due to their weight. This is where paper and plastics come in as they are cheap, elastic, and inert. Plastics also provide good tensile and barrier qualities with superior transparent and lightweight characteristics, as well as practical benefits such asheat sealability and moldability [4].

Plastics are by far the maximum often used materials for food packaging uses. They've been utilized extensively in both rigid and flexible packaging. Plastics, on the other hand, are nonbiodegradable in nature. A report published by the United States Environmental Protection Agency (US EPA), plastic is the main culprit reason to generate municipal waste generation (MSW), which is the most pollution-related issue in several countries. Moreover, due to the widespread usage of plastic packaging materials, 14.32 million tonnes of plastic garbage was discovered in 2014 US EPA [5]. Nicholas Appert found in 1809 that boiling a sealed container containing food extends the life of the food. This discovery paved the way for the use of metal containers for food packaging in food sterilizing. Cookies in sealed metal tins were first sold in 1830. Berger [2]; Risch [3] Plastic-based food packaging materials became available after the discovery of several types of synthetic/manmade polymers nearly in the 19th century. Today, bottles made of vinyl chloride and polyethylene terephthalate (PET) are used to package water and other liquids. PET can also be seen in food packaging and hot-fill items such as food jam [2, 3].

11.1.2 Characteristics of Typical Food Packaging Materials

- Able to protect food quality for a long time.
- Portable and convenient to use.
- Renewable and biodegradable.
- Providing no issue regarding no municipal solid waste accumulation.

11.1.3 Need for Biodegradable Food Packaging Materials

MSW creation is a major environmental concern since the beginning when plastics were primarily employed as food packaging material. This has prompted researchers to look at biopolymer-based food packaging technologies. These biopolymers have inherent qualities such as biodegradability, antibacterial activity, and antioxidant activity, which can help extend the shelf life of any food product.

Over time, customer and human demands have constantly increased with the increment of food types and complexities. The new advancement and varieties of food consistently have pushed the food industries to begin the search for bridgeable and low-cost biopolymers as food packaging materials. These biopolymers must have the capabilities to guarantee the accessibility of protected and quality food. The primary function of food packaging is to isolate food items from their surroundings and thus avoiding or preventing exposure to waste elements such as microbes, oxygen, temperature, and humidity. This also helps to avoid or postpone quality and nutritional loss and so lengthens the shelf life of the food items. On the other hand, food packaging systems play several other important crucial requirements such as providing comfort, marketing, and communicating with consumers by promoting the packed food product.

11.2 Chitin and Chitosan Chemical Structure

Deacetylation is used to remove the chitosan polymer from chitin. Chitin is a cellulosic biopolymer found in crustacean and insect exoskeletons, as well as fungal and yeast cell walls. Chitin is a structural material for the majority of sea animals, much like cellulose does in plant cells.

Chitin polymer consists primarily of (1–4)-linked 2-acetamido-2-deoxy-D-glucose monomers that are manufactured in massive quantities (1011 metric tonnes) every year [6]. However, only 150 000 tonnes of chitin are available for commercial use due to its solubility and flexible nature [7]. It is the most available biopolymer of animal origin and the second most plentiful biopolymer on the planet after cellulosic material. Chitin differs from other polysaccharides due to its nitrogen content. Chitosan is a deacetylated chitin derivative made up primarily of (1–4)-linked 2-amino-2-deoxy-D-glucose monomers. The degree of acetylation (DA), molecular size nitrogen/carbon (N/C) ratio, nitrogen content, and polydispersity of the chitin source influence the primary features of chitosan.

Deacetylation of chitin is the utmost common and cost-effective method for isolating chitosan. However, some fungus produces chitosan that can be extracted directly [8, 9]. Chitin is used as a structural substance by a wide variety of animal and fungal species.

11.3 Chitosan as a Potential Biodegradable Food Packaging Material

From the above explanation, it can be inferred that biodegradable alternatives to plastic polymers include polysaccharides such as starch, agar, cellulose, chitosan, and proteins such as casein. In the occurrence of adequate temperature and moisturized conditions (as well as the presence of oxygen) bioplastics decompose without leaving hazardous residues. Some biopolymers disintegrate in a matter of weeks and whereas manmade/synthetic polymers might take months to years to deteriorate, differing on the nature and origin of the polymer.

Chitin is the structural material of crustaceans, insects, and fungi. Deacetylation of chitin yields chitosan, a deacetylated derivative of chitin (Figure 11.1).

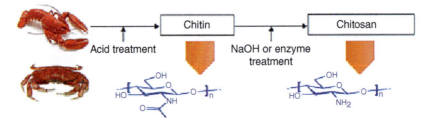

Figure 11.1 Preparation of chitosan. Source: Tripathi et al. [10] / Walter de Gruyter GmbH.

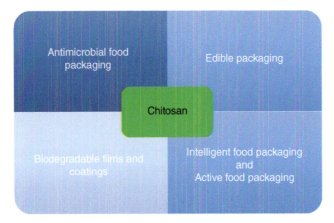

Figure 11.2 Various food packaging applications of chitosan materials.

The inclusion of amino groups is liable for the polymer's varied characteristics and that makes them functionally diverse biopolymer. It has a long history of industrial applications, but one of the most recent is as a biodegradable antimicrobial food packaging material. Chitin is also used in food packaging materials along with extracts nanoparticles, whiskers, nanofibres, edible food coatings, plasticizers, and cross-linking materials.

11.3.1 Chitosan as Food Packaging Material

Chitosan and Chitosan-based nanocomposites are used in the production and applications of food preservation and packaging. They are also equally used to preserve and packaging for organic food products such as vegetables, fruits, meat, and fish (as shown in Figure 11.2).

11.3.2 Chitosan Film in Food Packaging and Their Types

Chitosan can be found in food packaging in the form of flexible packaging films or coatings. The flexible films can be edible or inedible, but they are virtually always edible as they form a layer right on the food's top surface. Each style of packaging contains a variety of additives that confer various qualities. To avoid jeopardizing chitosan's food-grade certification, edible coatings are made with little or no additives [11, 12].

11.3.2.1 Chitosan-Based Films

Chitosan can be used in food packaging in the shape/form of packing films or coatings that are applied directly onto the food. Chitosan films have also been described in combination with other polymers/biopolymers such as proteins, polysaccharides, and synthetic plastics [13].

11.3.2.2 Flexible Packaging Films

Flexible chitosan films are prepared using the solvent casting method, in which chitosan is dissolved in suitable solvents (typically slightly acidified water) and then poured onto a flat surface to evaporate [14]. The characteristics of the films are governed by several factors such as the pH of the solvent, type of the acid used, and the molecular weight (MW). The tensile qualities of the films are critical from the standpoint of food packing. Tensile strength (TS) and elongation at break (EB) of chitosan films are equivalent to those of low-density polyethylene (LDPE), high-density polyethylene (HDPE), and cellophane [15]. However, because of its great sensitivity to moisture and humidity, chitosan's use in food packaging is limited. Furthermore, because chitosan films are not thermoplastic and deteriorate before reaching melting temperatures, they cannot be utilized as industrial processes such as molding or extrusion, nor can they be heat sealed or stretched [16, 17].

11.3.3 Chitosan Film in Food Packaging

Chitosan is used in food packaging by using as flexible or adaptable packaging films/coatings. In contrast to coatings, which are almost edible since they make a layer immediately on the food's top surface, flexible films can be either edible or inedible. Each style of packaging contains a variety of additives that confer various qualities. To avoid jeopardizing chitosan's food-grade certification, edible coatings are made with little or no additives (Figure 11.3).

11.3.4 Films Embedded with Nanomaterials

Copper, silver, zinc, titanium, and other metallic ions have been found in abundance in nature. In several eukaryotes, including humans, the majority of these ions are necessary minerals. Furthermore, below a certain concentration, compounds pose little hazard to eukaryotic cells. As a result, metal/metal oxide nanoparticles are used as fillers in food packaging polymers to improve photodegradation, thermal, and mechanical barrier qualities. Other features pertinent to food packaging are also addressed by these nanoparticles. They have remarkable antibacterial and antioxidant properties, resulting in increased product shelf life. Due to its noble characterizes or nature, silver is the most extensively used nanoparticle, and silver salts are commercially often used as antibacterial agents [18]. The inactivation of respiratory

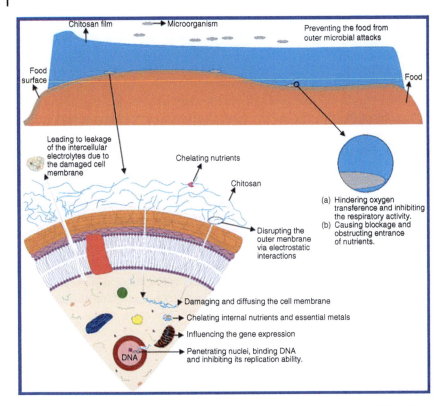

Figure 11.3 Antimicrobial mechanism of chitosan films. Source: Wang et al. [12] / with permission of American Chemical Society.

enzymes by silver ions results in the creation of reactive oxygen species and thus results in cell damage [19]. Tripathi et al. [20] investigated the antibacterial efficacy of chitosan films after adding silver nanoparticles. The silver nanoparticle-embedded chitosan films were found to have improved antibacterial activity against gram-positive bacteria *Bacillus subtilis* and *Staphylococcus aureus*, as well as gram-negative bacteria *Escherichia coli* and *Pseudomonas aeruginosa*. Vimala et al. [21] found that chitosan films embedded with silver nanoparticles had greater mechanical strength as well as improved antibacterial activity. Another major nanomaterial that is popular due to its safety and which is increasingly used in a variety of human use goods is zinc oxide. Zinc oxide nanoparticles were produced and used as fillers in the chitosan matrix [22]. Due to chitosan and zinc oxide, the films have synergistic antibacterial action against *B. subtilis* and *E. coli*. Furthermore, the reinforcing impact of zinc oxide nanoparticles results in a significant increase in mechanical characteristics.

11.3.5 Films Embedded with Clays

The exceptional feature of clays is to disperse smoothly into the polymer matrix, even with alower concentration, to fulfill the voids in the polymer matrix.

This provides a very compact structure made of polymer/clay composites which can be used as a strong food packaging material. The intercalation of clays in this tight structure creates a complex diffusion channel for gases and water vapor and thus making it difficult for them to pass from the atmosphere to the inside food products. The reinforcing effect of clays also improves the mechanical characteristics significantly. Clays of various sorts have been used as fillers to improve the qualities of plain chitosan and chitosan blend films. The nanocomposite films can improve oxygen permeability and thermal stability and as the clay concentration in the films increased, it makes them a good choice for food packaging purposes.

11.3.6 Films Embedded with Polysaccharide Particles, Fibres, and Whiskers

The use of fillers for reinforcement in polymeric films and sheets to increase mechanical performance is gaining popularity as a simple and practical method. Although numerous inorganic materials have been employed for this purpose such as clays and carbon nanotubes, but they are supposed to be less environmental friendly due to their nonbiodegradable or sometimes for their limited biodegradability. This resulted in the passage of polysaccharide-based biodegradable reinforcing resources such as chitin and cellulose in powders, crystals, and fiber forms. The possible antibacterial activity of these biodegradable reinforcing materials, unlike clays, is another rationale for their rising utilization. Many studies have employed these natural reinforcing elements to improve the characteristics of chitosan films for food packaging usage. According to their findings, the films gained not only improved mechanical qualities and thermal stability but also provides superior antioxidant and antibacterial properties, which made them the most suitable or viable active food packaging substance/material [1].

11.3.7 Films Embedded with Natural Oils and Extracts

As packaging experts' interest in active food packaging materials grows, greater emphasis is being placed on the search for ingredients that, in addition to improving film qualities, also impart biological action to the films. The development of biodegradable, biocompatible, nontoxic, and cost-effective biologically active chemicals is becoming increasingly crucial with the increased market rivalry to extend food shelf life while retaining environmental friendliness and economical aspects. Natural essential oils derived from renewable plant parts or agricultural wastes are proving to be a promising contenders for this application. They are made up of various plant metabolites that are both antibacterial and antioxidant, in addition to fatty acids and lipids. Furthermore, they are being nontoxic and also have no negative effects on human health when ingested in tiny amounts. Apart from providing antioxidant and antibacterial properties to the films, these essential oils can also provide which is a significant benefit by providing water and plasticizing resistance (Figure 11.4).

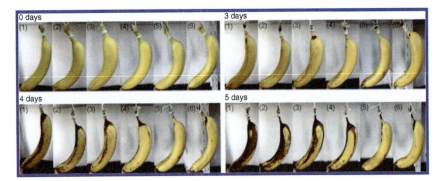

Figure 11.4 A comparative test of uncoated bananas (1), coated bananas by using CMC100 film (2), HTCC40/CMC60 film (3), HTCC70/CMC30 film (4), HTCC90/CMC10 film (5) and HTCC100 film (6). Carboxymethylcellulose, CMC. 2-N-Hydroxypropyl-3-trimethylammonium chloride chitosan, HTCC. Source: Reproduced with permission from Hu et al. [23] / Reproduced with permission from Elsevier.

Turmeric is another rhizome family member. Curcumin, an antioxidant and antibacterial ingredient found in turmeric extract, is a metabolite. The effect of curcumin dosage in chitosan films was investigated by Roy and Rhim [24]. The composite films were shown to have excellent antioxidant activity. Chitosan and curcumin have a synergistic effect, making them effective antibacterial packaging materials [24]. *Prunus armeniaca*, or bitter apricot which is another essential oil and agro-waste resource derived from its seeds, shows high antioxidants and antimicrobial properties. The antimicrobial and antioxidant properties of chitosan films infused with apricot kernel essential oils were investigated and it was discovered that the oil assisted in enhancing the antimicrobial and antioxidant properties of the films, resulting in an increase in the shelf life of packed bread slices by preventing mold growth for nearly 10 days [25]. Plant-based proteins, which are inexpensive and abundant, are emerging as promising active agents among plant extracts. Kidney bean protein isolates gave the chitosan films flexibility and a low surface free energy, resulting in a high hydrophobicity, which is critical for the packaging of dry, moisture-sensitive foods [26]. Rambabu et al. [27] have stated that the shelf life of packed cashew nuts is extended by 28 days from the normally packed cashew nut foods which are just a quite few days. It was discovered that adding mango leaf extract to the chitosan-based food film packet can increase the nuts oxidation resistance by 56% along with the superior increment in the moisture and mechanical barrier properties [27]. The color of an intelligent chitosan film can also able to preserve purple-fleshed sweet potato extract as fresh as all the pH varying from 3–10 (yellow to dark brown color) [28].

Quercetin (a flavonoid found in plants) is a complex similar to starch and can be used with the chitosan film in the food packaging area due to its long preservation days around 56 days capabilities of quercetin–starch-based films were also listed as the latest innovation. The chitosan films have improved antibacterial and antioxidant activity that builds them best for food packing [29].

11.4 Future Research Directions and Developments

11.4.1 Chitin/Chitosan Derivatives and Their Interactions with Microorganisms: A Comprehensive Review and Future Perspectives

Chitosan, which is made by deacetylating chitin, one of the most common naturally occurring polymers, has antibacterial effects against fungi and bacteria. It can also be used in various areas, including food, biomedicine, biotechnology, agriculture, and pharmaceuticals. According to research, the antibacterial activity of chitosan is dependent on various parameters, including pH, temperature, MW, ability to chelate metals, degree of deacetylation, chitosan source, and the type of bacterium involved. The review will focus on chitosan and its derivatives' antimicrobial characteristics in vitro and in vivo, as well as their mechanism of action in the treatment of infectious animal diseases and their usefulness in food safety. They wrap up with a list of the issues that come with using chitosan and its derivatives [30].

The following characteristics of future advances in chitosan-based films should be considered are as follows:

a) More chitosan derivatives and functional materials must be researched and manufactured to fulfill all the specific needs.
b) To move closer to industrial production, these existing technologies might be improved or coupled with additional favorable technologies. In addition, new technologies are still needed to mass-produce chitosan-based films with the requisite characteristics for various packaging applications.
c) To carry out a thorough toxicity study of chitosan-based films during they come into contact with food is required before they can be used commercially.
d) To investigate more research on the environmental effects of chitosan in the real-world-based food industry.

11.4.2 A Future Perspective in Crop Protection: Chitosan and its Oligosaccharides

Because of its biocompatibility, biodegradability, and bioactivity, chitosan and its oligosaccharides have generated a lot of interest for possible applications in agriculture, biomedicine, and biotechnology. Although synthetic bactericide treatment has long been the primary means of disease control, there is growing international concern about the indiscriminate application of synthetic compounds to crops due to the potential for harm to human health and the emergence of pathogen resistance to bactericides. As a result, there is a global tendency to investigate new options to limit the use of synthetic chemical agents. Based on the most recent research findings, this review focuses on the biological activities of chitosan and its oligosaccharides. Chitosan and chitooligosaccharides (COS) are two new families of biological macromolecules that are becoming more important in the study (COS). Due to their antibacterial properties, both become intriguing alternative treatments. They can be used in a variety of industries, including food, pharmaceuticals, and agriculture. COS has been shown to induce a variety of defense responses in plants in response to microbial infections, helping to minimize the harmful impact of diseases on agricultural yield and quality. Research summarizes the properties and applications of chitosan and its oligosaccharides, with an emphasis on plant bacterial diseases [31].

11.4.3 Chitosan in Molecularly-Imprinted Polymers: Current and Future Prospects

Because of its low-cost and high amount of amino and hydroxyl functional groups, chitosan is commonly employed in molecular imprinting Technology (MIT) as a functional monomer or supporting matrix. Chitosan is a prospective alternative to traditional functional monomers because of its several great qualities, including biodegradability, nontoxicity, biocompatibility, and suitable mechanical and physical properties. Chitosan molecularly imprinted polymers have recently received a lot of attention and have shown great promise in a variety of sectors, including pollution control, medicine, protein separation, and identification and chiral chemical separation [32].

11.4.4 Crosstalk Between Chitosan and Cell Signaling Pathways

In the current decade, the area of tissue engineering (TE) is at its most fascinating. Recent advances in TE have allowed it to be applied in clinical settings. TE uses biomaterial scaffolds based on chitosan and is easily able to establish suitable linkage between the backbone for its tissue regeneration and to regenerate damaged tissues. Chitosan is a popular biomaterial with several promising properties, including biocompatibility, antibacterial activity, and biodegradability. They explore the crosstalk between CS and several cell types in this review to offer a roadmap for more successful use of this polymer in TE and the production of regenerative medicine in the future [33].

11.4.5 Resorbable Chitosan Matrix – As a Promising Biomaterial for the Future

Chitosan is a naturally occurring polysaccharide with structural similarities to glycosaminoglycans that appears to be bioabsorbable and nontoxic. The chitosan matrix can be used in biomedical applications that include hemodialysis membranes, bilirubin removal, charcoal-encapsulated chitosan beads (ACCB), and drug delivery. As a result, chitosan is a feasible material that may be produced in a variety of shapes, such as granules, tablets, beads, and membranes, depending on the application. Among all of these biomedical applications, chitosan appears to be a promising future candidate for tissue interface anchoring [34].

11.5 Conclusions

In the field of packaging, biobased polymeric materials have a lot of promise for replacing petroleum-based packaging materials, which constitute a severe environmental and human health concern. Technology advancements are allowing the growth of chitosan-based films with a variety of qualities that could be utilized as food packaging materials in the future. Packaging applications such as barrier films, antibacterial films, and sensing films are the most applied fields for the usages of chitosan-based films because of the reason that chitosan is biocompatible, biodegradable, antibacterial and it also possesses low cytotoxicity. However, further research is required to improve chitosan-based films that can assemble all the practical needs of the food packaging industry.

References

1 Priyadarshi, R. and Rhim, J.W. (2020). Chitosan-based biodegradable functional films for food packaging applications. *Innovative Food Science & Emerging Technologies* 62: 102346.
2 Berger, K.R. (2003). A brief history of packaging. *EDIS* 2003 (17).
3 Risch, S.J. (2009). Food packaging history and innovations. *Journal of Agricultural and Food Chemistry* 57 (18): 8089–8092.
4 Marsh, K. and Bugusu, B. (2007). Food packaging - roles, materials and environmental issues: scientific status summary. *Journal of Food Science* 72 (3): R39–R55.
5 US EPA (2014). Municipal solid waste generation, recycling, and disposal in the United States: facts and figures for 2012. US Environmental Protection Agency, pp. 1–13.
6 Elieh-Ali-Komi, D. and Hamblin, M.R. (2016). Chitin and chitosan: production and application of versatile biomedical nanomaterials. *International Journal of Advanced Research* 4 (3): 411.
7 Guan, G., Azad, M., Kalam, A. et al. (2019). Biological effects and applications of chitosan and chitooligosaccharides. *Frontiers in Physiology* 10: 516.
8 Rane, K.D. and Hoover, D.G. (1993). Production of chitosan by fungi. *Food Biotechnology* 7 (1): 11–33.
9 White, S.A., Farina, P.R., and Fulton, I. (1979). Production and isolation of chitosan from Mucor rouxii. *Applied and Environmental Microbiology* 38 (2): 323–328.
10 Tripathi, S., Mehrotra, G.K., and Dutta, P.K. (2008). Chitosan-based antimicrobial films for food packaging applications. *E-Polymers* 8 (1): 1–7.
11 Priyadarshi, R., Riahi, Z., Rhim, J.W. et al. (2021). Sulfur quantum dots as fillers in gelatin/agar-based functional food packaging films. *ACS Applied Nano Materials* 4 (12): 14292–14302.
12 Wang, H., Qian, J., and Ding, F. (2018). Emerging chitosan-based films for food packaging applications. *Journal of Agricultural and Food Chemistry* 66 (2): 395–413.

13 Kumar, S., Mukherjee, A., and Dutta, J. (2020). Chitosan-based nanocomposite films and coatings: emerging antimicrobial food packaging alternatives. *Trends in Food Science & Technology* 97: 196–209.

14 Kim, K.M., Son, J.H., Kim, S.K. et al. (2006). Properties of chitosan films as a function of pH and solvent type. *Journal of Food Science* 71 (3): E119–E124.

15 Butler, B.L., Vergano, P.J., Testin, R.F. et al. (1996). Mechanical and barrier properties of edible chitosan films as affected by composition and storage. *Journal of Food Science* 61 (5): 953–956.

16 Pelissari, F.M., Yamashita, F., and Grossmann, M.V.E. (2011). Extrusion parameters related to starch/chitosan active films properties. *International Journal of Food Science & Technology* 46 (4): 702–710.

17 van den Broek, L.A., Knoop, R.J., Kappen, F.H., and Boeriu, C.G. (2015). Chitosan films and blends for packaging material. *Carbohydrate Polymers* 116: 237–242.

18 Sharma, R.K., Tahiliani, S., Jain, N. et al. (2013). Cynodon dactylon leaf extract assisted green synthesis of silver nanoparticles and their anti-microbial activity. *Advanced Science, Engineering and Medicine* 5 (8): 858–863.

19 Priyadarshi, R. and Negi, Y.S. (2019). Poly(vinyl pyrrolidone)-mediated synthesis of silver nanowires decorated with silver nanospheres and their antimicrobial activity. *Bulletin of Materials Science* 42 (3): 1–7.

20 Tripathi, S., Mehrotra, G.K., and Dutta, P.K. (2011). Chitosan–silver oxide nanocomposite film: preparation and antimicrobial activity. *Bulletin of Materials Science* 34 (1): 29–35.

21 Vimala, K., Yallapu, M.M., Varaprasad, K. et al. (2011). Fabrication of curcumin-encapsulated chitosan-PVA silver nanocomposite films for improved antimicrobial activity. *Journal of Biomaterials and Nanobiotechnology* 2 (01): 55.

22 Priyadarshi, R. and Negi, Y.S. (2017). Effect of varying filler concentration on zinc oxide nanoparticle embedded chitosan films as potential food packaging material. *Journal of Polymers and the Environment* 25 (4): 1087–1098.

23 Hu, D., Wang, H., and Wang, L. (2016). Physical properties and antibacterial activity of quaternized chitosan/carboxymethyl cellulose blend films. *LWT- Food Science and Technology* 65: 398–405.

24 Roy, S. and Rhim, J.W. (2020). Preparation of carbohydrate-based functional composite films incorporated with curcumin. *Food Hydrocolloids* 98: 105302.

25 Priyadarshi, R., Kumar, B., Deeba, F. et al. (2018). Chitosan films are incorporated with Apricot (*Prunus armeniaca*) kernel essential oil as active food packaging material. *Food Hydrocolloids* 85: 158–166.

26 Ma, W., Tang, C.H., Yang, X.Q., and Yin, S.W. (2013). Fabrication and characterization of kidney bean (*Phaseolus vulgaris* L.) protein isolate-chitosan composite films at acidic pH. *Food Hydrocolloids* 31 (2): 237–247.

27 Rambabu, K., Bharath, G., Banat, F. et al. (2019). Mango leaf extracts incorporated chitosan antioxidant film for active food packaging. *International Journal of Biological Macromolecules* 126: 1234–1243.

28 Yong, H., Wang, X., Bai, R. et al. (2019). Development of antioxidant and intelligent pH-sensing packaging films by incorporating purple-fleshed sweet potato extract into chitosan matrix. *Food Hydrocolloids* 90: 216–224.

29 Yadav, S., Mehrotra, G.K., Bhartiya, P. et al. (2020). Preparation, physicochemical and biological evaluation of quercetin-based chitosan-gelatin film for food packaging. *Carbohydrate Polymers* 227: 115348.

30 Riaz Rajoka, M.S., Mehwish, H.M., Wu, Y. et al. (2020). Chitin/chitosan derivatives and their interactions with microorganisms: a comprehensive review and future perspectives. *Critical Reviews in Biotechnology* 40 (3): 365–379.

31 Katiyar, D., Hemantaranjan, A., Singh, B., and Bhanu, A.N. (2014). A future perspective in crop protection: chitosan and its oligosaccharides. *Advances in Plants & Agriculture Research* 1 (1): 1–8.

32 Long, X., Yun-An, H., Qiu-Jin, Z., and Chun, Y. (2015). Chitosan in molecularly imprinted polymers; current and future prospect. *Open Access International Journal of Molecular Sciences* 16: 18328–18347.

33 Farhadihosseinabadi, B., Zarebkohan, A., Eftekhary, M. et al. (2019). Crosstalk between chitosan and cell signaling pathways. *Cellular and Molecular Life Sciences* 76 (14): 2697–2718.

34 Chandy, T. and Sharma, C.P. (1995). Resorbable chitosan matrix-a promising biomaterial for the future. In: *Proceedings of the 1995 Fourteenth Southern Biomedical Engineering Conference*, 282–285. IEEE.

12

Effect of Natural Materials on Thermal Properties of Food Packaging Film: An Overview

H. M. Prathibhani C. Kumarihami[1], Nishant Kumar[2], Pratibha[3], Anka T. Petkoska[4], and Neeraj[2]

[1] *University of Peradeniya, Department of Crop Science, Faculty of Agriculture, Peradeniya, 20400, Sri Lanka*
[2] *National Institute of Food Technology Entrepreneurship and Management (NIFTEM), Department of Food Science and Technology, Plot No. 97, Sector- 56, HSIIDC, Industrial Estate, Kundli, Sonepat, Haryana 131028, India*
[3] *National Institute of Technology, Department of Applied Sciences and Humanities & Management, Plot No. FA7, Zone, P1, GT Karnal Road, Delhi 110036, India*
[4] *St. Kliment Ohridski University-Bitola, Faculty of Technology and Technical Sciences, Dimitar Vlahov, 1400 Veles, Republic of North Macedonia*

Abbreviations

ΔH_m	melting enthalpy
AgNP	silver nanoparticle
BEO	basil leaf essential oil
CMC	carboxymethyl cellulose
CO_2	carbon dioxide
DEG	diethylene glycol
DSC	different scanning colorimetric
EG	ethylene glycol
EOs	essential oils
EVA	ethylene vinyl alcohol
FPI	fish protein isolate
FSG	fish skin gelatin
GEO	ginger essential oils
GRAS	Generally Recognized as Safe
HDPE	high-density polyethylene
LDPE	low-density polyethylene
MMT	montmorillonite
NC	nano clays

Natural Materials for Food Packaging Application, First Edition.
Edited by Jyotishkumar Parameswaranpillai, Aswathy Jayakumar,
E. K. Radhakrishnan, Suchart Siengchin, and Sabarish Radoor.
© 2023 WILEY-VCH GmbH. Published 2023 by WILEY-VCH GmbH.

NPs	nanoparticles
O_2	oxygen
PA	polyamide
PEG	poly(ethylene glycol)
PET	polyethylene terephthalate
PG	propylene glycol
PP	polypropylene
PS	polystyrene
PVA	polyvinyl alcohol
PVC	polyvinyl chloride
T_{cc}	crystallization temperature
T_g	glass transition
TGA	thermogravimetric analysis
T_m	melting temperature
T_o	oxidation temperature
XRD	X-ray powder diffraction
ZnONP	ZnO nanoparticles

12.1 Introduction

Food packaging system are played salient roles in the food processing; provides safety from physical, biological, and chemical hazards, helps in prolonging the shelf life of food products during the handling, transportation, and storage with ensuring food safety and quality [3, 47]. Four basic principles of food packaging, including containment, protection, convenience, and communication have been accepted widely [5, 64]. Containment is one of the fundamental functions of packaging of food products; it is crucial for the food products during transportation, storage, and distribution in addition to the fact that it also serves to determine portion sizes, stock keeping, and merchandizing. Moreover, the application of packaging has significant impacts on the shelf life of the food products by reducing the oxidation, barrier against water and gas transmission; minimize growth of microbial organism, physical damage, and accumulation of volatile compounds [18, 38, 39]. Food quality attributes such as color, texture, aroma, and flavor are significantly affected by the type of packaging that should also enable the food safety aspects [34, 86]. The mail functions of food packaging are graphically presented in Figure 12.1. Food packaging is necessary for providing convenience to the consumer through sizes, easy handling, easy opening and dispensing, resealability, and food preparation in the package [76]. Additionally, food packages display product information (communication), usually including the names and quantity of ingredients, nutritional data, food processor, manufacturing date, expiration date, country of origin, and a barcode for easy product traceability in the supply chain [64].

Different food products require a variety of food packaging materials. The primary functions of the package are determined by the physicochemical characteristics of both the packaging type as well as of food products. Various types of packaging

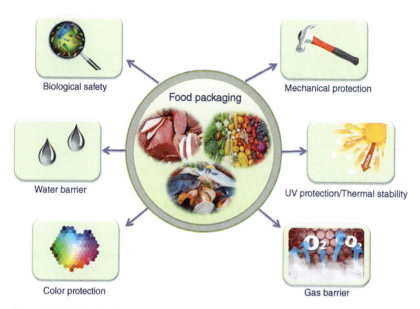

Figure 12.1 The fundamental functions of food packaging. Source: Burak Altan Alisan/Shutterstock; Bojsha / Adobe Stock; PublicDomainPictures/Pixabay; Amazon.com, Inc.

materials are used in the food packaging industry, including metals, glass, polymers, paper, plastics, and their composites. Examples of synthetic-based polymers are high-density polyethylene (HDPE), low-density polyethylene (LDPE), polyamide (PA), polypropylene (PP), polyethylene terephthalate (PET), polystyrene (PS), ethylene vinyl alcohol (EVA), and polyvinylchloride (PVC) are widely used in food packaging. However, these materials are nonbiodegradable and have serious impacts on the environment, climate changes, and biodiversity. Therefore, the need of materials from renewable resources to produce biodegradable packaging as an alternative to synthetic packaging is very important and should be our priority. Natural plant and animal biopolymers such as polysaccharides, proteins, lipids, and their composites have been used to develop environmentally friendly and nontoxic food packaging materials. In addition, the incorporation of active ingredients such as plant extracts, essential oils (EOs), and other bioactive compounds has influenced and enriched the properties of packaging materials [35, 96] preparing them as novel and active types of packaging.

12.2 Biodegradable Films: An Alternative for Food Packaging

Interdisciplinary research and development in the field of food packaging is promoted in many institution worldwide, especially in the area of biodegradable polymers intended for food packaging. They exhibit unique properties such as film and/or coating formation, suitable mechanical, barrier, or optical properties. Biodegradable films can be made from a different biopolymer sources such as polysaccharides, proteins, and lipids or their combinations. They are origin is from agricultural, marine, or animal sources, and present a suitable alternative to synthetic polymer-based materials [5]. In general, film is defined as a thin layer of biopolymer material that can be used as a film wrapping over food items or in some cases between food ingredients. Moreover, it possesses suitable barrier properties against moisture, gases (CO_2, O_2), volatile compounds, and certain mechanical properties. Biodegradable films can exhibit also suitable antibacterial properties. There are a lot of cases where biodegradable films have been created and applied in the packaging of various foods such as fruits, vegetables, cheese, meat, and their products. Today, the incorporation of food additives with active role(s) such as antioxidants, antibacterial agents, pigments, nutrients, and flavoring agents, into packaging is novel with aim to optimize the food quality and safety. In addition, plasticizers and other additives could be added for film formation that could improve the structure, flexibility, and/or elasticity of the film.

12.2.1 Biodegradable Polymers

Biopolymers such as polysaccharides, proteins, and lipids derived from natural sources are used to develop food packaging in form of films or coatings [5, 24, 26, 31, 76]. Natural polysaccharides (cellulose, starch, agar, gums, etc.) and their derivatives (pectin, chitosan, alginate, carrageenan, etc.) have been widely used in film-making due to their abundance and low toxicity [5, 34]. They are edible, naturally degradable, act as a barrier to gases (CO_2, O_2) and have excellent mechanical strength, but polysaccharide materials are very fragile and hydrophilic. On the other hand, proteins can be found in several sources at low cost and are among the most abundant polymers in nature, therefore protein-based films are widely used in film manufacturing due to their film-forming and nutritional properties. Films made from lipids have barrier properties against water vapor but lack sufficient structural integrity and durability. They are less polar, which results in poor gas barrier properties. Therefore, lipids can be incorporated into polysaccharides and proteins to obtain bilayer or multilayer hydrocolloid membranes. Waxes, fatty acids and alcohols, cocoa-based compounds, and their derivatives are commonly used representatives of lipid-based hydrophobic films. Additionally, natural antimicrobial and/or antioxidant compounds can be blended with biodegradable polymers to develop bioactive coatings; in fact biopolymers act as a carrier for various additives and active ingredients. Active packaging reduces the growth of microorganisms, acts against oxidants, preserves food safety, and/or could improve its quality. Biodegradable and natural substances are often most attractive and used in active packaging due to their positive benefits to human health as well as their eco-friendly nature [86, 87]. Incorporation of natural antioxidants and antimicrobials to biopolymer-based packages has been extensively studied by many researchers worldwide over the past decades. Addition of other types of additives such as anti-browning agents, antioxidants, prebiotics, probiotics, texture enhancers, colorants, and flavors in packaging materials can improve the functional and organoleptic properties of foods throughout long-term storage. The properties such as mechanical properties (tensile strength, elongation at break), thickness, gas barrier, water barrier, water solubility, and thermal properties are very important in the selection and application of biopolymer-based packages. These properties are affected and could be custom designed by addition of functional ingredients such as antimicrobial agents, antioxidants, plasticizers, cross-linkers, and texturizing agents [29].

12.3 Thermal Properties of Food Packaging

Thermal properties and stability are important properties of polymeric materials intended for food packaging. It can also be defined as the ability of a polymeric material to withstand heat and maintain the properties such as mechanical, barrier, and textural properties in food packaging [11, 71]. The pyrolysis profile of polymer-based packaging materials can be determined using various methods such as differential scanning colorimetric (DSC) and thermogravimetric methods [6, 37]. These methods offer the possibility to evaluate the temperature properties and corresponding phase change enthalpies of packaging materials. DSC analysis shows the melting temperature (T_m) as the endothermic peak of the DSC curve while the melting enthalpy (ΔH_m) can be determined from the area of the peak. Additionally, the glass transition temperature (T_g) can be viewed as the average derivative of the starting and ending temperatures and the thermal stability of the polymer film [14, 39]. Thermogravimetric analysis (TGA) is commonly applied to evaluate the weight change of polymer films under temperature conditions under a controlled nitrogen atmosphere [14]. Moisture loss, release of phenols, and other volatile compounds usually occur at increased temperatures. Therefore, TGA measurement is important and it usually ranges from 0 to 1000 °C. These tests should be performed under controlled conditions [54]. In general, the thermal stability of polymer films depends on variables such as crystallinity, chemical structure, and molecular weight of the tested materials [33]. Additionally, incorporating other additives and ingredients (plant extracts, EOs, colorants, flavors, nanomaterials, plasticizers, and emulsifiers) can positively or negatively affect the thermal properties of polymer films. These additives provide intermolecular and intermolecular interactions such as ionic, hydrogen covalent, and non-covalent [51]. Mohamed et al. [50] showed the decomposition step and mass loss (%) of thermal behavior of natural composite films of carboxymethyl cellulose (CMC) mixed with gelatin, shellac, and glycerol. Namely, 50 : 50% gelatin and CMC composite films were reported to have minimal weight loss up to 25% at 237 °C and 71% at 756 °C (71%), indicating that these films possess high thermal stability as compared to other composites. The thermal stability of gelatin/CMC composite films is improved by good distribution of gelatin in the matrix. TGA of carboxymethyl cellulose-grafted polyacrylamide (CMC-g-PAm) copolymer shows that thermal decomposition of the CMC segments occurs between 241 and 343 °C with a weight loss of 13%. In addition, pure CMC consumes two pyrolysis steps; a weight loss of 18% between 265 and 320 °C and a weight loss of 57% between 320 and 450 °C [50]. They perform higher thermal stability of natural composite films made of gelatin/CMC than CMC-g-PAm copolymer, styrene butadiene rubber (SBR)–CMC, and CMC-g-PDMAEMA due to better dispersion of gelatin up to 756 °C (71%). A composite film containing 20 wt% shellac: 80 wt% CMC showed an average mass loss (44%) at 230 °C and reached 78% at 382.5 °C. On the other hand, the 2 wt% shellac: 65 wt% gelatin: 33 wt%, indicated a higher thermal stability of the produced films. Addition of gelatin to shellac/CMC composites improved the heat stability of shellac. Shellac:CMC-containing gelatin composite films have improved thermal stability due to better dispersion of shellac in the presence of gelatin. Incorporation of gelatin in composites increases the hydrogen bonding interactions between gelatin/CMC and glycerol, and between shellac, gelatin, and glycerol and CMC, thus improving thermal stability. The formation of CMC–gelatin or CMC/shellac/gelatin complexes with multiple hydrogen bonds between numerous carbonyl groups (C=O) and (NH)-peptide groups of shellac and CMC, shellac, enhanced heat stability by gelatin. Therefore, gelatin lubricates and fills voids in CMC or shellac/CMC and improves the thermal stability of the resulting composite [50]. On other hand, Ganeson et al. [23] investigated the effects of cinnamon oil-based nano-emulsion on the gelatin-based edible films and reported that the 70:30 ratio of cinnamon oil nano-emulsion and gelatin was significantly increased the thermal stability of edible films. It showed the thermal degradation range between 121 and 129.7 °C of cinnamon oil-based nano-emulsions edible films, it was reported higher as compared to pure cinnamon oil (63–119 °C).

12.4 Effects of Natural Materials on the Thermal Stability of Food Packaging

12.4.1 Effects of Plant Extract

Generally, the addition of synthetic-based antioxidant agents can raise a concern about food safety and quality. The natural plant extract is considered an excellent antioxidant agent, which has different types of polyphenols, flavonoids, and anthocyanin compounds [36, 40, 70]. The incorporation of these compounds as antimicrobial and antioxidant agents can improve the biological activities of food packaging as well as the application of food products. They are safe and nontoxic [2, 19, 42]. The incorporation of plant extract as a natural antioxidant and antimicrobial agent could be responsible for improving the biological activities of food packaging but it affected the mechanical and thermal properties of food packaging as well [37, 46, 52, 95]. Therefore, the increasing concentration of natural plant extracts generally, affected the thermal stability of the food packaging due to disturbed crystallinity and structure of polymers; the molecular interaction between the natural additive and polymer also affected the thermal stability of the food and bio-based packaging [39]. Several researchers have investigated the effects of plant and other natural extracts on the thermal properties of food packaging for example, Sun et al. [77, 78]

reported that the incorporation of natural materials such as epigallocatechin gallate and mulberry anthocyanin extract improved the thermal stability of glucomannan/carboxymethyl chitosan biocomposite films and glucomannan/chitosan-based active packaging. The thermal stability of CMC [62] and chitosan-based biodegradable packaging [61] were investigated by DSC and TGA methods; they have found that thermal stability decreases with increasing concentration of *Allium tuberosum* as natural additives. This might be possible due to disturbed internal crystallinity and structure of materials [59]. A similar statement has been reported by Sun et al. [79]; they reported that the increasing polyphenol concentration (0.25%, 0.50%, 0.75%, and 1.0%) of thinned young apple is responsible for degrading the thermal stability (303.41, 300.99, 297.05, 296.65 °C) of the chitosan-based film due to the chemical etching, degradation of macromolecules because of the bond disruption [56]. The incorporation of protocatechuic acid grafted reduced the thermal stability of chitosan-based film with significantly higher weight loss [43]. The increased concentration of pomegranate peel extract [39] and pequi peel extract [12] showed the reducing thermal stability of chitosan-based film due to decreasing plasticity and crystallinity of the biopolymer. The slight reduction in thermal stability of thermoplastic-based food packaging after the inclusion of grape cane extract was reported by Díaz-Galindo et al. [22]. On the contrary, Wang et al. [92] reported that the addition of anthocyanin into chitosan hydrochloride and chitosan CMC-based noncomplex improved the thermal stability along with the mechanical and biological properties of the packaging material. On the other side, Matta Fakhouri et al. [49] reported that the increasing concentration of cranberry powder in arrowroot starch and gelatin-based edible packaging reduces the glass transition temperature of packaging due to reducing crystallinity confirmed by X-ray diffraction (XRD) analysis. The addition of curcumin (over 4%) [44] and rosemary extract [57] as a natural polyphenol in edible packaging was reported that the natural additives affected the thermal and mechanical strength of the packaging materials.

12.4.2 Effects of Essential Oils

Essential oils (EOs) are volatile oils extracted from various parts of aromatic plants (barks, seeds, flowers, skins, fruits, roots, leaves, trees, fruits, and whole plants) with multiple biological properties [63, 72]. EOs can be extracted by a variety of methods including hydro distillation, steam distillation, hydro diffusion, and solvent extraction [20, 63, 72]. *Azadirachta indica* (neem), *Lavandula angustifolia* (lavender), *Thymus vulgaris* (thyme), *Eucalyptus globulus* (eucalyptus), *Cinnamomum zeylanicum* (cinnamon), *Syzygium aromaticum* (clove), *Citrus limonum* (lemon), *Melaleuca alternifolia* (tea tree), *Ocimum basilicum* (basil), *Matricaria chamomile* (chamomile flower), *Eletaria cardamom* (cardamom seed), *Rosmarinus officinalis* (rosemary), *Brassica nigra* (mustard) are examples of different types of EOs [4, 58, 63, 72]. They are widely used as preservatives, flavorants, and stabilizers in the food and pharmaceutical industries due to their natural antimicrobial, antioxidant, or bio preservative properties [63]. Volatile and phenolic compounds in EOs play a role in inhibiting microbial growth, but antioxidant and antibacterial properties are also enhanced by

EO, helping to extend the shelf life of foods [63, 75]. In recent years, EOs have been widely used as a stabilizer, antibacterial agent, and antioxidant and incorporated into food and packaging materials. Various methods including emulsification or homogenization can be used to incorporate EOs into edible films and coatings. Low thermal stability and high volatility are the main limitations of EOs as an antimicrobial agent incorporated into polymer matrices [20]. A film with an antimicrobial effect was produced based on linear low-density polyethylene (LLDPE) and EO-thymol-incorporated montmorillonite (MMT). The thermal stability of EOs during melt processing was greatly enhanced by the addition of MMT incorporated together with the blowing agent [20]. The thermal properties of the three antimicrobial films were evaluated by Efrati et al. [20] uses DSC to evaluate the effects of his EO and/or nanoclays (NC) on the thermal properties of films compared to the pure polymer (LLDPE). Therefore, this technology allows the fabrication of antimicrobial nanocomposite packages with high concentrations of volatiles as active ingredients. Syafiq et al. [80] investigated effect of various levels of cinnamon essential oil (0–2% CEO) on flammability, thermal stability, and morphological properties of starch-biopolymer-based composites edible films reinforced with nano cellulose fibers. In these CEO-containing composite films, the number and size of microporous holes tended to increase with increasing CEO concentration. Subsequently, the loading of CEO significantly improved the thermal stability compared to the pure composite. This could be attributed to the increase in the crystallinity index with increasing CEO concentration. It has been reported that the thermal stability is proportional to the crystallinity, and the higher crystallinity resulted thermal stability. Furthermore, it has been reported that the linear burning rate of composites decreased with increasing CEO loading [80]. Ma et al. [45] also successfully prepared gelatin–olive oil composite films and concluded that the inclusion of olive oil in the film matrix increased the melting point (T_m) of the helix-coil transition of gelatin. Olive oil formed an immiscible emulsified phase within the film [45]. Noshirvani et al. [53] also studied the effects of cinnamon and ginger oils on the physical, mechanical, thermal, and water vapor transmission properties of oleic acid-emulsified chitosan–CMC films. They reported that film crystallinity and thermal stability decreased with increasing cinnamon oil concentration. However, the lowest concentration of ginger essential oil (GEO) resulted in low thermal stability, whereas the higher added amount resulted in higher thermal stability [53]. Addition of high concentrations of GEO may reorganize the polymer structure. The structure of the polymer can be modified with minimal GEO while adding EO results in a more uniform structure and improved thermal stability. Therefore, as the concentration of GEO increases, the crystallinity index increases. Furthermore, various properties of films, such as physical, mechanical, thermal, and water vapor permeability, depending on the essential oil composition [53].

12.4.3 Effects of Color Agent

Intelligent and smart packaging materials developed based on natural dyes and biodegradable films have the potential to replace traditional food packaging [41]. Natural pigments such as anthocyanins, curcumin, betacyanin, carminic acid, carotenoids, chlorophyll, alizarin, betalain, and shikonin are extracted from plant and vegetable/fruit sources [10, 66, 67]. They are economical, safe, nontoxic, and delicate natural pigments that act as quality indicators for smart packaging and facilitate recycling of food waste in packaging systems. These dyes change color under the influence of external factors such as pH, temperature, UV light, and humidity, giving the material smart packaging properties. Color change indicates material durability or quality [41, 66, 67]. These packaging films can therefore be optimized and commercialized as active and intelligent packaging for the visual quality assessment of fresh produce. Latos-Brozio and Masek [41] made biodegradable polymers (polylactide and polyhydroxybutyrate) with added natural colorants (β-carotene, chlorophyll, curcumin, lutein) commonly used in the food industry. They reported that the addition of natural dyes did not change the temperature range of the samples, such as T_g, crystallization temperature (T_{cc}), and melting temperature (T_m). However, a significant increase in oxidation temperature (T_o) was observed. This may be due to the addition of stabilizing substances to the colorant, which are not indicated in the composition of the product. In addition, the antioxidant pigments lead to higher oxidation temperatures, increased material stability, and improved resistance to oxidation. However, the addition of food coloring did not significantly alter the thermal stability of the polymer material [41]. The addition of the natural dyes anthocyanins and shikonins slightly increased the mechanical strength and improved the water vapor barrier properties of CMC/agar-based halochromic films without altering the thermal stability of the films [67]. Similar results were previously reported by Roy et al. [66] in CMC/agar-based films combined with shikonin and cellulose nanocrystals. Addition of shikonin provided functional properties, including pH-dependent color change, and antioxidant and antimicrobial properties, without significantly altering the film's water vapor transmission rate and thermal stability. Liu et al. [44] prepared a colorimetric film based on κ-carrageenan, a natural hydrocolloid extracted from red algae and containing curcumin, to monitor the freshness of pork and shrimp. κ-carrageenan and curcumin interact through hydrogen bonding, and a small amount of curcumin ($\leq 3\%$) can be well dispersed in the κ-carrageenan matrix. These properties significantly improved the barrier, mechanical and thermal stability of smart films. Incorporated curcumin may improve the thermal stability of the κ-carrageenan matrix due to the presence of hydrogen bonding. Some of the curcumin incorporated into the κ-carrageenan matrix was also present in crystalline form. When the anthocyanin-rich purple sweet potato extract was incorporated into the chitosan matrix, the thickness of the chitosan film, water solubility, UV–vis light barrier properties and thermal stability may be significantly improved [97]. Similar results were reported for antioxidant and pH-sensitive films developed by incorporating anthocyanin-rich black soybean seed hull extract into chitosan films [92]. TGA showed excellent thermal stability of chitosan/polyvinyl alcohol (PVA) films containing bentonite and anthocyanins [32].

12.4.4 Effects of Nanomaterials

The great potential of nanotechnology has attracted attention to develop promising and desirable materials in food packaging systems. Nanocomposite has been developed to improve the functional properties of food packaging, such as barrier performance, mechanical strength, and thermal stability. Additionally, nanomaterials can include bacteriostatic, antioxidants, plant extracts, and enzymes to extend the shelf life of foods [27]. Applications of nanocomposite materials in the food industry include enhancement of the original properties of polymer-based packaging system properties such as mechanical, barrier properties, and stability under various temperature and humidity conditions [27]. Nanomaterials used in food packaging include inorganic materials (metals and metal oxides, clay nanoparticles [NPs], and nanofibers embedded in bio nano-composite films) and organic materials such as chitosan, phenols, quaternary ammonium salts, natural polysaccharides and proteins-based polymers [27, 28]. Bionanocomposites have been introduced as a new generation of nanocomposite films made from biopolymers [8, 9, 28, 87]. They consist of a natural polymer matrix and organic/inorganic fillers with at least one dimension in the nanometer range [8, 9]. The excellent interfacial interaction of nanofillers on the polymer branches greatly improved the thermal, mechanical, and water barrier properties of the polymer due to the large specific surface area and high surface energy of the polymer [8, 9, 28]. The size of nanofillers used and the uniformity of distribution and dispersion of nanoparticles (NPs) within the matrix has a significant impact on the performance of nanocomposite films and packaging [28].

Active nanocomposite films based on fish protein isolate (FPI) and fish skin gelatin (FSG) incorporating basil leaf essential oil (BEO) and ZnO nanoparticles (ZnONP) significantly affected the thermal stability of the packaging films [9]. Bio nanocomposite films developed based on FSG and bimetallic Ag–Cu nanoparticles (Ag–Cu NPs) have significantly improved thermal stability, mechanical properties, and antibacterial activity against both gram-positive and gram-negative bacteria have been shown to be effective. Ag–Cu NPs within the FSG matrix may act as a thermal insulator, leading to retarded pyrolysis and enhanced thermal stability of the composites [8]. CMC/ZnO nanocomposite films with high concentrations of ZnO NPs showed improved thermal stability compared to control films [21]. Films containing ZnO NPs incorporated directly into solutions of chitosan (CS) and PVA polymers with varying concentrations of the plasticizer Tween 80 (T80) showed better thermal stability compared to CS/PVA [90]. The improved thermal stability of biopolymer films with ZnO is attributed to the reduction of interatomic distances, so more energy is required to degrade these films [90]. Yu et al. [99] also reported that CMC–sodium nanocomposites-containing ZnO NPs exhibited better thermal stability than control CMC films due to the interaction between ZnO and CMC. A gelatin/MMT hybrid nanocomposite prepared by Zheng et al. [100] and reported that the MMT significantly improve the thermal properties of gelatin-based composite. The thermal stability of nanocomposite films was mainly attributed to the thermal stability of nanomaterials in the gelatin matrix. On other side, silver nanoparticles (AgNPs) are considered most attractive antibacterial agents, and PVA film-coated with AgNPs silk using natural silk as a reducing agent and stabilizer has excellent thermal stability, mechanical, and antimicrobial activity [81]. These results provided evidence that the application of biodegradable nanocomposite films leads to more efficient thermal properties.

12.4.5 Effects of Plasticizers

The thermal properties of the polymers-based food packaging are generally depending on the nature of the material. The thermal degradation of the food packaging is caused by shear force and lower molecular interactions which lead to decreased crystallinity of the polymers [55]. To overcome the problem of thermal degradation of food packaging, plasticizers such as glycerol, sorbitol, poly(ethylene glycol) (PEG), citrate, polyols, triacetine, oligomeric ester amides, and others, i.e. fatty acid derivatives are incorporated with the matrix to improve the thermal stability of the packaging [82, 91]. Plasticizers are low molecular weight nonvolatile substances, potential to reduce the biopolymer chain interaction which resulting enhances the thermal stability of food packaging [1]. Plasticizers are widely used in food packaging to improve their thermoplastic properties by retrogradation process; this process helps in the prevention of amide groups of polymers into plasticization system and forms a stable hydrogen bond with bio-based polymers [55]. The compatibility of the plasticizers with polymer is a major issue for the effective plasticization of food packaging materials [83]. The plasticizers are incorporated with polysaccharide, protein, and lipid-based polymers to improve their properties. Additionally, glycerol and sorbitol plasticizers are commonly used in starch-based food packaging due to their molecular interaction and formation of strong inter molecule bonds [68, 91]. Many researchers have focused on the use of plasticizers such as glycerol, ethyl glycol, propylene glycol (PG), mannitol, xylitol, surfactants, and other fatty acids that enhance the thermal, mechanical, and other characteristics of edible or biodegradable food packaging [91]. Sanyang et al. [69] reported that the incorporation of glycerol with sugar palm starch improves the thermal stability of packaging material. Requena et al. [60] also increased the heat resistance of poly[(3-hydroxybutyrate)-co-(3-hydroxyvalerate)]-based film after incorporation of PEG 1000 and PEG 4000 of plasticizers. On the other hand, Martino et al. [48] and Courgneau et al. [17] were reported that the incorporation of plasticizers in poly(lactic acid) film significantly decreased the thermal resistance of packaging. Similarly, Jouki et al. [30] reported the increasing concentration of glycerol as a plasticizer in cress seed carbohydrate gum reduces the thermal resistance of packaging due to lower interaction between polymer and plasticizer. On the contrary, Yousefnia Pasha et al. [98] investigated the thermal stability of poly(lactic acid) film and reported that the incorporation of triacetin and PEG improved the thermal stability of the film. The polarity, bonding of hydrogen, solubility, constant, and other parameters is depending on the compatibility between the plasticizer and polymers; these properties are responsible for thermal resistance and degradation [15, 89]. The thermal stability of the poly (3-hydroxybutyrate-*co*-3-hydroxyvalerate) based film was improved after the addition of soyabean oil, epoxidized soybean oil, dibutyl phthalate, and triethyl citrate as plasticizers [15]. The polyols plasticizers such as glycerol, PG, diethylene glycol (DEG), and EG were used to enhance the thermal properties of pigskin gelatin-based film by Van de Velde and Kiekens [88]. On the other side, Cao et al. [13] reported the increasing concentration of glycerol and sorbitol improves the thermal resistance of cassia gum-based edible film. Tarique et al. [82] investigate the effects of different concentrations (15%, 30%, and 45%) on the thermal, barrier, and other characteristics of arrowroot starch-based packaging; they reported the higher thermal resistance of food packaging with glycerol. The higher homogeneity was recorded with a higher concentration of glycerol. Based on the previous investigations and their scientific evidence can be concluded that the amount and concentration of plasticizers, types of polymer, and their chemical structure desired for the thermal and other properties of packaging materials. The specific plasticizers act in different manners according to the matrix polymers.

12.4.6 Effects of Emulsifiers

The emulsifiers/surfactants are surface-active ingredients; they are the potential to protect oil–water droplets against immediate recoalescence. They are considered amphiphilic substances and able to improve the functional properties of food packaging due to the hydrophobic nature of lipids and oil substances [7, 16, 85, 94]. Several types of emulsifiers such as soy lecithin, gum arabic, agar, topica pearl, polysorbate/tween (20 and 80), carrageenan, egg lecithin, palm oil, canola oil, and mustard lecithin are used in the food processing sector to develop edible packaging. The most common surfactant, i.e. lecithin is widely used due to its higher emulsifying ability and attribute to spontaneous agglomeration [73]. The other natural emulsifier agents such as saponins, glycosides of steroid, and polycyclic terpenes are also used to decrease the surface tension of water [74]. In food packaging, the use of emulsifiers produced an electrostatic complex in the protein and polysaccharide-based emulsion. Several researchers have been evidentially reported that the incorporation of emulsifiers/surfactants is the potential to improve the functional as well as thermal properties of edible and food packaging. For example, Ghiasi et al. [25] reported that the incorporation of structure oil nanoparticles in polysaccharide matrix could help to improve the thermal and other functional properties of edible film. Generally, the thermal stability, (T_g) of the edible packaging increased by the increasing amount of bond and chain stiffness, cross-linking between the matrix, and their crystallinity [65]. Tongnuanchan et al. [85] improved the thermal resistance of the FSG-based film using palm oil as additives. They reported the addition of palm oil directly affected the thermal properties of fish gelatin-based film as compared to control. On the other side, Tongnuanchan et al. [84] also reported that the incorporation of Tween 80 and soy lecithin with EOs affected the thermal behaviors and reported thermal degradation between 301.29 and 334.48 °C of the fish gelatin film. On the basis of the previous literature, it can be concluded that the thermal behavior of the edible film and packaging is depending on the types of polymers and used emulsifiers/surfactants.

12.5 Conclusions

The application of food packaging plays a key role in extending the shelf life of food while preserving quality attributes such as color, taste, and sensory quality by minimizing the risk of oxidation, water transpiration, and microbial load. The use of naturally derived biopolymers in packaging can be used as an alternative to synthetic polymers for environmental protection and consumer health safety. Natural biopolymers such as polysaccharides, proteins, lipids, and their compositions are commonly used to design nontoxic and biodegradable packaging for food.

These biopolymers are biodegradable, inherently nontoxic, and readily available at low cost. Additionally, the incorporation of active ingredients such as EOs, emulsifiers, colorants, plasticizers, and nanomaterials into packaging materials is one of the emerging approaches to improving packaging properties. Thermal properties and stability are important properties of polymeric materials and food packaging, which indicates the heat resistance of packaging materials to maintain quality attributes of products. In the present work, most of the possible additions to biodegradable matrices intended for food packaging are evaluated; some of them showed positive effects on thermal properties of examined compositions and others did not affect the thermal properties in great content (even showed negative effects), but improved other (functional) properties of the polymer composites that suit particular food packaging applications.

References

1 Abdorreza, M.N., Cheng, L.H., and Karim, A.A. (2011). Effects of plasticizers on thermal properties and heat sealability of sago starch films. *Food Hydrocolloids* 25 (1): 56–60.
2 Akhtar, M.J., Jacquot, M., Jasniewski, J. et al. (2012). Antioxidant capacity and light-aging study of HPMC films functionalized with natural plant extract. *Carbohydrate Polymers* 89 (4): 1150–1158.
3 Alamri, M.S., Qasem, A.A., Mohamed, A.A. et al. (2021). Food packaging's materials: a food safety perspective. *Saudi Journal of Biological Sciences* 28 (8): 4490–4499.
4 Al-Hilifi, S.A., Al-Ali, R.M., and Petkoska, A.T. (2022). Ginger essential oil as an active addition to composite chitosan films: development and characterization. *Gels* 8 (6): 327.
5 Alizadeh-Sani, M., Mohammadian, E., Rhim, J.W., and Jafari, S.M. (2020). pH-sensitive (halochromic) smart packaging films based on natural food colorants for the monitoring of food quality. *Trends in Food Science & Technology* 105: 93–144.
6 Andrade Martins, Y.A., Ferreira, S.V., Silva, N.M. et al. (2020). Edible films of whey and Cassava starch: physical, thermal, and microstructural characterization. *Coatings* 10 (11): 1059.
7 Andreuccetti, C., Carvalho, R.A., Galicia-García, T. et al. (2011). Effect of surfactants on the functional properties of gelatin-based edible films. *Journal of Food Engineering* 103 (2): 129–136.
8 Arfat, Y.A., Ahmed, J., Hiremath, N. et al. (2017). Thermo-mechanical, rheological, structural and antimicrobial properties of bionanocomposite films based on fish skin gelatin and silver-copper nanoparticles. *Food Hydrocolloids* 62: 191–202.
9 Arfat, Y.A., Benjakul, S., Prodpran, T. et al. (2014). Properties and antimicrobial activity of fish protein isolate/fish skin gelatin film containing basil leaf essential oil and zinc oxide nanoparticles. *Food Hydrocolloids* 41: 265–273.

10 Balbinot-Alfaro, E., Craveiro, D.V., Lima, K.O. et al. (2019). Intelligent packaging with pH indicator potential. *Food Engineering Reviews* 11 (4): 235–244.

11 Begum, S. A., Rane, A. V., & Kanny, K. (2020). Applications of compatibilized polymer blends in automobile industry. In Compatibilization of Polymer Blends (pp. 563–593). Elsevier. Ajitha A. R. and Sabu Thomas

12 Breda, C.A., Morgado, D.L., Assis, O.B.G., and Duarte, M.C.T. (2017). Processing and characterization of chitosan films with incorporation of ethanolic extract from "pequi" peels. *Macromolecular Research* 25 (11): 1049–1056.

13 Cao, L., Liu, W., and Wang, L. (2018). Developing a green and edible film from Cassia gum: the effects of glycerol and sorbitol. *Journal of Cleaner Production* 175: 276–282.

14 Chakravartula, S.S.N., Soccio, M., Lotti, N. et al. (2019). Characterization of composite edible films based on pectin/alginate/whey protein concentrate. *Materials* 12 (15): 2454.

15 Choi, J.S. and Park, W.H. (2004). Effect of biodegradable plasticizers on thermal and mechanical properties of poly(3-hydroxybutyrate). *Polymer Testing* 23 (4): 455–460.

16 Corredig, M. (2009). Molecular understanding of the interaction of dairy proteins with other food biopolymers. In: *Dairy-Derived Ingredients* (ed. M. Corredig), 371–393. Woodhead Publishing.

17 Courgneau, C., Domenek, S., Guinault, A. et al. (2011). Analysis of the structure properties relationships of different multiphase systems based on plasticized poly(lactic acid). *Journal of Polymers and the Environment* 19: 362–371.

18 Daniloski, D., Petkoska, A.T., Lee, N.A. et al. (2021). Active edible packaging based on milk proteins: a route to carry and deliver nutraceuticals. *Trends in Food Science & Technology* 111: 688–705.

19 De'Nobili, M.D., Pérez, C.D., Navarro, D.A. et al. (2013). Hydrolytic stability of L-(+)-ascorbic acid in low methoxyl pectin films with potential antioxidant activity at food interfaces. *Food and Bioprocess Technology* 6 (1): 186–197.

20 Efrati, R., Natan, M., Pelah, A. et al. (2014). The combined effect of additives and processing on the thermal stability and controlled release of essential oils in antimicrobial films. *Journal of Applied Polymer Science 131* (15): https://doi.org/10.1002/app.40564.

21 Espitia, P.J.P., Soares, N.D.F.F., Teófilo, R.F. et al. (2013). Physical–mechanical and antimicrobial properties of nanocomposite films with pediocin and ZnO nanoparticles. *Carbohydrate Polymers* 94 (1): 199–208.

22 Díaz-Galindo, E.P., Nesic, A., Cabrera-Barjas, G. et al. (2020). Physico-Chemical and Antiadhesive Properties of Poly(Lactic Acid)/Grapevine Cane Extract Films against Food Pathogenic Microorganisms. *Polymers* 12: 2967.

23 Ganeson, K., Razifah, M.R., Mubarak, A. et al. (2022). Improved functionality of cinnamon oil emulsion-based gelatin films as potential edible packaging film for wax apple. *Food Bioscience* 47: 101638.

24 Gheorghita, R., Gutt, G., and Amariei, S. (2020). The use of edible films based on sodium alginate in meat product packaging: an eco-friendly alternative to conventional plastic materials. *Coatings* 10 (2): 166.

25 Ghiasi, F., Golmakani, M.T., Eskandari, M.H., and Hosseini, S.M.H. (2020). A new approach in the hydrophobic modification of polysaccharide-based edible films using structured oil nanoparticles. *Industrial Crops and Products* 154: 112679.

26 Gupta, V., Biswas, D., and Roy, S.A. (2022). Comprehensive review of biodegradable polymer-based films and coatings and their food packaging applications. *Materials* 15: 5899.

27 Huang, Y., Mei, L., Chen, X., and Wang, Q. (2018). Recent developments in food packaging based on nanomaterials. *Nanomaterials* 8 (10): 830.

28 Jafarzadeh, S. and Jafari, S.M. (2021). Impact of metal nanoparticles on the mechanical, barrier, optical and thermal properties of biodegradable food packaging materials. *Critical Reviews in Food Science and Nutrition* 61 (16): 2640–2658.

29 Jakubowska, E., Gierszewska, M., Szydłowska-Czerniak, A. et al. (2022). Development and characterization of active packaging films based on chitosan, plasticizer, and quercetin for repassed oil storage. *Food Chemistry* 399: 133934.

30 Jouki, M., Khazaei, N., Ghasemlou, M., and HadiNezhad, M. (2013). Effect of glycerol concentration on edible film production from cress seed carbohydrate gum. *Carbohydrate Polymers* 96 (1): 39–46.

31 Kocira, A., Kozłowicz, K., Panasiewicz, K. et al. (2021). Polysaccharides as edible films and coatings: characteristics and influence on fruit and vegetable quality—a review. *Agronomy* 11: 813.

32 Koosha, M. and Hamedi, S. (2019). Intelligent chitosan/PVA nanocomposite films containing black carrot anthocyanin and bentonite nanoclays with improved mechanical, thermal and antibacterial properties. *Progress in Organic Coatings* 127: 338–347.

33 Król-Morkisz, K. and Pielichowska, K. (2019). Thermal decomposition of polymer nanocomposites with functionalized nanoparticles. In: *Polymer Composites with Functionalized Nanoparticles* (ed. P. Krzysztof and M.M. Tomasz), 405–435. Elsevier.

34 Kumar, N. and Neeraj (2019). Polysaccharide-based component and their relevance in edible film/coating: a review. *Nutrition & Food Science* 49 (5): 793–823.

35 Kumar, N., Daniloski, D., D'cunha, N.M. et al. (2022). Pomegranate peel extract – a natural bioactive addition to novel active edible packaging. *Food Research International* 156: 111378.

36 Kumar, N., Neeraj, Pratibha, and Singla, M. (2020). Enhancement of storage life and quality maintenance of litchi (*Litchi chinensis* Sonn.) fruit using chitosan: pullulan blend antimicrobial edible coating. *International Journal of Fruit Science* 20 (3): S1662–S1680.

37 Kumar, N., Ojha, A., and Singh, R. (2019). Preparation and characterization of chitosan-pullulan blended edible films enrich with pomegranate peel extract. *Reactive and Functional Polymers* 144: 104350.

38 Kumar, N., Petkoska, A.T., Al-Hilifi, S.A., and Fawole, O.A. (2021a). Effect of chitosan–pullulan composite edible coating functionalized with pomegranate peel extract on the shelf life of mango (*Mangifera indica*). *Coatings* 11 (7): 764.

39 Kumar, N., Trajkovska Petkoska, A., Khojah, E. et al. (2021b). Chitosan edible films enhanced with pomegranate peel extract: study on physical, biological, thermal, and barrier properties. *Materials* 14 (12): 3305.

40 Kumar, N., Ojha, A., Upadhyay, A. et al. (2021c). Effect of active chitosan-pullulan composite edible coating enrich with pomegranate peel extract on the storage quality of green bell pepper. *LWT – Food Science and Technology* 138: 110435.

41 Latos-Brozio, M. and Masek, A. (2020). The application of natural food colorants as indicator substances in intelligent biodegradable packaging materials. *Food and Chemical Toxicology* 135: 110975.

42 Li, J.H., Miao, J., Wu, J.L. et al. (2014). Preparation and characterization of active gelatin-based films incorporated with natural antioxidants. *Food Hydrocolloids* 37: 166–173.

43 Liu, J., Meng, C.G., Liu, S. et al. (2017). Preparation and characterization of protocatechuic acid grafted chitosan films with antioxidant activity. *Food Hydrocolloids* 63: 457–466.

44 Liu, J., Wang, H., Wang, P. et al. (2018). Films based on κ-carrageenan incorporated with curcumin for freshness monitoring. *Food Hydrocolloids* 83: 134–142.

45 Ma, W., Tang, C.H., Yin, S.W. et al. (2012). Characterization of gelatin-based edible films incorporated with olive oil. *Food Research International* 49 (1): 572–579.

46 Mahcene, Z., Khelil, A., Hasni, S. et al. (2020). Development and characterization of sodium alginate based active edible films incorporated with essential oils of some medicinal plants. *International Journal of Biological Macromolecules* 145: 124–132.

47 Marsh, K. and Bugusu, B. (2007). Food packaging—roles, materials, and environmental issues. *Journal of Food Science* 72 (3): R39–R55.

48 Martino, V.P., Ruseckaite, R.A., and Jiménez, A. (2009). Ageing of poly(lactic acid) films plasticized with commercial polyadipates. *Polymer International* 58 (4): 437–444.

49 Matta Fakhouri, F., Nogueira, G.F., de Oliveira, R.A., and Velasco, J.I. (2019). Bioactive edible films based on arrowroot starch incorporated with cranberry powder: microstructure, thermal properties, ascorbic acid content and sensory analysis. *Polymers* 11 (10): 1650.

50 Mohamed, S.A., El-Sakhawy, M., Nashy, E.S.H., and Othman, A.M. (2019). Novel natural composite films as packaging materials with enhanced properties. *International Journal of Biological Macromolecules* 136: 774–784.

51 Munir, S., Javed, M., Hu, Y. et al. (2020). The effect of acidic and alkaline pH on the physico-mechanical properties of surimi-based edible films incorporated with green tea extract. *Polymers* 12 (10): 2281.

52 Nogueira, G.F., Soares, C.T., Cavasini, R. et al. (2019). Bioactive films of arrowroot starch and blackberry pulp: physical, mechanical and barrier properties and stability to pH and sterilization. *Food Chemistry* 275: 417–425.

53 Noshirvani, N., Ghanbarzadeh, B., Gardrat, C. et al. (2017). Cinnamon and ginger essential oils to improve antifungal, physical and mechanical properties of chitosan-carboxymethyl cellulose films. *Food Hydrocolloids* 70: 36–45.

54 Ogale, A.A., Cunningham, P., Dawson, P.L., and Acton, J.C. (2000). Viscoelastic, thermal, and microstructural characterization of soy protein isolate films. *Journal of Food Science* 65 (4): 672–679.

55 Paluch, M., Ostrowska, J., Tyński, P. et al. (2022). Structural and thermal properties of starch plasticized with glycerol/urea mixture. *Journal of Polymers and the Environment* 30 (2): 728–740.

56 Pankaj, S.K., Bueno-Ferrer, C., Misra, N.N. et al. (2014). Physicochemical characterization of plasma-treated sodium caseinate film. *Food Research International* 66: 438–444.

57 Piñeros-Hernandez, D., Medina-Jaramillo, C., López-Córdoba, A., and Goyanes, S. (2017). Edible cassava starch films carrying rosemary antioxidant extracts for potential use as active food packaging. *Food Hydrocolloids* 63: 488–495.

58 Puscaselu, R.G., Lobiuc, A., and Gutt, G. (2022). The future packaging of the food industry: the development and characterization of innovative biobased materials with essential oils added. *Gels* 8: 505.

59 Qiao, C., Ma, X., Zhang, J., and Yao, J. (2017). Molecular interactions in gelatin/chitosan composite films. *Food Chemistry* 235: 45–50.

60 Requena, R., Jiménez, A., Vargas, M., and Chiralt, A. (2016). Effect of plasticizers on thermal and physical properties of compression-moulded poly [(3-hydroxybutyrate)-*co*-(3-hydroxyvalerate)] films. *Polymer Testing* 56: 45–53.

61 Riaz, A., Lagnika, C., Luo, H. et al. (2020b). Chitosan-based biodegradable active food packaging film containing Chinese chive (*Allium tuberosum*) root extract for food application. *International Journal of Biological Macromolecules* 150: 595–604.

62 Riaz, A., Lagnika, C., Luo, H. et al. (2020a). Effect of Chinese chives (*Allium tuberosum*) addition to carboxymethyl cellulose based food packaging films. *Carbohydrate Polymers* 235: 115944.

63 Ribeiro-Santos, R., Andrade, M., de Melo, N.R., and Sanches-Silva, A. (2017). Use of essential oils in active food packaging: recent advances and future trends. *Trends in Food Science & Technology* 61: 132–140.

64 Robertson, G.L. (2005). *Food Packaging: Principles and Practice*. CRC Press.

65 Rogers, C.E. (1985). Permeation of gases and vapours in polymers. In: *Polymer Permeability* (ed. J. Comyn). Dordrecht: Springer https://doi.org/10.1007/978-94-009-4858-7_2.

66 Roy, S., Kim, H.J., and Rhim, J.W. (2021a). Synthesis of carboxymethyl cellulose and agar-based multifunctional films reinforced with cellulose nanocrystals and shikonin. *ACS Applied Polymer Materials* 3 (2): 1060–1069.

67 Roy, S., Kim, H.J., and Rhim, J.W. (2021b). Effect of blended colorants of anthocyanin and shikonin on carboxymethyl cellulose/agar-based smart packaging film. *International Journal of Biological Macromolecules* 183: 305–315.

68 Sanyang, M.L., Sapuan, S.M., Jawaid, M. et al. (2015a). Effect of plasticizer type and concentration on tensile, thermal and barrier properties of biodegradable films based on sugar palm (*Arenga pinnata*) starch. *Polymers* 7 (6): 1106–1124.

69 Sanyang, M.L., Sapuan, S.M., Jawaid, M. et al. (2015b). Effect of glycerol and sorbitol plasticizers on physical and thermal properties of sugar palm starch based films. In: *Proceedings of the 13th International Conference on Environment, Ecosystems and Development*, 157–162. WSEAS Press.

70 Shahbaz, M.U., Arshad, M., Mukhtar, K. et al. (2022). Natural plant extracts: an update about novel spraying as an alternative of chemical pesticides to extend the postharvest shelf life of fruits and vegetables. *Molecules* 27: 5152.

71 Shaikh, S., Yaqoob, M., and Aggarwal, P. (2021). An overview of biodegradable packaging in food industry. *Current Research in Food Science* 4: 503–520.

72 Sharma, S., Barkauskaite, S., Jaiswal, A.K., and Jaiswal, S. (2021). Essential oils as additives in active food packaging. *Food Chemistry* 343: 128403.

73 Shchipunov, Y.A. and Schmiedel, P. (1996). Phase behavior of lecithin at the oil/water interface. *Langmuir* 12 (26): 6443–6445.

74 Simões, C.M.O., Schenkel, E.P., Gosmann, G. et al. (2007). *Farmacognosia: da planta ao medicamento*, 711–740. Porto Alegre: UFRGS Editora.

75 Singla, M., Pareek, S., Kumar, N. et al. (2022). Chitosan-cinnamon oil coating maintains quality and extends shelf life of ready-to-use pomegranate arils under low-temperature storage. *Journal of Food Quality* https://doi.org/10.1155/2022/3404691.

76 Suhag, R., Kumar, N., Petkoska, A.T., and Upadhyay, A. (2020). Film formation and deposition methods of edible coating on food products: a review. *Food Research International* 136: 109582.

77 Sun, J., Jiang, H., Li, M. et al. (2020a). Preparation and characterization of multifunctional konjac glucomannan/carboxymethyl chitosan biocomposite films incorporated with epigallocatechin gallate. *Food Hydrocolloids* 105: 105756.

78 Sun, J., Jiang, H., Wu, H. et al. (2020b). Multifunctional bionanocomposite films based on konjac glucomannan/chitosan with nano-ZnO and mulberry anthocyanin extract for active food packaging. *Food Hydrocolloids* 107: 105942.

79 Sun, L., Sun, J., Chen, L. et al. (2017). Preparation and characterization of chitosan film incorporated with thinned young apple polyphenols as an active packaging material. *Carbohydrate Polymers* 163: 81–91.

80 Syafiq, R.M., Sapuan, S.M., and Zuhri, M.R. (2020). Effect of cinnamon essential oil on morphological, flammability and thermal properties of nanocellulose fibre-reinforced starch biopolymer composites. *Nanotechnology Reviews* 9 (1): 1147–1159.

81 Tao, G., Cai, R., Wang, Y. et al. (2017). Biosynthesis and characterization of AgNPs–silk/PVA film for potential packaging application. *Materials* 10 (6): 667.

82 Tarique, J., Sapuan, S.M., and Khalina, A. (2021). Effect of glycerol plasticizer loading on the physical, mechanical, thermal, and barrier properties of arrowroot (*Maranta arundinacea*) starch biopolymers. *Scientific Reports* 11 (1): 1–17.

83 Tavera-Quiroz, M.J., Urriza, M., Pinotti, A., and Bertola, N. (2012). Plasticized methylcellulose coating for reducing oil uptake in potato chips. *Journal of the Science of Food and Agriculture* 92 (7): 1346–1353.

84 Tongnuanchan, P., Benjakul, S., and Prodpran, T. (2014). Structural, morphological and thermal behaviour characterisations of fish gelatin film incorporated with basil and citronella essential oils as affected by surfactants. *Food Hydrocolloids* 41: 33–43.

85 Tongnuanchan, P., Benjakul, S., Prodpran, T., and Nilsuwan, K. (2015). Emulsion film based on fish skin gelatin and palm oil: physical, structural and thermal properties. *Food Hydrocolloids* 48: 248–259.

86 Trajkovska Petkoska, A., Daniloski, D., Kumar, N., and Broach, A.T. (2021a). Active edible packaging: a sustainable way to deliver functional bioactive compounds and nutraceuticals. In: *Sustainable Packaging* (ed. S.S. Muthu), 225–264. Singapore: Springer.

87 Trajkovska Petkoska, A., Daniloski, D., Kumar, N., and Broach, A.T. (2021b). Biobased materials as a sustainable potential for edible packaging. In: *Sustainable Packaging* (ed. S.S. Muthu), 111–135. Singapore: Springer.

88 Van de Velde, K. and Kiekens, P. (2002). Biopolymers: overview of several properties and consequences on their applications. *Polymer Testing* 21 (4): 433–442.

89 Van Oosterhout, J.T. and Gilbert, M. (2003). Interactions between PVC and binary or ternary blends of plasticizers. Part I. PVC/plasticizer compatibility. *Polymer* 44 (26): 8081–8094.

90 Vicentini, D.S., Smania, A. Jr., and Laranjeira, M.C. (2010). Chitosan/poly(vinyl alcohol) films containing ZnO nanoparticles and plasticizers. *Materials Science and Engineering: C* 30 (4): 503–508.

91 Vieira, M.G.A., da Silva, M.A., dos Santos, L.O., and Beppu, M.M. (2011). Natural-based plasticizers and biopolymer films: a review. *European Polymer Journal* 47 (3): 254–263.

92 Wang, S., Xia, P., Wang, S. et al. (2019a). Packaging films formulated with gelatin and anthocyanins nanocomplexes: physical properties, antioxidant activity and its application for olive oil protection. *Food Hydrocolloids* 96: 617–624.

93 Wang, X., Yong, H., Gao, L. et al. (2019b). Preparation and characterization of antioxidant and pH-sensitive films based on chitosan and black soybean seed coat extract. *Food Hydrocolloids* 89: 56–66.

94 Winwood, R.J. (2015). Algal oils: properties and processing for use in foods and supplements. In: *Specialty Oils and Fats in Food and Nutrition* (ed. G. Talbot), 159–172. Woodhead Publishing.

95 Wu, H., Lei, Y., Zhu, R. et al. (2019). Preparation and characterization of bioactive edible packaging films based on pomelo peel flours incorporating tea polyphenol. *Food Hydrocolloids* 90: 41–49.

96 Yadav, A., Kumar, N., Upadhyay, A. et al. (2022). Effect of mango kernel seed starch-based active edible coating functionalized with lemongrass essential oil on the shelf-life of guava fruit. *Quality Assurance & Safety of Crops and Food* 14 (3): 103–115.

97 Yong, H., Wang, X., Bai, R. et al. (2019). Development of antioxidant and intelligent pH-sensing packaging films by incorporating purple-fleshed sweet potato extract into chitosan matrix. *Food Hydrocolloids* 90: 216–224.

98 Yousefnia Pasha, H., Mohtasebi, S.S., Tabatabaeekoloor, R. et al. (2021). Preparation and characterization of the plasticized polylactic acid films produced by the solvent-casting method for food packaging applications. *Journal of Food Processing and Preservation* 45 (12): e16089.

99 Yu, J., Yang, J., Liu, B., and Ma, X. (2009). Preparation and characterization of glycerol plasticized-pea starch/ZnO–carboxymethylcellulose sodium nanocomposites. *Bioresource Technology* 100 (11): 2832–2841.

100 Zheng, J.P., Li, P., Ma, Y.L., and Yao, K.D. (2002). Gelatin/montmorillonite hybrid nanocomposite. I. Preparation and properties. *Journal of Applied Polymer Science* 86 (5): 1189–1194.

13

Mechanical Properties of Natural Material-Based Packaging Films: Current Scenario

Johnsy George[1], Muhammed Navaf[2], Aksalamol P. Raju[1], Ranganathan Kumar[1], and Kappat V. Sunooj[2]

[1] *Defence Food Research Laboratory, Food Engineering and Packaging Technology Division, Mysore, 570011, India*
[2] *Pondicherry University, Department of Food Science and Technology, Puducherry, 605014, India*

13.1 Introduction

The application of natural polymer-based films in packaging is generating a lot of interest among the public as they minimize the negative ecological impact, sustainability, government emphasis on environmentally friendly packaging, and rising consumer awareness [1]. The global packaging industry mainly includes packaging markets for food, beverages, cosmetics, and healthcare; however, food packaging has the lion's share in the total packaging market. Food packaging material has a share of around 50% of the total plastic products manufactured in the packaging industry and can directly impact the environment [2]. The exponential growth of packaging requirements coupled with the nonbiodegradability of commodity plastics created a major threat to our environment. Considering the environmental impact of plastic waste materials, the development of packaging films from natural polymers or other bio-based polymers is important as they considerably reduce the negative impact of solid waste on the environment [3].

The global market of biodegradable polymers for packaging films is expected to rise further by several folds in the next decade. Currently, many industries are developing degradable food packaging films. But the popularity of biodegradable packaging films is not high compared to conventional packaging materials. Bioplastics currently account for less than 1% of the 368 million tons of plastic produced each year, according to European Bioplastics [4]. However, as demand grows and more specialized biopolymers, applications, and products emerge, the bioplastics industry

---

Natural Materials for Food Packaging Application, First Edition.
Edited by Jyotishkumar Parameswaranpillai, Aswathy Jayakumar,
E. K. Radhakrishnan, Suchart Siengchin, and Sabarish Radoor.
© 2023 WILEY-VCH GmbH. Published 2023 by WILEY-VCH GmbH.

continues to expand and diversify. The increased cost, limited mechanical and barrier properties, limited market availability, and lack of understanding prevent the general public away from utilizing it. A good packaging film should possess enough strength, elongation, transparency, flexibility, abrasion and puncture resistance, and barrier properties [5]. Due to the poor processing characteristics, such as low thermal and chemical stability, conventional methods for manufacturing films are not easily applicable to degradable materials.

Food packaging materials should meet some key design requirements, such as providing physical protection and creating conditions that are essential for achieving desired shelf life. In packaging films, mechanical characteristics are vital for ensuring the mechanical strength of the film and integrity during shipment, handling, and storage of foods packaged in them. The mechanical properties of packaging films are related to the material behavior when it is subjected to external forces such as tension and compression. In food packaging, evaluation of mechanical characteristics is very important because the packaging helps to protect each type of foodstuff from external damage. In many applications, packaging materials will be subjected to mechanical damage during the handling, transportation, and distribution of food packets. To evaluate the suitability of natural materials for packaging applications, the following mechanical properties such as tensile strength (TS), percentage elongation, modulus, seal strength, tear resistance, and puncture resistance need to be checked. This chapter deals with the current scenario of mechanical properties of natural material-based packaging films.

13.2 Mechanical Properties of Packaging Films

Mechanical properties of a packaging material indicate the response of such materials to the applied loads. Mechanical properties are determined to evaluate the behavior of a particular material, which is subjected to external loads when used for certain applications. Determination of mechanical properties is very important in food packaging, as it helps in material selection as well as predicting the behavior of packaging material during the storage and distribution process. The mechanical properties of packaging films, such as tensile properties, seal strength, tear strength, and puncture resistance, are studied to ensure the specific applicability of packaging materials. These values are important to get general information about the suitability of such materials in different food packaging applications. Various factors such as polymer architecture, spatial arrangement of macromolecules, chemical structure, and crystallinity play an important role in the mechanical properties of polymer films. The mechanical properties of packaging films are considered essential criteria for selecting a packaging film, and each application has different mechanical property requirements.

Tensile properties are the most important mechanical properties of packaging film and are widely practiced testing methods of packaging films. Determining the tensile properties of film will be helpful to identify and utilize the suitable applications of the packaging film. These data can be used for engineering design as well as quality control of packaging materials. Tensile properties include TS, Young's

Figure 13.1 Universal testing machine.

modulus, and elongation at break (EB). The values of tensile parameters depend on various factors such as the film thickness, preparation methods, testing speed, grips used, way of measuring extension, and test conditions. The film's tensile properties were measured according to the standard test method for tensile properties of thin plastic sheeting, ASTM D-882 [6]. The universal testing machine (Figure 13.1) is mainly used to determine the tensile properties of the packaging films.

Some of the important mechanical properties that determine the suitability of a polymer film for packaging applications are as follows.

13.2.1 Tensile Strength (TS)

TS is one of the standard parameters used to detect the strength of a polymer. It measures the maximum stress that a material can withstand without fracture before its failure, which is measured as force per unit area.

$$\text{Tensile Strength}(\sigma) = \frac{\text{Force}(F)}{\text{Area}(A)}$$

The TS depends on several factors such as the type of polymer, processing conditions, and modifications made on polymer. It may change with temperature and storage.

13.2.2 Young's Modulus (Y)

Young's modulus is also known as Modulus of Elasticity or Tensile Modulus, and it is the ratio of stress to the strain in the elastic region. It is an index of the stiffness of the material.

$$\text{Young's modulus} = \frac{\text{Tensile stress}(\sigma)}{\text{Tensile strain}(\varepsilon)}$$

The tensile energy required to break the sample, which is also referred to as the toughness, is used to evaluate packaging materials that may be subjected to transport and other related handling abuses.

13.2.3 Elongation at Break (EB)

EB indicates to what extent a material can stretch without a break, and it is defined as the ratio of the change in length to initial length. The percentage EB is estimated by dividing the extension at the time of sample rupture by the initial gauge length of the specimen multiplied by 100.

$$\text{Percentage elongation} = \frac{\text{Change in length}}{\text{Original length}} \times 100$$

These values indicate how ductile a polymer is to form different shapes. It measures in percentage, and a polymer with a higher value indicates a stronger polymer.

13.2.4 Seal Strength

The average seal strength is defined as the force per unit width of the seal, which is required to separate a flexible material from another rigid or flexible material that is sealed together [7]. It is important to determine the opening force needed to open a seal and thereby can be used to verify the package integrity. Seal strength at some minimum level is a very important packaging requirement, and sometimes it is also needed to limit the seal strength so as to facilitate easy opening.

In the heat-sealing process, the two layers of polymer films are kept between two hot metal jaws that provide sufficient heat to melt the polymers, resulting in an interfacial interaction between melted layers. Seal strength is defined as the force required to separate the two layers of the seal under particular conditions. It is the maximum force required per unit width of seal to separate a flexible material progressively from another material. The seal strength of packaging films can be determined by using the ASTM F 88M-09 method. Seal strength is a quantitative measure for use in process validation, process control, and capability. It also measures the ability of the packaging process to produce a consistent seal.

13.2.5 Tear Resistance

Tear resistance is the force necessary to propagate a tear in the plastic film. The force needed to initiate tearing across a specific geometry of a film can be calculated from load–time or load–displacement data [8, 9]. Tear resistance is a complex function of its resistance to rupturing. There is no direct linear relationship between tear strength and the thickness of the testing film. Different testing methods, such as ASTM D1922 and ISO 6383-2, are used to study the tear resistance of packaging films.

13.2.6 Puncture Resistance

Puncture resistance is the ability of packaging material to inhibit the intrusion of a foreign object, which can be measured by using a slow controlled penetration by a probe [10]. Puncture resistance of a material is the measure of the maximum force required to penetrate a material.

13.2.7 Impact Resistance

Impact testing is used to assess the ability of packaging material to withstand the conditions of rapid deformation, such as when a packet is dropped or when something is dropped on the package. This testing is mainly done by clamping a packaging material to a pneumatic ring, and a dart is released to drop on the center of the test specimen [11].

13.2.8 Burst Strength

Burst strength of film material is defined as the resistance offered by the sample to an externally applied pressure (at right angles) on its surface under certain defined conditions. The bursting strength of a sample is measured as the pressure exerted at the time of its failure. This test also indirectly gives an idea about the capacity of a film to absorb energy.

13.3 Mechanical Properties of Natural Polymer-Based Packaging Films

Natural polymers or other bio-based polymer materials can be classified into three categories such as polymers found in nature or can be extracted from biomass, polymers synthesized from natural/bioderived monomers, and polymers that can be synthesized from microorganisms. The first category of naturally occurring polymers consists of polysaccharides, proteins, etc., which are abundant in nature. Polysaccharides are the most abundant carbohydrate polymer found in nature, such as cellulose, starch, and chitosan. Protein-based film-forming natural polymers consist of gelatin, soy protein, etc. The second category consists of polymeric materials that are synthesized from naturally occurring monomers. This class of biopolymers includes polymers such as poly(lactic acid) (PLA)) made by the condensation of lactic acid, polyethylene furanoate (PEF) made by the poly condensation of 2,5-furandicarboxylic acid and ethylene glycol, poly(butylene succinate) (PBS) made by the esterification reaction of succinic acid and 1,4 butanediol, poly(butylene adipate-*co*-terephthalate) (PBAT), which is a random copolymer of adipic acid, 1,4 butanediol, and terephthalate. The third category consists of polymers such as Polyhydroxyalkanoates (PHAs), bacterial cellulose, and other microbial polysaccharides.

13.3.1 Naturally Occurring Polymers

This group consists of polymers found in nature or polymers that can be extracted from biomass. Polysaccharides such as starch, cellulose, chitosan, proteins, and gums belong to this category. Polysaccharide-based films have low cost, high availability, renewable nature, and biodegradability, enhancing their use as raw material for packaging films. In general, polysaccharide-based films have lower mechanical properties than conventional plastic materials.

13.3.1.1 Starch

Starch is one of the most widely studied biological polymers for synthesizing packaging film due to its abundant and cheap availability, biodegradable, and edible characteristics. It is an agricultural biopolymer that can be obtained from various sources such as cereals, fruits, tubers, and pulses [12]. The composition and granular structure of starch may vary with the starch source. Starch is insoluble in cold water. However, on heating, the starch granules break, and the degradation of the crystalline region will facilitate the hydrogen bonding between hydroxyl groups in starch with water molecules. It results in partial solubilization of starch granules.

Generally, starch films are synthesized by wet and dry methods. The edible films are synthesized by employing the wet process, in which the polymer is solubilized, and the solution is dried. The dry method is carried out by extrusion, which is based on the thermoplastic properties of starch. To convert starch into thermoplastic, the starch is plasticized with glycerol, sorbitol, etc., and heated above glass transition temperature under restricted moisture content. Due to the high ordered hydrogen-bonded network, the starch-based film has high oxygen barrier property, whereas the high brittleness of starch film makes it poor in mechanical strength [13].

The majority of starch used for film synthesis comes from cassava, potato, corn, and rice [14]. Starch sources play a significant role in film features due to the ratio of amylose and amylopectin and the different structural properties of macromolecules. Granule swelling is the primary process involved in starch gelatinization. It depends on the granule size, shape, amylose amylopectin interactions, etc. It is reported that the film synthesized from potato starch exhibited a better barrier and weaker mechanical properties than the film synthesized from corn and wheat. It may be due to the difference in granule size between them. A high amylopectin content in starch restricts the solubility and leads to aggregation of micro starch granules, which results in lower mechanical properties of the film [15]. Other factors such as processing temperature, amount and type of plasticizer used, and presence of copolymers will also affect the mechanical and barrier properties of starch-based films. The temperature-induced degradation of the crystalline region facilitates the phase transformation of the ordered molecular structure, and it is observed that the gelatinization temperature of most of the starches is around 90 °C. Likewise, plasticizer plays a vital role in maintaining the integrity of packaging films and help in reducing the brittleness. The film synthesized without plasticizer are brittle and exhibits high TS and low EB [15]. The mechanical properties of starch-based films are shown in Table 13.1.

Table 13.1 Mechanical properties of starch-based films.

Film	Plasticizer/ acid used	Tensile strength (MPa)	Young's modulus (MPa)	Elongation at break (%)	References
Sago starch	Glycerol	9.87	$6.17 \times 10^7 \, N/m^2$	17.11	[16]
Sago starch	Sorbitol	25.03	$4.33 \times 10^7 \, N/m^2$	59.35	[16]
Tapioca starch	Glycerol	2.17	—	5.17	[17]
Potato starch	Glycerol	1.656	0.193 MPa	15.903	[18]
Tapioca starch	Glycerol	2.67	24.60 MPa	23	[19]
Tapioca starch/ microcrystalline cellulose	Glycerol	2.36	33.61 MPa	16	[19]
Chitosan	Acetic acid	15.33	0.55 MPa	63.96	[20]
Potato starch/ chitosan	Glycerol	9.72	—	5.70	[21]
Potato starch/ chitosan/citric acid (10%)	Glycerol	12.53	—	7.13	[21]

The blending of polymers is a common way to change the characteristics of biodegradable films. It offers the benefit of being able to use technology at a minimal cost. The typical goal of creating a unique blend of two or more polymers is to maximize the blend's performance rather than substantially changing the characteristics of the components. Nowadays, the properties of the starch-based film are improved mainly by adding other biodegradable polymers such as PLA and PBAT. Moreover, the blending of starch with natural polymers such as cellulose will improve the properties of the starch film very efficiently.

Ghanbarzadeh et al. [22] have discussed the effect of carboxymethyl cellulose on corn starch film. The addition of 15% and 20% carboxymethyl cellulose caused a two and threefold increase in the TS of corn starch film, respectively. Interestingly they noted that the elongation at the break did not decrease considerably. However, the addition of carboxymethyl cellulose to cassava starch significantly reduced the EB, and a similar effect was observed in many starches [23]. The increased strength of starch film by blending with cellulose is due to the intermolecular interaction between the hydroxyl group of starch with the carboxyl group of cellulose. During the film drying process, the initial hydrogen bonds in the starch are replaced by a newly formed hydrogen bond between starch and cellulose, which gives a more compact structure to the starch film [23].

Similarly, the blending of starch with proteins will also affect the mechanical properties, depending on the type of starch used. Hassan and Norziah [16] studied the impact of gelatin in sago starch film; glycerol and sorbitol were used as plasticizers. The study reported that the sago starch blending with gelatin significantly reduced the TS in both cases, and the EB was increased with gelatin content. The mixing of sago starch with gelatin results in interaction between hydroxyl groups of starch and protein, which reduces the interaction between starch chains and leads to lower TS. Contrary to this, many researchers reported an increase in the TS of the starch film by blending with gelatin. According to the study conducted by Tongdeesoontorn et al. [24], adding gelatin to cassava starch boosted TS and lowered EB. It's because the starch hydroxyl group and the protein NH_3 group form intermolecular hydrogen bonds. Biodegradable films based on nonconventional starch such as Kitul (*Caryota urens*) were also reported. Various surface treatment techniques such as gamma irradiation and energetic neutral nitrogen atom treatment were found to have efficacy in improving the mechanical properties [25, 26].

The inclusion of nanosized particles is another way to improve the mechanical characteristics of starch-based films. Different types of nanomaterials such as nanofiber, nanoparticles, and metal and metal oxide nanoparticles are used to optimize the mechanical properties of the starch-based film. The form, number, and alignment of nanoparticles, as well as the ability of the polymer matrix to transmit stress, determine the mechanical characteristics of biocomposites [27]. Nanoparticles provide a large surface area; it fills the voids and thus provides strong interaction between starch matrix and nanoparticles, resulting in a more compact structure and enhanced mechanical strength [17]. Many researchers studied the impact of nanoparticles in starch-based film. In general, the reinforcement with nanoparticles increased the TS and Young's modulus but decreased the EB. Adding ZnO nanoparticles to cassava starch film improved the TS at lower concentrations; however, it was found to decrease at higher concentrations (1%). Likewise, the percentage elongation was also noted to decrease with increased nanoparticle concentration. This is because the particles need more space in the matrix at higher concentrations, thus negatively affecting the TS [28]. A similar result was also observed in nano cellulose-reinforced starch film [27]. Othman et al. [17] analyzed the effect of chitosan nanoparticles in cassava starch film at different temperatures and relative humidity. At a lower temperature up to 30 °C, the addition of chitosan nanoparticles significantly increased the TS; however, nanocomposite showed a lower TS than cassava starch film at a higher temperature. Nanocomposite showed lower EB at lower temperature and increased at a higher temperature. Starch nanocrystals can be synthesized from starch by acid hydrolysis, which results in nanoscale platelet morphology and higher crystallinity [29]. These nanocrystals were also reported to reinforce starch-based films and exhibited good enhancement effects on the physicochemical properties due to the self-reinforcing effect [30, 31].

Starch-based films have a low tear strength as compared to synthetic polymers. Petersen et al. [32] reported that the film synthesized from wheat starch has a tear strength of 21 and 25 N/mm in the machine direction and transverse direction, respectively. Similarly, sago starch film synthesized with a 10% plasticizer exhibited a tear strength of 6.51 N/mm, and further addition of plasticizer to the film has a negative impact on its tear strength [33]. The reinforcement of starch film with cellulose would increase its tear strength. Ismail et al. [34] studied the effects of reinforcement of tapioca starch with cellulose prior to ultrasonication for different periods.

They observed that the starch blended with five hours of ultrasonicated cellulose exhibited a 241% increase in tear strength. The homogeneous blending of starch with fillers facilitates its penetration into the matrix and thus restricts the slippage of macromolecules, resulting in the film's increased tear strength.

Starch films formed from nanocomposite show higher burst strength than pure starch films. Many researchers studied the impact of starch nanoparticles on the burst strength of the composite films. Gujral et al. [35] synthesized potato starch nanocomposite film and observed that the burst strength of the pure starch film was 387.50 g. And the burst strength of nanofilm increased with an increase in nano starch concentration. Higher burst strength of 1058.30 g was observed in a film containing 10% nanoparticles. Similarly, Sharma et al. [36] also reported that the burst strength of kidney bean starch film increased from 1255.7 to 2777.7 g by formulating a 10% nano starch composite film. A similar enhancement in burst strength was reported in starch nanoparticle blended water chestnut and mung bean starches.

Xu et al. [37] synthesized corn starch nanocomposite film using cellulose nanocrystals (CNCs). They observed that the reinforcement with CNCs increased the burst strength of corn starch film from 1437.79 to 2477.91 g. The burst strength of cellulose films will increase with an increase in water activity. Cazón et al. [38] studied the impact of water activity (0.1–0.9) on the burst strength of cellulose film and reported that burst strength ranges from 145.03 to −338.10 g.

The heat sealability of starch-based films depends on the type of starch and the plasticizer used. The film synthesized from sago and potato starches exhibited a good sealing strength of near 500 N/m and is comparable to synthetic polymers with a seal strength of nearly 600 N/m. However, the film synthesized from mung bean starch exhibits very low seal strength. The addition of nanoparticles to starch film was found to have improved the seal strength of the film. The increasing hydrogen and covalent bonds involving C—O—H and N—C may be the main forces responsible for the sealed joint formation of the films [39]. Nouri and Nafchi [40] reported an increase in seal strength of sago starch by the addition of betel leaf extract at a concentration of 20%. Moreover, the addition of compounds containing nitrogen or oxygen functional groups will increase chain interaction and thus would increase the seal strength of starch films [41].

13.3.1.2 Cellulose

Cellulose is the most abundant natural polymer available on the earth and is one of the broadly exploited biopolymers for developing biodegradable films [42]. Its renewability, low cost, biodegradability, nontoxicity, and chemical stability make it favorable for film synthesis. Cellulose is produced from different sources such as wood, cotton, hemp, plant-based materials, and as well as from microbial sources. Cellulose is composed of β-1,4 linked glucopyranose units that form a linear homopolymer [43]. Cellulose is insoluble in polar solvents and is soluble in a few solvents with no similar chemical properties. The insolubility of cellulose is the challenging focus during film development. Several methods are adopted to improve the solubility of cellulose. The solvents such as *N*-methylmorphin N-oxide, ionic liquids, and an aqueous solution of NaOH are used to improve the solubility of cellulose. The water solubility of cellulose is increased by producing its derivatives such as carboxymethyl cellulose, methylcellulose, and hydroxypropyl cellulose. These derivatives are synthesized by soaking cellulose in an alkali solution followed by a reaction with chloroacetic acid, methyl chloride, or propylene oxide. This derived cellulose is water-soluble and is used as raw material for cellulose film synthesis. The sealing strength of methylcellulose film and hydroxypropyl methylcellulose (HPMC) films highly depends on sealing temperature. The film containing methylcellulose and HPMC exhibited a high heat seal of 0.211 and 0.385 N/mm, respectively, at a temperature of 166 °C [44]. The film formed from methylcellulose and hydroxypropyl cellulose possesses high oxygen and lipid barrier properties but has low water vapor barrier properties. Among the cellulose derivatives, hydroxyl propyl methylcellulose is a good film-forming material with good flexibility, transparency, and oil resistance [45].

Cellulose-based films have poor mechanical qualities, which can be enhanced using various techniques such as mixing with other biodegradable polymers or adding plasticizers [46, 47]. The addition of plasticizers to cellulose film increases the free volume of polymer structure or increases molecular mobility. Mahadevaiah et al. [48] reported the effect of different plasticizers, i.e. glycerol and polyethylene glycol, on the various mechanical properties of HPMC. They observed that the addition of plasticizers decreased TS and Young's modulus and is proportional to plasticizer concentration. Compared to polyethylene glycol, glycerol significantly impacts film's TS. The film synthesized by glycerol is weaker, more stretchable, and more flexible than polyethylene glycol film. The EB increased with glycerol concentration, decreasing with polyethylene glycol concentration. Hybrid HPMC nanocomposites with a combination of different nanocrystals were reported to have better mechanical properties [49].

13.3.1.3 Chitosan

Compared to starch-based films, chitosan film possesses high mechanical properties. The mechanical properties of chitosan-based films directly depend on the molecular weight and deacetylation degree. Chitosan with higher molecular weight and lower deacetylation degree shows better TS and elongation. The increased intermolecular bonding can explain it during film formation, and longer chains with minimum branching facilities have better packing efficiency [50]. Moreover, researchers employed a few techniques, such as the addition of cross-linking agents, irradiation and ultrasonic treatments, and nanocomposite development to improve the mechanical properties of chitosan film [51]. Chitosan film synthesized with acetic acid as solvent exhibited higher TS than glycerol. The use of glycerol decreased the TS of chitosan film, whereas acetic acid reduced the EB, and it increased in glycerol film [52]. The reinforcement of chitosan film with 10% nanocrystalline cellulose caused a 24% increase in TS and a 5% reduction in EB [53]. Many researchers have also reported similar findings when chitosan is reinforced with nanosized particles. The addition of nanoparticles to the chitosan matrix facilitates strong interaction between fillers and matrix, restricting the matrix movement and decreasing EB. García et al. [54] studied the effect of gamma irradiation on molecular weight reduction and mechanical properties of chitosan film with 2% lactic acid and tween 80. The irradiation increased the mechanical strength, while EB decreased. Contrary, the irradiation of chitosan film containing 2% acetic acid decreased the TS and elongation at the break [20]. The incorporation of chitin whiskers in chitosan films was reported to have considerably increased the tensile properties [55].

Chitosan films have very low tear strength as compared to synthetic polymers. The tearing strength of chitosan films may vary with the plasticizer used. The film synthesized using 1% acetic acid exhibited a tear strength of 1.91 and 2.50 N/mm in the machine and transverse direction, respectively [56]. However, the cross-linked chitosan films showed a lower value [57]. Similarly, the film synthesized with 1% acetic acid exhibited lower tear strength of 0.26 N. The addition of 40% glycerol and 20% sorbitol improved the tear strength of chitosan film [58]. Prateepchanachai et al. [59] reported that the pure chitosan film exhibited a seal strength of 5.5 N/m and was increased by reinforcement with glycerol and glycerol–gelatin to a value of 76.5 and 110.7 N/m, respectively.

13.3.1.4 Alginates

Alginate is an anionic exopolysaccharide found in marine brown algae (Phaeophyceae). It is made up of 1,4-linked hexuronic acid residues with varying ratios, which are β-D-mannuronic acid (M) and α-L-guluronic acid (G) [60]. The most crucial property of alginates is their ability to react with tri and divalent cations to form a film. Alginate films can be synthesized by the casting method of glycerin as a plasticizer [61]. The multivalent metal ions like calcium can react with alginate and form a cooperative association between M and G blocks, forming a 3D network where they may pack and coordinate [62]. Alginate film exhibits a hydrophilic nature. Hence, the cross-linking of alginate with multivalent metal ions will enhance their water barrier properties, mechanical resistance, cohesiveness, and rigidity [5].

13.3.1.5 Pectin

Pectin is a group of acidic water-soluble hygroscopic polysaccharides in vegetables and fruits and rich in citrus peel. It is commonly present in the intercellular region of primary cells of higher plants. Pectin is an ionic polymer with a linear structure and is made up of α-1,4-linked D-galactopyranosyluronic acid units [63]. Pectin is negatively charged due to ionized carboxylic acid groups along its backbone that have pK_a values of about 3.6 [64]. It exists either with methoxylated carboxyl groups or with the amidated carboxyl group. The former is obtained by esterifying the carboxyl group of galacturonic acid with methanol. Latter is obtained by converting galacturonic acid with ammonia to a carboxylic acid amide. Pectin is classified as high methoxyl pectin and low methoxyl pectin based on the degree of esterification with methanol. Pectin is given the generally recognized as safe (GRAS) status by the US FDA and is exploited as a gelling agent in various food [65].

Development of pectin-based edible film can be carried out by different methods such as casting, extrusion, spraying, and knife-coating. An aqueous solution of pectin can be cast into a transparent film, and it exhibits low elongation and resistance to water permeability. However, it shows excellent mechanical properties and barriers to oil and oxygen. In general, solvents like water and ethanol are used to synthesize pectin film [66]. The casting method of pectin can be classified into solution casting and continuous casting methods. In both, the preparation of the casting medium consists of mixing pectin and plasticizer in a suitable solvent.

The pectin is usually blended with other polysaccharides, mainly starch, and plasticized with glycerol in the extrusion method. The film obtained through extrusion has advanced microstructure and thermal dynamic mechanical properties than the film synthesized by the solution casting method. The spray film formation is a wet process in which the pectin is dissolved into the liquid phase and sprayed to a nonsticky surface, giving a uniform film. The knife coating method is broadly practiced in industrial applications. Knives spread a precise solution layer onto a surface, moving under the knife. This method has excellent control over film thickness, and hence, the film formed by this method exhibits uniform thickness [65]. The use of additives like plasticizers such as glycerol, acetylated monoglycerides, polyethylene glycol, and cross-linking agents make pectin film more flexible and mechanically strong [5].

Pectin films have higher puncture resistance as compared to starch films, and the pectin film synthesized with 15% glycerol by the continuous casting method exhibited a puncture resistance of 326 N/mm. However, the addition of spent coffee powders to the film matrix decreases the puncture resistance [67]. Moreover, the plasticizer and pH significantly affect the puncture resistance of pectin films. Oliveira et al. [68] synthesized high methoxyl pectin film using casting and extrusion methods at different pH levels. And they noticed that casted film synthesized at a pH of 2 exhibited high puncture strength. However, extrusion film synthesized at pH 7 exhibited high puncture resistance.

13.3.1.6 Casein

Casein is a milk protein that comprises 82% of total milk protein. Casein exists in four subunits, κ-casein, β-casein, α s1-casein, and α s2-casein, making up 13%, 36%, 38%, and 10% of the casein composition, respectively. From milk, casein can be obtained by the precipitation of milk at a pH of 4.6. It can be obtained from other sources by using electrophoresis, chromatographic, enzymatic, and membrane filtration processes [69]. Due to its high nutritional and functional properties, casein protein is widely used as food additive in bakery applications, beverages, milk products, snack foods, etc. Casein's biodegradability, high thermal stability, nontoxicity, capability to bind tiny molecules and ions, and micelle production potential make it ideal for creating an edible biofilm. Casein's four subunits serve an important role in the protein's ability to form films. Due to the high interchain cohesion induced by their random-coil structure and substantially high intermolecular interactions, it may swiftly form films from aqueous solutions without further processing [70]. Casein films include multiple polar functional groups such as hydroxyl and amino groups, and they have high barrier properties. Casein film's unique feature allows it to be used in active packaging. It may be paired with other packaging materials to protect items prone to oxidation or moisture [71]. However, casein films have low elasticity, and they will have affected by the plasticizer used during film formation.

Casein-based films are susceptible to moisture and are soluble in water, which will affect the film's mechanical and barrier properties [72].

13.3.1.7 Whey Protein

Whey is a milk protein that is acquired as a by-product of the cheese-making process and is defined as the leftover substance in the milk serum after casein coagulation at pH 4.6 and 20 °C. It is produced in large quantities, has admirable functional properties, and can develop edible film [73]. Whey proteins consist of different globular proteins such as β-lactoglobulin (57%), α-lactalbumin (19%), bovine serum albumin (7%), several immunoglobulins (13%), and the polypeptides protease–peptone (4%) [74]. The whey-based films can be developed from either whey protein concentrate or whey protein isolate, and the portion content varies with the product. Whey protein products are rich in sulfur-containing amino acids such as cysteine and methionine [70].

The film-forming ability of the whey proteins primarily depends on the unfolding and dissociation of an individual protein in an aqueous medium. Whey proteins are globular proteins, which should be unfolded, and their internal hydrophobic and sulfhydryl groups need to be exposed during filmmaking. Changes in temperature, pH, shear pressures, or the addition of organic solvents or salts can all aid in the unfolding and dissociation of globular proteins. The hydrophobic groups of unfolded protein molecules are further oriented and arranged to form a 3D chemical network, which is stabilized by disulfide linkage and or by physical interactions such as van der Waals interactions, hydrogen bonding, and electrostatic and hydrophobic interactions [74].

Many authors reported the potential use of whey protein film in packaging, and these are highly flexible, transparent, and glossy. It also offers a good barrier property against aroma, fat, and oxygen but provides less barrier to moisture. The incorporation of plasticizer helps avoid the intrinsic brittleness formation in the whey coatings. It is because the plasticizers will facilitate the mobility of the polymer chain. Increasing intermolecular spacing allows films with enough flexibility to keep their integrity once applied and subsequently formed [75]. The physical and chemical treatments like the exposure of the film-forming solution to ultraviolet radiation, ultrasound, or alkalization treatments will improve the properties such as mechanical properties and color of the film [70].

A study reported by Socaciu et al. [76] noted that pure whey protein film has a puncture resistance of 6.2 N/mm, and the heat treatment of film-forming solution decreased the value. However, the addition of tarragon essential oil has considerably increased the puncture resistance of film formed from heat-treated solution compared to untreated film. Other methods such as ultrasound treatment, UV radiation, and gamma irradiation are also performed to improve the puncture resistance of protein films. Banerjee et al. [77] used ultrasound treatment to enhance the film's mechanical properties and observed a 28% and 120% increase in puncture resistance by treatment of 30 minutes and 1 hour, respectively.

13.3.1.8 Collagen

Collagen is a naturally abundant protein in the animal body, and it comprises approximately 30% of total protein. It is fibrous and found in connective tissues.

Collagen possesses a triple helix structure of three parallel chains of the left-handed helix. Two of these are homologous chains (α-1) and one supplementary chain that varies slightly in its chemical composition. Collagen has great TS and forms stable fibrils. Collagen is versatile and contains various amino acids in different proportions. It is distributed along with macromolecular structure, making it a highly functional biomaterial and suitable for film formation [78]. Collagen films are developed by air-drying cast collagen solution with a thickness of 0.01–0.5 mm [79].

Collagen produced under mild conditions exists in its native form without any changes in its triple helix form, resulting in fibrillary structure film with excellent mechanical properties. Due to the extensive interaction between protein chains by hydrogen bonding, electrostatic and hydrophobic interaction, the film made without the plasticizer are brittle in appearance. Collagen films are commonly employed as barrier membranes to protect food items from oxygen and solutes while also providing structural integrity and vapor permeability. The collagen films are also impervious to polar compounds [80]. Inter and intramolecular cross-linking using aldehydes such as glyoxal, formaldehyde, and others can improve the functional qualities of collagen films, such as water resistance [81].

13.3.1.9 Gelatin

Gelatin is a protein derived by the hydrolysis of collagen protein present in the bones and skin of the animal. It is a water-soluble and odorless protein that forms and randomizes polypeptide chains in an aqueous solution. Its structure consists of rigid bar-like molecules arranged in fibers interconnected by covalent bonds. Gelatin is a heterogeneous polypeptide mixture of α, β, and γ-chains [82]. Among the protein, the wide availability, low cost, biodegradability, and good film-forming property of gelatin make it an essential material for packaging films [83].

Gelatin films can be synthesized either by casting gelatin in an aqueous solution, extrusion, or blown film extrusion methods. The cast film is done at a temperature less than room temperature, known as cold cast film. The film made at a temperature above 35 °C is termed as hot cast film, and the conformational structure of gelatin films depends on the temperature [70]. Gelatin films are transparent, and their melting temperature is close to room temperature [84]. When gelatin comes into contact with moisture, it swells or dissolves due to its highly hygroscopic nature. As a result, adding cross-linking agents, plasticizers, and strengthening agents to collagen films can improve their functional qualities. These additives may reduce the intermolecular force between the protein chains. The action of molecular structure reduces the hydrophilic nature or facilitates the formation of strong covalent bonds in the protein matrix [82].

Ciannamea et al. [12] synthesized a gelatin film containing 40% glycerol that had a tear strength of 1.17 N, and the value increased by replacing a part of glycerol with soybean oil. Likewise, 8% gelatin, sodium caseinate, and whey protein films exhibited a tear strength of 0.30, 0.13, and 0.14 N, respectively [85]. The tearing strength of collagen films could be enhanced by developing blended film with nanoparticles. At optimum concentration, nanoparticles cause molecular rearrangement in the polymer matrix. Voloskova et al. [86] reported that the addition of nanosized corundum to collagen film resulted in a 25% increase in tear strength.

13.3.1.10 Zein Protein

Zein is the major protein found in corn, and it accounts for 45–50% of the gluten portion. It is rich in nonpolar amino acids. Hence it is hydrophobic and is soluble in 70–80% alcohol solution. Biodegradable films and coatings made from zein proteins gain attention because of their film-forming property and high barrier capacity to moisture [84].

Zein-based films can be synthesized by extrusion, solvent casting, or spin-casting methods. Firstly, in solvent casting, the zein protein and plasticizers such as polyols or fatty acids are dissolved in suitable solvents; ethanol is commonly used as a solvent. Then the aqueous solution is cast on a flat and nonstick surface [70]. During the drying process, the hydrophobic hydrogen and disulfide bond, to some extent, are formed between zein chains [87]. Compression molding or extrusion of zein resin at a high temperature can also be used to make zein films, which results in higher plasticization efficiency [46]. A spin coater may also be used to make zein films, which include four steps: putting a polymer solution to the substrate, speeding spinning to dispense the polymer solution, spinning at a steady speed, and solvent evaporation [88]. Due to many nonpolar amino acids, zein-based films have excellent water vapor barrier properties. However, the film formed from pure zein is brittle. Hence the plasticizer plays a significant role in controlling the brittle formation in the zein film, making the film more flexible [89]. Treatment such as exposure to UV and gamma radiations improves zein films' functionality. Gao et al. [90] reported that the use of glycerol as a plasticizer in zein film resulted in a gradual decrease in TS with an increase in glycerol concentration; however, the elongation at the break did not cause much difference at a level of 20% and exhibited sharp increase above 20% glycerol.

The mechanical characteristics of protein-nanocomposite films will influence the intrinsic features of nanomaterials, such as modulus, aspect ratio, matrix nature, dispersibility, and compatibility between the nanoparticle and the matrix [91]. Qu et al. [92] reported that the reinforcement of TiO_2 nanoparticles up to concentration of 0.15 wt% to zein/chitosan blend film significantly enhanced the TS. Higher concertation of nanoparticles negatively affected the TS. Similarly, nano cellulose fiber and nano-silica in soy protein-based composite film resulted in a 41.55% and 20.92% increment, respectively, in TS. However, the EB decreased from 153.54% to 117.45%

Cross-linking of protein will also facilitate an improved mechanical property of the film. Physical, chemical, or enzymatic approaches can be used to generate stronger intermolecular covalent connections, tighter molecule packing, and lower polymer mobility. Chemicals such as glutaraldehyde, formaldehyde, dialdehyde starch, carbonyldiimidazole, and mild alkali treatments are employed to cross-link protein films. In general, the cross-linking of protein with chemicals increases the TS, but it lowers the EB [93]. Wihodo and Moraru [94] reported an alkali treatment of soy protein isolate with 0.1 M NaOH resulted in a 35% increase in percentage elongation; however, it did not cause any changes in TS. Fan et al. [95] studied the effect of glutaraldehyde cross-linking on the mechanical properties of gelatin and zein films. They observed that the cross-linking considerably increased the TS, whereas the EB was decreased. Gao et al. [90] reported that adding more than 1% glutaraldehyde to zein protein reduces TS. The intermolecular interaction between $-NH_2$ and $-CHO$ is responsible for cross-linking in zein protein. However, when a higher concentration of glutaraldehyde is added, cross-linking primarily happens between amino acid molecules.

Enzymes are another tool to produce cross-linked proteins. One of the most prevalent enzymes involved in protein cross-linking is transglutaminase. It can catalyze an acyl transfer process in which the γ-carboxyamide group of glutamine acts as the acyl donor, and the e-amino group of lysine acts as the acyl acceptor, resulting in the synthesis of ε-(γ-glutamyl)-lysine cross-links in proteins. As a result, intramolecular and intermolecular covalent cross-linkages are formed inside protein molecules [94]. Other enzymes like horseradish peroxidase and tyrosinase are also used for the protein cross-linking process. Yildrim and Hettiarachchy [96] cross-linked whey protein isolate and soy 11S globulin with transglutaminase and observed that the film made from cross-linked protein is two times stronger than non-cross-linked film. Cheng et al. [97] conducted a similar study in collagen fibers and reported that the enzyme cross-linking increased the mechanical strength and elongation at the break of film.

The oxidation of amino acids, breaking of covalent bonds, production of protein-free radicals, and recombination and polymerization events may all occur when proteins are exposed to ionizing radiation. As a result, irradiating proteins causes cross-linking or disintegration. The TS of gluten film increased after an irradiation dose of 10 kGy. However, the EB decreased [98]. Similar findings were also observed by Sabato et al. [99] when working with gamma-irradiated whey and soy proteins. Similarly, UV light at a dose of $51.8 J/m^2$ for 24 hours increased the TS of wheat gluten, maize zein, and egg albumin films significantly [94]. The mechanical properties of a few protein-based films are summarized in Table 13.2.

13.3.1.11 Soy Protein

Soy proteins are globular proteins isolated from soybeans, and it is made up of a mixture of albumins and globulins. 90% of which function as storage protein, and it consists mainly of 7S (b-conglycinin) and 11S (glycinin) globulins [103]. Soy protein films are primarily prepared from soy protein isolates. It contains 90% of protein and is synthesized from finely ground, low-heat-treated, defatted soy flakes in which a great proportion of nonprotein substances are removed [104]. Soy protein film can also be made from soy flour, soymilk, and fractioned proteins. Its functionality depends on the type of protein used. The TS and elongation of the soy protein films increase with an increase in the molecular weight of the protein.

Several methods such as heating, extruding, spinning, casting, and thermally compacting are adopted to synthesize soy protein films. However, the casting method is widely accepted. The heating method of film formation was practiced and is used to form soy protein–lipid films. During the extrusion film manufacturing process, soy

Table 13.2 Mechanical properties of few protein-based films.

Protein film	Tensile strength (MPa)	Young's modulus (MPa)	Elongation at break (%)	References
Whey protein	4.68	—	114	[93]
Whey protein/glutaraldehyde	8.30		107	[93]
Whey protein/UV	6.40		110	[93]
Whey protein	5.64			[96]
Whey protein/transglutaminase	12.53			[96]
Gluten	1.7	13	501	[98]
Gluten/formaldehyde	8.1	70	190	[98]
Soy protein isolate	3.49	145.45	153.54	[100]
Soy protein isolate/cellulose nanoparticles	4.94	191.66	117.45	[100]
Soy protein isolate/nano silica	4.22	139.48	139.48	[100]
Zein protein	3.72	157	29.5	[101]
Casein/glycerol	3.4	5.5	55.3	[70]
Casein/sorbitol	4.6	6.9	66.1	[70]
Collagen	25.3	181.8	14.7	[78]
Collagen/chitosan	13.5	55.3	24.7	[78]
Gelatin	2.17	92.2	82.6	[102]
Gelatin/chitosan	13.08	439.3	44.5	[102]

proteins, polyethylene oxide, and low-density polyethylene are extruded. Spinning was done using an alkaline dope solution that included water, soy protein isolate, and sodium hydroxide. The dope was extruded via a spinning nozzle and placed into acetate buffer (pH 4.7), where it coagulated and formed a wet film [70].

In the casting method, the film-forming solution can be either acidic or alkaline; however, the property of the film depends on the pH of the casting medium. The film formed in alkaline solution exhibits better functionality than film processed in acidic solution. The functional properties of soy protein films can be improved by additives such as sodium dodecyl sulfate and carboxymethyl cellulose, which improve extensibility, moisture barrier property, and oxygen barrier property, respectively [99].

13.3.1.12 Gluten Protein

Wheat protein gluten is a hydrophobic globular protein that combines polypeptide molecules. It comprises two wheat flour proteins, prolamine, and glutelin. Wheat protein is separated into gliadin and glutenin based on their solubility in aqueous alcohol.

The former is soluble in 70% ethanol, but the latter is not. Wheat gluten may be dissolved in aqueous high or low-pH solutions at low ionic strength, but it is insoluble in water. Gluten has qualities that make it an excellent material for film production, such as cohesion and flexibility [84].

Two methods are generally employed for gluten film synthesis. One is the casting of an aqueous protein solution, and the second is the boiling of protein solution and collecting the layer formed on the surface of the solution. The functionality of the film is varied with the method employed for development. The film formed in the casting method exhibits higher elongation properties; however, the film formed by the boiling method shows stronger rupture resistance [70]. During the drying process, the disulfide bonds broken by the heating of film-forming solutions are reformed, and hydrogen and hydrophobic bonds are formed [105]. The purity of gluten plays a significant role in film appearance and its mechanical properties, and plasticizers and glycerin enhance the flexibility of gluten film [84].

13.3.2 Polymers Synthesized from Natural/Bioderived Monomers

This group consists of polymers synthesized from at least one monomer produced from renewable raw materials. These polymers have a lower carbon footprint and are sustainable. Some of the commonly used bio-based polymers are PLA, PEF, PBS, Poly(butylene succinate-*co*-butylene adipate) (PBSA), PBAT, etc.

13.3.2.1 Polylactic Acid (PLA)

PLA is a thermoplastic aliphatic polyester obtained from natural renewable materials. It is considered a biocompatible material and exhibits good thermomechanical properties. The synthesis of PLA starts from producing lactic acid monomers by the fermentation process. Agricultural goods such as maize and wheat and waste products from agriculture and the food industry such as molasses, whey, and green juice are examples of carbohydrate feedstocks that can be used as a source of lactic acid. The lactic acid monomers synthesized by fermentation are then polymerized to produce PLA. Lactic acid's molecular structure includes an asymmetric carbon atom, resulting in two optically active forms known as the L- and D-isomers. Some critical macromolecular characteristics, such as crystallinity, melting temperature, and processing ease, are influenced by the ratio of L- to D-monomer units [106].

Two methods are mainly used to produce PLA from lactic acid monomers. The first approach includes polycondensing lactic acid by removing water molecules with a solvent in a high vacuum and high-temperature environment. Lactide ring-opening polymerization (ROP) is another technique that produces high molecular weight PLA. Jacobsen et al. [107] established a continuous one-stage method employing reactive extrusion technology to make PLA manufacture commercially viable. This approach demands that the bulk polymerization be completed within a short period (five to seven minutes), as dictated by the extrusion system's residence time. PLA stability is sufficient at the processing temperature.

PLA has been listed as GRAS by the US Food and Drug Administration and authorized by the European Commission to be used in contact with food. PLA has high strength, stiffness, excellent organoleptic characteristics, and thermoplasticity. Its TS and elastic modulus are comparable to that of conventional petroleum-based plastic-like poly(ethylene terephthalate) (PET) and polystyrene (PS) [108]. However, it has low impact strength and disadvantages such as bitterness and poor barrier to gas and water. PLA has a glass transition temperature of 55 °C and a melting temperature of 175 °C. High molecular weight PLA is colorless, glossy, and stiff, and it has properties comparable to polystyrene [109]. The mechanical properties of some of the PLA-based and other bio-derived polymers are summarized in Table 13.3.

Its mechanical properties depend on additives used, such as plasticizer and presence of nanoparticles. The molecular weight and stereochemical composition of PLA's backbone determine its mechanical characteristics. PLA films have high TS and Young modulus, but their EB is relatively low due to their crystallinity and brittleness [87]. The remarkable mechanical strength of PLA is mostly due to individual lactic acid units. PLA films have an elastic modulus of 3000–4000 MPa and a TS of 50–70 MPa. On the other hand, the elongation values during break are lower, ranging from 1% to 7%. The TS of extruded PLA film in the longitudinal direction rose as the number of rotations increased, according to Sikora et al. [117].

Shankar et al. [111] reinforced ZnO nanoparticles to pure PLA film and observed that adding ZnO nanoparticles increases the TS; however, it decreased the elasticity

Table 13.3 Mechanical properties of a few films synthesized from bio-derived monomers.

Polymer	Tensile strength (MPa)	Young's modulus (MPa)	Elongation at break (%)	References
PLA	47.78	3027.79	5.35	[110]
PLA/cinnamaldehyde/ZnO NPs	18.16	2604.31	32.22	[110]
PLA	42.5	2490	4.9	[111]
PLA/ZnO NP	54.8	2460	4.4	[111]
PLA	52.1	1130	32.1	[112]
PLA/cellulose nano particles	49.5	1172	5.5	[112]
PLA/cellulose nano particles/ maleic anhydride	50.6	1154	13.2	[112]
PLA	49.6	3600	2.4	[113]
PLA/micro crystalline cellulose	38.2	4100	1.8	[113]
PLA	65.78	1680	8.91	[114]
PLA/triallyl isocyanurate	67.86	1950	3.87	[114]
PEF	85	2800	5	[115]
PPE	31	1363	3	[116]
PPTF	12	1419	2	[116]

modulus and EB. Similarly, Li et al. [110] studied the effect of ZnO nanoparticles in PLA/cinnamaldehyde blended film. They reported that the addition of ZnO nanoparticles causes a reduction in TS, whereas the elasticity modulus and EB were increased. However, the TS value of PLA/cinnamaldehyde is much less than that of pure PLA film. Kumar et al. [118] reported that cellulose nanofiber composite PLA film exhibited increased TS than pure PLA film, showing relatively low EB. Highly rigid nanofillers and the affinity between nanofillers and biopolymers at the interface are credited with improving nanocomposites' mechanical and thermal properties [119].

Another technique for improving the mechanical characteristics of PLA film is in situ cross-linking. Zhou et al. [112] synthesized PLA film in situ cross-linked with maleic anhydride-modified CNCs by casting method. They reported that the pure PLA film has a TS of 52.1 MPa, Young's modulus of 1130 MPa, and a 32.1% EB. However, the addition of pure CNCs decreased the mechanical parameters. Contrary, the use of maleic anhydride-modified cellulose nanoparticles with PLA using dicumyl peroxide increased the TS and Young's modulus of film. Many researchers observed that the direct incorporation of cellulose nanoparticles into PLA film did not cause a significant impact on mechanical properties. The dispersibility of nanoparticles, nanoparticle orientation in film matrix, and the interfacial interaction between them play an essential role in the mechanical properties of PLA nanocomposite films.

The film synthesized from pure PLA exhibited lower tear strength of 162.0 and 168.0 KN/m in the machine and transverse directions, respectively, due to the brittleness [120]. Ai et al. [121] reported that the PLA film containing 0.1% bis (tertbutyl dioxy isopropyl) benzene exhibited a tear strength of 95.7 and 75.4 kN/m in the transverse and machine direction, respectively. Similarly, Li et al. [120] reported PLA film containing 0.15% chain extender exhibited a tear strength of 122.8 and 144.8 KN/m in the machine and transverse direction, respectively. Further, adding PBAT to the previous PLA blended film increased the tear strength. However, the latter PLA composite film decreased the tear strength of the film.

PLA films have very low seal strength as compared to other polymers. PLA films exhibit a seal strength of around 2.75 N/1.5 mm. The seal strength of PLA films can be enhanced by forming a blended film. Ye et al. [122] studied the impact of tea polyphenol and chitosan in the tear strength of PLA film. They observed that the blending of PLA with a 1:9 proportion of tea polyphenol and chitosan exhibited higher seal strength of 9.54 N/15 mm. Contrary to this, Fathima et al. [123] reported that the addition of nano chitosan to PLA film significantly decreased the seal strength of the film. However, the addition of chain extenders to the PLA film will enhance the seal strength of the film.

Studies conducted on the burst strength of PLA film reported that the pure PLA films have a burst strength of 23.50 MPa. PLA blended film with a low concentration (0.025%) of fungal melanin increased the burst strength to 27.45 MPa, whereas higher content (0.2%) of fungal melanin decreased the burst strength of the film [124]. However, Hou et al. [125] reported that the burst strength of lignocellulosic film considerably increased by adding 0.5–4% PLA into it. Kumar et al. [126]

studied the effect of chemically modified nano zirconia in a PLA-based polymer blend and observed that nanoparticle addition to the blended film increased its burst strength.

Herrera et al. [127] synthesized PLA film plasticized with glycerol triacetate by blown film forming method and reported that the film exhibited a puncture resistance of 77 N/mm. Moreover, the addition of chitosan nanoparticles (5%) to the film increased the puncture resistance by 300%. However, the addition of essential oils like nisaplin, clove essential oil, and peppermint to PLA film decreased the puncture resistance of PLA films [128]. The processing conditions will also influence the puncture resistance of PLA films. Sikora et al. [117] reported that the rotational speed of the extruder has a significant impact on the puncture resistance of PLA-blown extruded films. An increase in rpm from 300 to 500 resulted in a 46% increase in puncture resistance of the film.

13.3.2.2 Polyethylene Furanoate (PEF)

PEF is considered a promising renewable resource-based solution to its fossil-based counterpart PET. It is a purely bio-based polymer that is generally synthesized by the condensation polymerization of 2,5-furandicarboxylic acid ($C_6H_4O_5$) (FDCA) and monoethylene glycol (MEG). FDCA is a relatively novel chemical made from the C5–C6 sugars of biomass by dehydration into hydroxymethylfurfural (HMF), with its subsequent catalytic oxidation [129]. Corn, grain, sugar cane, agricultural raw materials, and paper wastes can be used as a source of sugar in biofermentation, which can be replaced by oil-based terephthalic acid (TA) in PET to produce PEF [130].

PEF has a strong barrier and thermal qualities that are similar to PET. It has a water vapor barrier that is twice as effective as PET. PEF has a glass transition temperature of 86 °C, while PET has a glass transition temperature of 74 °C. PEF melts at 235 °C, which is lower than PET's 365 °C melting point, making it easier to process. Its strength modulus is almost 1.5 times higher than PET [131]. Condensation polymerization of PEF is a time-consuming process, and it may cause degradation and undesirable discoloration. As a result, a quick synthesis technique for bottle-grade PEF was established using ROP from cyclic PEF oligomers in minutes, preventing product deterioration and discoloration. As a result, the cyclic oligomer mixture's melting point is about 370 °C [132].

13.3.2.3 Polybutylene Succinate (PBS)

PBS is generally synthesized from 1,4 Butanediol and succinic acid. Succinic acid can be produced by microbial fermentation of several wild-type strains such as *Actinobacillus succinogenes* [133]. Succinic acid is considered as one of the major bio-based monomer for the synthesis of bio-based polymers. The ester linkages present in PBS make it biodegradable, and at the same time, it possesses excellent mechanical properties. The PBS can be blended with thermoplastic starch or other polymers such as poly ethylene succinate to make it more economical and suitable for many packaging applications. This type of blending helps in making biodegradable polyesters with similar properties to PET [134].

13.3.2.4 Poly(Butylene Adipate-*co*-Terephthalate)
PBAT is a copolyester made up of two comonomers consisting of a rigid part made by the polymerization of 1,4 butanediol with an aromatic TA and a flexible unit made by the polymerization of 1,4 butanediol with an aliphatic adipic acid. PBAT exhibits better mechanical properties. PBAT can be used to blend with PLA so that fully biodegradable blends with good processability, toughness, and impact resistance can be manufactured. The properties of PLA and PBAT are complimentary in nature, and hence PLA/PBAT formulations are developed as a sustainable alternative for conventional nondegradable polymers [135].

13.3.2.5 Bio-based Polyethylene
Polyethylene is the most exploited polymer globally, and it is produced from the monomer ethylene, which is also used in the production of other polymers such as PVC and PS. Commercially ethylene is synthesized from petroleum feedstock by distillation. Recently, great interest has been seen in developing these polymers from biological resources. Ethylene monomers can be synthesized by dehydration of ethanol produced by glucose fermentation from bio-based resources. Various bioresources feedstock such as sugarcane, sugar beet, starch crops coming from maize, wheat, or other grains, and lignocellulosic materials can be used to produce bioethanol. The process involves the anaerobic fermentation of glucose present in the feedstock to ethanol and followed by ethanol distillation to remove water, resulting in an azeotrope mixture of hydrous ethanol and by-product vinasse. Further polymerization of these ethylene monomers is similar to petroleum-based ethylene monomers [63]. The properties of polyethylene obtained by the bioprocess are identical to that of petroleum-based polymer in terms of chemical, physical, and mechanical properties [136].

13.3.2.6 Bio-Based Polypropylene (Bio-PP)
After ethylene, propylene is the most explored building block in the production of polymers. Bio-based polypropylene (Bio-PP) can be synthesized from bioresources rich in glucose by butylene dehydration of bio-isobutanol followed by polymerization. As compared to bio-polyethylene (Bio-PE), Bio-PP is not explored commercially.

13.4 Mechanical Properties of Natural Polymers Synthesized from Microorganisms-Based Packaging Films

13.4.1 Polymer Processed from Microorganisms

13.4.1.1 Polyhydroxyalkanoate (PHA)
PHAs are renewable, biodegradable, biocompatible, and optically active polyester. These are linear polyesters formed as intracellular particles by many bacterial species, and they serve as an energy and carbon reserve material. More than 150 different monomers of hydroxyalkanoate have been discovered, which can be combined to yield polymers with various properties. One of the factors that determine PHA

properties is monomer composition. It can be controlled by choosing different substrates and bacterial strains to synthesize the polyesters [137]. Poly 3-hydroxybutyrate (PHB) and polyhydroxybutyrate-*co*-valerate (PHBV) are two PHAs that are commercially available and received most of the industrial attention [129].

Many PHA-producing bacteria are identified, which consume renewable feedstock such as lignocellulose, vegetable oils, and fatty acids. The microbial sources of PHAs and their carbon sources are summarized in Table 13.4. In general, the microbial production of PHA includes several steps such as fermentation, where the feedstock is introduced into the reactor until the bacterial growth, followed by isolation and purification. Once the PHA accumulation is completed, the PHA is extracted using an organic solvent. Finally, the PHA polymer is isolated and purified by precipitation in alcohol. Adjusting the substrate and fermentation conditions can change the characteristics of PHA.

The nutrients in the fermentation medium will aid in managing the chemical composition of PHAs, either as a mixture of monomer subunits or as a single monomer subunit. Pure bacterial culture and refined feedstocks are used in the industrial manufacturing process, making them costly. Mixed culture and low-cost feedstock, such as sewage water, were used as substrates in certain research. Homopolymers, heteropolymers, and both can be used to create PHAs. Depending on the purpose, PHA

Table 13.4 Microbial sources of PHAs and their carbon sources.

Polymer	Microorganism	Carbon source	References
PHB	*Ralstonia eutropha* H16	Glucose	[138]
	Escherichia coli	Glucose	[139]
	Alcaligenes eutrophus	Glucose	[140]
	R. eutropha H16	Fructose	[141]
	Bacillus pasteurii and *Micrococcus luteus*	2% glucose, sucrose, and fructose	[142]
	Rhodococcus pyridinivorans BSRT1-1	Sucrose	[143]
	Bacillus pumilus (E10)	*Arthrospira platensis* biomass	[144]
PHA	*Alphaproteobacteria Betaproteobacteria*	Crude glycerol	[145]
PHBV	*Azotobacter vinelandii*	Sucrose	[146]
	Cupriavidus necator	Fructose and propionic acid	[147]
	C. necator	Glycerol	[148]
	Rhodospirillum rubrum	Fructose	[149]
Poly(3-hydroxybutyrate-*co*-3-hydroxyhexanoate)	*R. eutropha* Re2133/pCB81	Food waste	[150]

can be made as high or low-molecular-weight polymers. It may be used as raw material, a mix, or an additive to change other polymers, including PLA, PBS, and polycaprolactone. [130]

The mechanical qualities of PHA blended films are determined by the copolymer's composition and molecular structure. PHB is a rigid, highly crystalline thermoplastic polymer with mechanical qualities similar to isotactic PP. Poly (3-hydroxybutyrate-*co*-3-hydroxyvalerate), commonly known as PHBV, is another PHA-type which is a biodegradable plastic produced naturally by bacteria and a good alternative for many nonbiodegradable synthetic polymers.

Due to its characteristics and melting temperature of 50–180 °C, PHA films have been employed in various applications [151]. Pérez-Arauz et al. [141] synthesized PHA by using peanut oil as a source of carbon, and they reported that the film synthesized by using this PHA exhibited a TS of 75.8 MPa. Similarly, the film's Young's and elastic modulus was 2.7 MPa and 25.7, respectively. However, these TS and Young's modulus are much lower than the standard reference value of PHA film. It has standard values of 820 and 20 MPa of TS and Young's modulus.

Further, to increase the mechanical properties of PHA film, plasticizers are used, which increase the flexibility, ductility, and extensibility of the final film by reducing the intermolecular attraction between monomeric chains. An addition of 5% epoxidized broccoli oil into PHA film as a plasticizer reduced the Young's modulus by 55%, enabling maximum elongation of the film [152]. Similarly, an increase in the elongation at the break of PHB film was also reported by different plasticizers.

Jin et al. [153] studied the tear strength of PHBV film and reported that the pure PHBV film has a tear strength of 24.08 kN/m, and he observed that the blending of PHBV with polyamidoamine increased the tear strength of the film. A high proportion of polyamidoamine resulted in a more than 50% increase in the tear strength of the composite film. Likewise, the blending of PHBV with other polymers such as cellulose acetate butyrate, and starch also increased the tear strength of the composite film.

13.4.1.2 Bacterial Cellulose

Bacterial cellulose is produced by various species of gram-negative and gram-positive bacteria such as *Gluconacetobacter xylinus*, *Agrobacterium*, *Aerobacter*, *Achromobacter*, *Azotobacter*, *Rhizobium*, *Sarcina*, *Salmonella*, *Alcaligenes*, *Aerobacter*, *Pseudomonas*, *Dickeya*, and *Rhodobacter*. Among these, *G. xylinus* is the most effective source for bacterial cellulose production due to its high productivity [154]. Bacterial cellulose, an exopolysaccharide obtained in a ribbon-like structure from pellicles of *Acetobacter xylinum* (*G. xylinus*), is a unique biopolymer in terms of its molecular structure, mechanical strength, and chemical stability [155]. It gives unique structural and biochemical properties such as ultrafine nanofiber network structure, inadaptability, inert, biodegradability, hypoallergenity, bioconsistency, crystallinity, high water holding capacity, and chemical stability [156, 157].

Bacterial cellulose has a basic fibril structure with the chemical formula $(C_6H_{10}O_5)n$, consisting of glucose units joined by a 1–4% glucan bond. Inter and intra hydrogen bonding holds the glucan chains of bacterial cellulose together. It's fiber network comprises a well-arranged three-dimensional nanofibres, resulting in a hydrogel sheet with a large surface area and porosity [158]. Bacterial cellulose has an advantage over plant cellulose, and it is free from hemicellulose, pectin, and lignin that is found in plant cellulose [159]. Bacterial cellulose's uniform and ultrafine structure provide improved mechanical properties, including high TS, elastic modulus, and high wet strength [160]. This polymer is expected to play a major role as a food additive, as a scaffold in tissue engineering, food packaging, and preparation of composite materials [161].

Bacterial cellulose consists of highly ordered, crystalline regions along with some amorphous regions in varying proportions. When the cellulose fibers are subjected to various mechanical, chemical, and enzyme treatments, the amorphous region can be removed and highly crystalline regions of the cellulose microfibrils can be extracted in the form of CNCs. These nanocrystals possessed exceptionally high mechanical properties. The theoretical TS of CNCs was found to be in the range of 7.5–7.7 GPa, and the elastic modulus of the CNC was found to be around 150 GPa [162, 163]. CNC can be used as the load-bearing constituent in many polymer nanocomposite systems as it significantly improves the mechanical properties of polymeric materials even at very low volume fractions. Several natural polymers, such as starch, chitosan, natural rubber, cellulose acetate butyrate, carboxymethyl cellulose, HPMC, gelatin, and soy protein, were reinforced with these nanocrystals [43].

13.4.1.3 Xanthan

The bacteria *Xanthomonas campestris* produces xanthan, which is an exopolysaccharide. It is made up of five distinct sugar units, the most common of which are D-glucose and D-mannose, as well as D-glucuronic acid and pyruvic acid [164]. It is manufactured by submerged aerobic fermentation of a pure *X. campestris* culture [165]. Xanthan features a 1,4-linked-D-glucose structure with a trisaccharide side chain connected to an alternating D-glucose residue. Xanthans are listed under GRAS. It can be readily soluble in hot and cold water with minimal effect on viscosity from either temperature or pH [5].

13.4.1.4 Pullulan

Pullulan is a nonionic exopolysaccharide produced by the fungus-like yeast *Aureobasidium pullulans*. It is a linear, unbranched, water-soluble polysaccharide made up of maltotriose units interconnected by [1, 6] glycosidic units [166]. Pullulans are highly soluble in hot and cold water and dilute alkali; however, they are insoluble in alcohol and other organic solvents except for dimethylsulfoxide and formamide [167]. Pullulan films are colorless, odorless, tasteless, water permeable, heat sealable, and transparent. It also provides a barrier to oxygen and oil [168]. Combined with glutathione and chitooligosaccharides, pullulan film offers extended shelf life to various fruits [169]. Similarly, pullulan coating extended the shelf life of strawberries and non-climacteric fruits by altering the internal atmosphere resulting in reduced respiration [168]. It also reported that pullulan would inhibit the fungal growth in food [167].

In general, the film synthesized from pullulan is brittle. To overcome this brittleness, plasticizers are used, reducing the intermolecular attraction and increasing polymer chains' mobility. The use of sorbitol and sucrose fatty acid ester as a plasticizer in pullulan film significantly affected the mechanical properties, and the film changed its brittle nature to ductile [168]. Glycerol plasticized pullulan film exhibited a TS of 10.60 MPa [170]. However, the addition of nanoparticles to pullulan film makes it more brittle. The TS of nano TiO_2 pullulan film was higher than that of pure film, and the EB exhibited an opposite tendency. The film synthesized by adding 0.04 g nanoparticles showed a TS of 15.99 MPa [171]. It is because nanoparticle increases the film's cohesion force and reduce flexibility. The edible film synthesized from gallan is more brittle than pullulan films.

13.4.1.5 Gellan

Gellan is an exopolysaccharide synthesized by *Sphingomonas elodea* (*Pseudomonas elodea*). It is a tetracyclic repeating unit made of one rhamnose, one glucuronic acid, and two glucose units replaced with acyl groups to form *O*-glycosidically-linked esters. It has a unique colloidal structure and excellent coating and gelling properties. It is widely utilized in the food industry as a gelling and texturizing agent [172]. Gellan seems to offer a lot of promise as a film-forming agent. Gellan-based coatings are applied to fresh-cut veggies to enhance quality and increase shelf life [173]. Zhu et al. [170] studied the mechanical properties of gellan and their blended film with pullulan. Gellan film has a TS and elongation at a break of 27.65 MPa, and 8.7%, the addition of pullulan to the gellan decreased the TS. In contrast, the EB increased with the increase in pullulan concentration.

13.4.1.6 Levan

In the food sector, levan is one of the widely studied hydrocolloids. It is a homopolysaccharide composed primarily of fructose units, linked by (2→6)-glycosidic bonds, with some (2→1)-linked branch chains, and a D-glucosyl residue at the end of the chain [174]. It possesses biomedical antioxidant, anti-inflammatory, anticarcinogenic, anti-AIDS, and hyperglycemic inhibitor effects, among others [175]. Extracellular levansucrase is used to produce levan by microorganisms such as *Erwinia herbicola*, *Zymomonas mobilis*, *Microbacterium laevaniformans*, *Bacillus subtilis*, and *Bacillus methylotrophicus* [176], which utilize sucrose rich substrates. Levan has a large molecular weight and widespread branching [174]. Levan can be used to make film and coatings in the food industry as a prospective and functional biopolymer. Levan-based films are transparent, flexible, and robust, with high oxygen barrier properties, making them ideal for food packaging [177].

13.5 Conclusion

The nonbiodegradability and environmental impact of petroleum-based packaging materials lead to the exploration of new sources of biopolymers or packaging films, which are biodegradable and biocompatible in nature. The biodegradable packaging materials from various sources such as synthesized from naturally occurring raw materials. It includes polymers from bioderived monomers such as PLA, polymer synthesized from biomass such as starch, protein, and polymers derived from microbial sources such as PHAs. The mechanical properties of packaging film are essential in determining the specific application. Compared to synthetic petroleum-based plastic packaging, the mechanical properties such as tensile properties, seal, tear, and burst strength of bio-based packaging films are limited to a certain extent. Bio-based packaging materials exhibit an excellent oxygen barrier; however, they are weak in water vapor barrier properties. However, various methods are adopted to improve the mechanical properties of bio-based packaging materials. The formation of blended films of different polymers, reinforcement with nanoparticles, use of additives such as cross-linking agents and plasticizers are a few techniques used to improve the mechanical properties of bio-based packaging films.

References

1 Ali, A. and Ahmed, S. (2018). Recent advances in edible polymer based hydrogels as a sustainable alternative to conventional polymers. *Journal of Agricultural and Food Chemistry* 66 (27): 6940–6967.
2 Jacob, J., Lawal, U., Thomas, S., and Valapa, R.B. (2020). Biobased polymer composite from poly(lactic acid): processing, fabrication, and characterization for food packaging. In: *Processing and Development of Polysaccharide-Based Biopolymers for Packaging Applications* (ed. Y.E. Zhang), 97–115. Amsterdam, The Netherlands: Elsevier.
3 George, J., Kumar, R., Jayaprahash, C. et al. (2006). Rice bran-filled biodegradable low-density polyethylene films: development and characterization for packaging applications. *Journal of Applied Polymer Science* 102 (5): 4514–4522.
4 Europe P (2020). Bioplastics market development update 2020. European Bioplastics. pp. 2019–20.
5 Mohamed, S.A.A., El-Sakhawy, M., and El-Sakhawy, M.A.M. (2020). Polysaccharides, protein and lipid-based natural edible films in food packaging: a review. *Carbohydrate Polymers* 238: 116178. https://doi.org/10.1016/j.carbpol.2020.116178.
6 Standard Test Method for Tensile Properties of Thin Plastic Sheeting, ASTM D 882.
7 Standard Test Method for Seal Strength of Flexible Barrier Materials, ASTM F 88.
8 Tear-Propagation Resistance (Trouser Tear) of Plastic Film and Thin Sheeting by a Single-Tear Method ASTM D 1938.
9 Standard Test Method for Tear Resistance (Graves Tear) of Plastic Film and Sheeting1 ASTM D 1004.
10 Standard Test Method for Slow Rate Penetration Resistance of Flexible Barrier Films and Laminates, ASTM F1306.
11 Standard Test Methods for Impact Resistance of Plastic Film by the Free- Falling Dart Method, ASTM D1709.
12 Ciannamea, E.M., Castillo, L.A., Barbosa, S.E., and De Angelis, M.G. (2018). Barrier properties and mechanical strength of bio-renewable, heat-sealable films based on gelatin, glycerol and soybean oil for sustainable food packaging. *Reactive and Functional Polymers* 125: 29–36.

13 Cazón, P., Velazquez, G., Ramírez, J.A., and Vázquez, M. (2017). Polysaccharide-based films and coatings for food packaging: a review. *Food Hydrocolloids* 68: 136–148.

14 Molavi, H., Behfar, S., Ali Shariati, M. et al. (2015). A review on biodegradable starch based film. *Journal of Microbiology, Biotechnology and Food Sciences* 04 (5): 456–461.

15 Thakur, R., Pristijono, P., Scarlett, C.J. et al. (2019). Starch-based films: major factors affecting their properties. *International Journal of Biological Macromolecules* 132: 1079–1089.

16 Al-Hassan, A.A. and Norziah, M.H. (2012). Starch-gelatin edible films: water vapor permeability and mechanical properties as affected by plasticizers. *Food Hydrocolloids* 26 (1): 108–117.

17 Othman, S.H., Kechik, N.R.A., Shapi'i, R.A. et al. (2019). Water sorption and mechanical properties of starch/chitosan nanoparticle films. *Journal of Nanomaterials* 2019: 3843949.

18 Nisa, I.U., Ashwar, B.A., Shah, A. et al. (2015). Development of potato starch based active packaging films loaded with antioxidants and its effect on shelf life of beef. *Journal of Food Science and Technology* 52 (11): 7245–7253.

19 Othman, S.H., Majid, N.A., Tawakkal, I.S.M.A. et al. (2019). Tapioca starch films reinforced with microcrystalline cellulose for potential food packaging application. *Food Science and Technology* 39 (3): 605–612.

20 Zainol, I., Akil, H.M., and Mastor, A. (2009). Effect of γ-irradiation on the physical and mechanical properties of chitosan powder. *Materials Science and Engineering: C* 29 (1): 292–297.

21 Wu, H., Lei, Y., Lu, J. et al. (2019). Effect of citric acid induced crosslinking on the structure and properties of potato starch/chitosan composite films. *Food Hydrocolloids* 97: 105208.

22 Ghanbarzadeh, B., Almasi, H., and Entezami, A.A. (2010). Physical properties of edible modified starch/carboxymethyl cellulose films. *Innovative Food Science and Emerging Technologies* 11 (4): 697–702.

23 Tongdeesoontorn, W., Mauer, L.J., Wongruong, S. et al. (2011). Effect of carboxymethyl cellulose concentration on physical properties of biodegradable cassava starch-based films. *Chemistry Central Journal* 5 (1): 1–8.

24 Tongdeesoontorn, W., Mauer, L.J., Wongruong, S. et al. (2012). Mechanical and physical properties of cassava starch-gelatin composite films. *International Journal of Polymeric Materials and Polymeric Biomaterials* 61 (10): 778–792.

25 Sudheesh, C., Sunooj, K.V., Jamsheer, V. et al. (2021). Development of bioplastic films from γ – irradiated kithul (*Caryota urens*) starch; morphological, crystalline, barrier, and mechanical characterization. *Starch/Staerke* 73 (5–6): 1–7.

26 Sudheesh, C., Sunooj, K.V., Sasidharan, A. et al. (2020). Energetic neutral N_2 atoms treatment on the kithul (*Caryota urens*) starch biodegradable film: physico-chemical characterization. *Food Hydrocolloids* 103: 105650.

27 Bangar, S.P. and Whiteside, W.S. (2021). Nano-cellulose reinforced starch bio composite films – a review on green composites. *International Journal of Biological Macromolecules* 185: 849–860.

28 Harunsyah, Yunus, M., and Fauzan, R. (2017). Mechanical properties of bioplastics cassava starch film with zinc oxide nanofiller as reinforcement. *IOP Conference Series: Materials Science and Engineering* 210 (1): 12015.

29 George, J., Nair, S.G., Kumar, R. et al. (2021). A new insight into the effect of starch nanocrystals in the retrogradation properties of starch. *Food Hydrocolloids for Health* 1: 100009.

30 Dai, L., Yu, H., Zhang, J., and Cheng, F. (2021). Preparation and characterization of cross-linked starch nanocrystals and self-reinforced starch-based nanocomposite films. *International Journal of Biological Macromolecules* 181: 868–876.

31 Kumar, S.V., George, J., and Sajeevkumar, V.A. (2018). PVA based ternary nanocomposites with enhanced properties prepared by using a combination of rice starch nanocrystals and silver nanoparticles. *Journal of Polymers and the Environment* 26 (7): 3117–3127.

32 Petersen, K., Nielsen, P.V., and Olsen, M.B. (2001). Physical and mechanical properties of biobased materials – starch, polylactate and polyhydroxybutyrate. *Starch/Staerke* 53 (8): 356–361.

33 Halimatul, M.J., Sapuan, S.M., Jawaid, M. et al. (2019). Effect of sago starch and plasticizer content on the properties of thermoplastic films: mechanical testing and cyclic soaking-drying. *Polimery/Polymers* 64 (6): 422–431.

34 Ismail, I., Osman, A.F., and Leong Ping, T. (2019). Effects of ultrasonication process on crystallinity and tear strength of thermoplastic starch/cellulose biocomposites. *IOP Conference Series: Materials Science and Engineering* 701 (1): 012045.

35 Gujral, H., Sinhmar, A., Nehra, M. et al. (2021). Synthesis, characterization, and utilization of potato starch nanoparticles as a filler in nanocomposite films. *International Journal of Biological Macromolecules* 186: 155–162.

36 Sharma, I., Sinhmar, A., Thory, R. et al. (2021). Synthesis and characterization of nano starch-based composite films from kidney bean (*Phaseolus vulgaris*). *Journal of Food Science and Technology* 58 (6): 2178–2185.

37 Xu, Y., Scales, A., Jordan, K. et al. (2017). Starch nanocomposite films incorporating grape pomace extract and cellulose nanocrystal. *Journal of Applied Polymer Science* 134 (6): 44438.

38 Cazón, P., Velázquez, G., and Vázquez, M. (2020). Bacterial cellulose films: evaluation of the water interaction. *Food Packaging and Shelf Life* 25: 100526.

39 Nafchi, A.M. and Alias, A.K. (2013). Mechanical, barrier, physicochemical, and heat seal properties of starch films filled with nanoparticles. *Journal of Nano Research* 25: 90–100.

40 Nouri, L. and Mohammadi Nafchi, A. (2014). Antibacterial, mechanical, and barrier properties of sago starch film incorporated with betel leaves extract. *International Journal of Biological Macromolecules* 66: 254–259.

41 López, O.V., Lecot, C.J., Zaritzky, N.E., and García, M.A. (2011). Biodegradable packages development from starch based heat sealable films. *Journal of Food Engineering* 105 (2): 254–263.

42 George, J., Bawa, A.S., and Hatna, S. (2010). Synthesis and characterization of bacterial cellulose nanocrystals and their PVA nanocomposites. *Advances in Materials Research* 123–125: 383–386.

43 George, J. and Sabapathi, S.N. (2015). Cellulose nanocrystals: synthesis, functional properties, and applications. *Nanotechnology, Science and Applications* 8: 45–54.

44 Das, M. and Chowdhury, T. (2016). Heat sealing property of starch based self-supporting edible films. *Food Packaging and Shelf Life* 9: 64–68.

45 George, J., Kumar, R., Sajeevkumar, V.A. et al. (2014). Amine functionalised nanoclay incorporated hydroxypropyl methyl cellulose nanocomposites: synthesis and characterisation. *International Journal of Plastics Technology* 18 (2): 252–262.

46 George, J., Sabapathi, S.N., and Hatna, S. (2015). Water soluble polymer-based nanocomposites containing cellulose nanocrystals. In: *Eco-friendly polymer nanocomposites Advanced Structured Materials* (ed. V.K. Thakur and M.K. Thakur), 259–293. New Delhi: Springer.

47 Liu, Y., Ahmed, S., Sameen, D.E. et al. (2021). A review of cellulose and its derivatives in biopolymer-based for food packaging application. *Trends in Food Science and Technology* 112: 532–546.

48 Dasaiah, M., Shivakumara, L.R., Demappa, T., and Vasudev, S. (2017). Mechanical and barrier properties of hydroxy propyl methyl cellulose edible polymer films with plasticizer combinations. *Journal of Food Processing & Preservation* 41 (4): 13020.

49 George, J., Kumar, R., Sajeevkumar, V.A. et al. (2014). Hybrid HPMC nanocomposites containing bacterial cellulose nanocrystals and silver nanoparticles. *Carbohydrate Polymers* 105 (1): 285–292.

50 Moura, J.M., Farias, B.S., Cadaval, T.R.S., and Pinto, L.A.A. (2020). Chitin/Chitosan based films for packaging applications. In: *Bio-BasedPackaging: Material, Environmental and Economic Aspects* (ed. S.M. Sapuan and R.A. Ilyas), 69.

51 Elsabee, M.Z. and Abdou, E.S. (2013). Chitosan based edible films and coatings: a review. *Materials Science and Engineering: C* 33 (4): 1819–1841.

52 Adila, S.N., Suyatma, N.E., Firlieyanti, A.S., and Bujang, A. (2013). Antimicrobial and physical properties of chitosan film as affected by solvent types and glycerol as plasticizer. *Advances in Materials Research* 748: 155–159.

53 Khan, A., Khan, R.A., Salmieri, S. et al. (2012). Mechanical and barrier properties of nanocrystalline cellulose reinforced chitosan based nanocomposite films. *Carbohydrate Polymers* 90 (4): 1601–1608.

54 García, M.A., Pérez, L., De La Paz, N. et al. (2015). Effect of molecular weight reduction by gamma irradiation on chitosan film properties. *Materials Science and Engineering: C* 55: 174–180.

55 George, J., Aaliya, B., Sunooj, K.V., and Kumar, R. (2022). An overview of higher barrier packaging using nanoadditives. In: *Nanotechnology-Enhanced Food Packaging* (ed. J. Parameswaranpillai, R.E. Krishnankutty, A. Jayakumar, et al.), 235–264. Wiley.

56 Kittur, F.S., Kumar, K.R., and Tharanathan, R.N. (1998). Functional packaging properties of chitosan films. *European Food Research and Technology* 206 (1): 44–47.

57 Singh, K., Suri, R., Tiwary, A.K., and Rana, V. (2012). Chitosan films: crosslinking with EDTA modifies physicochemical and mechanical properties. *Journal of Materials Science. Materials in Medicine* 23 (3): 687–695.

58 Rodríguez-Núñez, J.R., Madera-Santana, T.J., Sánchez-Machado, D.I. et al. (2014). Chitosan/Hydrophilic plasticizer-based films: preparation, physicochemical and antimicrobial properties. *Journal of Polymers and the Environment* 22 (1): 41–51.

59 Prateepchanachai, S., Thakhiew, W., Devahastin, S., and Soponronnarit, S. (2017). Improvement of mechanical and heat sealing properties of chitosan films via the use of glycerol and gelatin blends in film-forming solution. *18th National Conference and the 10th TSAE International Conference TSAE 2017*, IMPACT Exhibition center, Bangkok, Thailand (7-9 September 2017). pp. 30–35.

60 Abdullah, N.A.S., Mohamad, Z., Khan, Z.I. et al. (2021). Alginate based sustainable films and composites for packaging: a review. *Chemical Engineering Transactions* 83: 271–276.

61 Rhim, J.W. (2004). Physical and mechanical properties of water resistant sodium alginate films. *LWT – Food Science and Technology* 37 (3): 323–330.

62 Benavides, S., Villalobos-Carvajal, R., and Reyes, J.E. (2012). Physical, mechanical and antibacterial properties of alginate film: effect of the crosslinking degree and oregano essential oil concentration. *Journal of Food Engineering* 110 (2): 232–239.

63 Arfin, T. and Sonawan, K. (2018). Bio-based materials: past to future. In: *Bio-based Materials for Food Packaging* (ed. S. Ahmed). Springer. 11 pp.

64 Bayarri, M., Oulahal, N., Degraeve, P., and Gharsallaoui, A. (2014). Properties of lysozyme/low methoxyl (LM) pectin complexes for antimicrobial edible food packaging. *Journal of Food Engineering* 131: 18–25.

65 Espitia, P.J.P., Du, W.X., De Avena-Bustillos, R., J. et al. (2014). Edible films from pectin: physical-mechanical and antimicrobial properties – a review. *Food Hydrocolloids* 35: 287–296.

66 Campos, C.A., Gerschenson, L.N., and Flores, S.K. (2011). Development of edible films and coatings with antimicrobial activity. *Food and Bioprocess Technology* 4 (6): 849–875.

67 Mendes, J.F., Martins, J.T., Manrich, A. et al. (2019). Development and physical–chemical properties of pectin film reinforced with spent coffee grounds by continuous casting. *Carbohydrate Polymers* 210: 92–99.

68 de Oliveira, A.C.S., Ferreira, L.F., de Oliveira Begali, D. et al. (2021). Thermoplasticized pectin by extrusion/thermo-compression for film industrial application. *Journal of Polymers and the Environment* 29 (8): 2546–2556.

69 Wusigale, Liang, L., and Luo, Y. (2020). Casein and pectin: structures, interactions, and applications. *Trends in Food Science and Technology* 97: 391–403.

70 Chen, H., Wang, J., Cheng, Y. et al. (2019). Application of protein-based films and coatings for food packaging: a review. *Polymers (Basel)* 11 (12): 1–32.

71 Said, N.S. and Sarbon, N.M. (2019). Protein-based active film as antimicrobial food packaging: a review. In: *Active Antimicrobial Food Packaging* (ed. I. var and S. Uzunlu), 3–19.

72 Bonnaillie, L.M., Zhang, H., Akkurt, S. et al. (2014). Casein films: the effects of formulation, environmental conditions and the addition of citric pectin on the structure and mechanical properties. *Polymers (Basel)* 6 (7): 2018–2036.

73 Javanmard, M. (2009). Biodegradable whey protein edible films as a new biomaterials for food and drung packaging. *Iranian Journal of Pharmaceutical Sciences* 5 (3): 129–134.

74 Schmid, M. and Müller, K. (2018). Whey protein-based packaging films and coatings. In: *Whey Proteins: From Milk to Medicine* (ed. H.C. Deeth and N. Bansal), 407–437.

75 Schmid, M., Dallmann, K., Bugnicourt, E. et al. (2012). Properties of whey-protein-coated films and laminates as novel recyclable food packaging materials with excellent barrier properties. *International Journal of Polymer Science* 2012: 5–7.

76 Socaciu, M.I., Fogarasi, M., Semeniuc, C.A. et al. (2020). Formulation and characterization of antimicrobial edible films based on whey protein isolate and tarragon essential oil. *Polymers (Basel)* 12 (8): 1–21.

77 Banerjee, R., Chen, H., and Wu, J. (1996). Milk protein-based edible film mechanical strength changes due to ultrasound process. *Journal of Food Science* 61 (4): 824–828.

78 Ahmad, M., Nirmal, N.P., Danish, M. et al. (2016). Characterisation of composite films fabricated from collagen/chitosan and collagen/soy protein isolate for food packaging applications. *RSC Advances* 6 (85): 82191–82204.

79 Sionkowska, A., Skrzyński, S., Śmiechowski, K., and Kołodziejczak, A. (2017). The review of versatile application of collagen. *Polymers for Advanced Technologies* 28 (1): 4–9.

80 Hashim, P., Mohd Ridzwan, M.S., Bakar, J., and Mat Hashim, D. (2015). Collagen in food and beverage industries. *International Food Research Journal* 22 (1): 1–8.

81 Sommer, I. and Kunz, P.M. (2012). Improving the water resistance of biodegradable collagen films. *Journal of Applied Polymer Science* 125 (5): E27–E41.

82 Ramos, M., Valdés, A., Beltrán, A., and Garrigós, M. (2016). Gelatin-based films and coatings for food packaging applications. *Coatings* 6 (4): 41.

83 George, J., Sabapathi, S.N., and Siddaramaiah (2017). Water soluble polymer based hybrid nanocomposites. In: *Hybrid Polymer Composite Materials: Processing* (ed. V.K. Thakur, M.K. Thakur and R.K. Gupta), 71–88. Woodhead Publishing.

84 Hassan, B., Chatha, S.A.S., Hussain, A.I. et al. (2018). Recent advances on polysaccharides, lipids and protein based edible films and coatings: a review. *International Journal of Biological Macromolecules* 109: 1095–1107.

85 Wang, L.Z., Liu, L., Holmes, J. et al. (2007). Assessment of film-forming potential and properties of protein and polysaccharide-based biopolymer films. *International Journal of Food Science and Technology* 42 (9): 1128–1138.

86 Voloskova, E.V., Baikina, L.K., and Poluboyarov, V.A. (2014). Effect of nanosized corundum on the structure and properties of molecular collagen. *Russian Journal of Physical Chemistry A* 88 (2): 301–304.

87 Zhang, Y., Cui, L., Che, X. et al. (2015). Zein-based films and their usage for controlled delivery: origin, classes and current landscape. *Journal of Controlled Release* 206 (2699): 206–219.

88 Shi, K., Kokini, J.L., and Huang, Q. (2009). Engineering zein films with controlled surface morphology and hydrophilicity. *Journal of Agricultural and Food Chemistry* 57 (6): 2186–2192.

89 Wang, Y., Rakotonirainy, A.M., and Padua, G.W. (2003). Thermal behavior of zein-based biodegradable films. *Starch/Starke* 55: 25–29.

90 Gao, Y., Zheng, H., Wang, J. et al. (2021). Physicochemical properties of zein films cross-linked with glutaraldehyde. *Polymer Bulletin* 79 (7): 4647–4665.

91 Tian, H., Weng, Y., Kumar, R. et al. (2021). Protein-based biodegradable polymer: from sources to innovative sustainable materials for packaging applications. In: *Bio-based Packaging: Material, Environmental and Economic Aspects* (ed. S.M. Sapuan and R.A. Ilyas), 51–67.

92 Liangfan, Q., Chen, G., Dong, S. et al. (2019). Improved mechanical and antimicrobial properties of zein_chitosan films by adding highly dispersed nano-TiO$_2$. Elsevier Enhanced Reader.pdf. *Industrial Crops and Products* 130: 450–458.

93 Ustunol, Z. and Mert, B. (2004). Water solubility, mechanical, barrier, and thermal properties of cross-linked whey protein isolate-based films. *Journal of Food Science* 69: FEP129–FEP133.

94 Wihodo, M. and Moraru, C.I. (2013). Physical and chemical methods used to enhance the structure and mechanical properties of protein films: a review. *Journal of Food Engineering* 114 (3): 292–302.

95 Fan, H.Y., Duquette, D., Dumont, M.J., and Simpson, B.K. (2018). Salmon skin gelatin-corn zein composite films produced via crosslinking with glutaraldehyde: optimization using response surface methodology and characterization. *International Journal of Biological Macromolecules* 120: 263–273.

96 Yildirim, M. and Hettiarachchy, N.S. (1998). Properties of films produced by cross-linking whey proteins and 11s globulin using transglutaminase. *Journal of Food Science* 63 (2): 248–252.

97 Cheng, S., Wang, W., Li, Y. et al. (2019). Cross-linking and film-forming properties of transglutaminase-modified collagen fibers tailored by denaturation temperature. *Food Chemistry* 271: 527–535.

98 Micard, V., Belamri, R., Morel, M.H., and Guilbert, S. (2000). Properties of chemically and physically treated wheat gluten films. *Journal of Agricultural and Food Chemistry* 48 (7): 2948–2953.

99 Sabato, S.F., Ouattara, B., Yu, H. et al. (2001). Mechanical and barrier properties of cross-linked soy and whey. *Journal of Agricultural and Food Chemistry* 49: 1397–1403.

100 Qin, Z., Mo, L., Liao, M. et al. (2019). Preparation and characterization of soy protein isolate-based nanocomposite films with cellulose nanofibers and nano-silica via silane grafting. *Polymers (Basel)* 11 (11): 1835.

101 Wang, Q. and Padua, G.W. (2005). Properties of zein films coated with drying oils. *Journal of Agricultural and Food Chemistry* 53 (9): 3444–3448.

102 Fakhreddin Hosseini, S., Rezaei, M., Zandi, M., and Ghavi, F.F. (2013). Preparation and functional properties of fish gelatin-chitosan blend edible films. *Food Chemistry* 136: 1490–1495.

103 Guerrero, P. and De La Caba, K. (2010). Thermal and mechanical properties of soy protein films processed at different pH by compression. *Journal of Food Engineering* 100 (2): 261–269.

104 Kunte, L.A., Hanna, M.A., and Weller, C.L. (1997). Cast films from soy protein isolates and fractions. *Cereal Chemistry* 74: 115–118.

105 Gennadios, A., Weller, C.L., and Gooding, C.H. (1994). Measurement errors in water vapor permeability of highly permeable, hydrophilic edible films. *Journal of Food Engineering* 21 (4): 395–409.

106 Tawakkal, I.S.M.A., Cran, M.J., Miltz, J., and Bigger, S.W. (2014). A review of poly(lactic acid)-based materials for antimicrobial packaging. *Journal of Food Science* 79: 1477–1490.

107 Jacobsen, S., Fritz, H.-G., Degée, P. et al. (2000). New developments on the ring opening polymerisation ofpolylactide. *Industrial Crops and Products* 11: 265–275.

108 Zhao, X., Hu, H., Wang, X. et al. (2020). Super tough poly(lactic acid) blends: a comprehensive review. *RSC Advances* 10 (22): 13316–13368.

109 Garlotta, D. (2019). A literature review of poly(lactic acid) a literature review of poly (lactic acid). *Journal of Polymers and the Environment* 9 (2): 63–84.

110 Li, W., Li, L., Cao, Y. et al. (2017). Effects of PLA film incorporated with ZnO nanoparticle on the quality attributes of fresh-cut apple. *Nanomaterials* 7 (8): 207.

111 Shankar, S., Wang, L.F., and Rhim, J.W. (2018). Incorporation of zinc oxide nanoparticles improved the mechanical, water vapor barrier, UV-light barrier, and antibacterial properties of PLA-based nanocomposite films. *Materials Science and Engineering: C* 93: 289–298.

112 Zhou, L., He, H., Chun, L.M. et al. (2018). Enhancing mechanical properties of poly(lactic acid) through its in-situ crosslinking with maleic anhydride-modified cellulose nanocrystals from cottonseed hulls. *Industrial Crops and Products* 112: 449–459.

113 Mathew, A.P., Oksman, K., and Sain, M. (2005). Mechanical properties of biodegradable composites from poly lactic acid (PLA) and microcrystalline cellulose (MCC). *Journal of Applied Polymer Science* 97 (5): 2014–2025.

114 Lin, Y.S., Wu, Z.H., Yang, W., and Yang, M.B. (2008). Thermal and mechanical properties of chemical crosslinked polylactide (PLA). *Polymer Testing* 27 (8): 957–963.

115 Wang, J., Liu, X., Jia, Z. et al. (2017). Synthesis of bio-based poly(ethylene 2,5-furandicarboxylate) copolyesters: higher glass transition temperature, better transparency, and good barrier properties. *Journal of Polymer Science, Part A: Polymer Chemistry* 55 (19): 3298–3307.

116 Guidotti, G., Soccio, M., Lotti, N. et al. (2018). Poly(propylene 2,5-thiophenedicarboxylate) vs. poly(propylene 2,5-furandicarboxylate): two examples of high gas barrier bio-based polyesters. *Polymers (Basel)* 10 (7): 785.

117 Sikora, J.W., Majewski, Ł., and Puszka, A. (2021). Modern biodegradable plastics – processing and properties part II. *Materials (Basel)* 14 (10): 1–28.

118 Kumar, R., Kumari, S., Rai, B. et al. (2019). Effect of nano-cellulosic fiber on mechanical and barrier properties of polylactic acid (PLA) green nanocomposite film. *Materials Research Express* 6 (12): 125108.

119 Chawla, R., Sivakumar, S., and Kaur, H. (2021). Antimicrobial edible films in food packaging: current scenario and recent nanotechnological advancements – a review. *Carbohydrate Polymer Technologies and Applications* 2: 100024.

120 Li, X., Ai, X., Pan, H. et al. (2018). The morphological, mechanical, rheological, and thermal properties of PLA/PBAT blown films with chain extender. *Polymers for Advanced Technologies* 29 (6): 1706–1717.

121 Ai, X., Li, X., Yu, Y. et al. (2019). The mechanical, thermal, rheological and morphological properties of PLA/PBAT blown films by using bis(*tert*-butyl dioxy isopropyl) benzene as crosslinking agent. *Polymer Engineering and Science* 59: E227–E236.

122 Ye, J., Wang, S., Lan, W. et al. (2018). Preparation and properties of polylactic acid-tea polyphenol-chitosan composite membranes. *International Journal of Biological Macromolecules* 117: 632–639.

123 Fathima, P.E., Panda, S.K., Ashraf, P.M. et al. (2018). Polylactic acid_chitosan films for packaging of Indian white prawn (*Fenneropenaeus indicus*). *International Journal of Biological Macromolecules* 117: 1002–1010.

124 Łopusiewicz, Ł., Jedra, F., and Mizieińska, M. (2018). New poly(lactic acid) active packaging composite films incorporated with fungal melanin. *Polymers (Basel)* 10 (4): 5–8.

125 Hou, Q.X., Chai, X.S., Yang, R. et al. (2006). Characterization of lignocellulosic-poly(lactic acid) reinforced composites. *Journal of Applied Polymer Science* 99 (4): 1346–1349.

126 Kumar, R., Upadhyaya, P., and Chand, N. (2014). Effect of chemically modified nano zirconia addition on properties of LLDPE/LDPE/PLA/MA-g-PE bio-nanocomposite blown films for packaging applications. *International Journal of Chemical and Physical Sciences* 3: 2319–6602.

127 Herrera, N., Roch, H., Salaberria, A.M. et al. (2016). Functionalized blown films of plasticized polylactic acid/chitin nanocomposite: preparation and characterization. *Materials and Design* 92: 846–852.

128 Czaja-jagielska, N., Praiss, A., and Sankowska, N. (2020). Biodegradable packaging based on PLA with antimicrobial properties. *Scientific Journal of Logistics* 16 (2): 279–286.

129 Meraldo, A. (2016). Chapter 4 – Introduction to bio-based polymers. In: *Multilayer Flexible Packaging* (ed. R. John and J. Wagner), 47–52. Elsevier.

130 Reichert, C.L., Bugnicourt, E., Coltelli, M.B. et al. (2020). Bio-based packaging: materials, modifications, industrial applications and sustainability. *Polymers (Basel)* 12 (7): 1–35.

131 Kernitskiy, V.I. (2018). Bio-polyesters. *Materials Science Forum* 935: 89–93.

132 Rosenboom, J.-G., Hohl, D.K., Fleckenstein, P. et al. (2018). Bottle-grade polyethylene furanoate from ring-opening polymerisation of cyclic oligomers. *Nature Communications* 9: 1–7.

133 Rydz, J., Musioł, M., Zawidlak-w, B., and Sikorska, W. (2018). Present and future of biodegradable polymers for food packaging applications. In: *Biopolymers for Food Design* (ed. A.M. Grumezescu and A.M. Holban), 431–467.

134 Beauprez, J.J., De Mey, M., and Soetaert, W.K. (2010). Microbial succinic acid production: natural versus metabolic engineered producers. *Process Biochemistry* 45 (7): 1103–1114.

135 Farias da Silva, J.M. and Soares, B.G. (2021). Epoxidized cardanol-based prepolymer as promising biobased compatibilizing agent for PLA/PBAT blends. *Polymer Testing* 93: 106889.

136 Siracusa, V. and Blanco, I. (2020). Bio-polyethylene (Bio-PE), bio-polypropylene (Bio-PP) and bio-poly(ethylene terephthalate) (Bio-PET): recent developments in bio-based polymers analogous to petroleum-derived ones for packaging and engineering applications. *Polymers (Basel)* 12: 1–17.

137 Weber, C.J., Haugaard, V., Festersen, R., and Bertelsen, G. (2002). Production and applications of biobased packaging materials for the food industry production and applications of biobased packaging. *Food Additives and Contaminants* 19: 172–177.

138 Du, G., Chen, J., Yu, J., and Lun, S. (2001). Continuous production of poly-3-hydroxybutyrate by *Ralstonia eutropha* in a two-stage culture system. *Journal of Biotechnology* 88 (1): 59–65.

139 Wu, H., Chen, J., and Chen, G.Q. (2016). Engineering the growth pattern and cell morphology for enhanced PHB production by *Escherichia coli*. *Applied Microbiology and Biotechnology* 100 (23): 9907–9916.

140 Salgado, P.R., D'Amico, D.A., Seoane, I.T. et al. (2021). Improvement of water barrier properties of soybean protein isolate films by poly(3-hydroxybutyrate) thin coating. *Journal of Applied Polymer Science* 138 (5): 49758.

141 Pérez-Arauz, A.O., Aguilar-Rabiela, A.E., Vargas-Torres, A. et al. (2019). Production and characterization of biodegradable films of a novel polyhydroxyalkanoate (PHA) synthesized from peanut oil. *Food Packaging and Shelf Life* 20: 100297.

142 Thapa, C., Shakya, P., Shrestha, R. et al. (2019). Isolation of polyhydroxybutyrate (PHB) producing bacteria, optimization of culture conditions for PHB production, extraction and characterization of PHB. *Nepal Journal of Biotechnology* 6 (1): 62–68.

143 Trakunjae C., Boondaeng A., Apiwatanapiwat W., Kosugi A., Arai T., Sudesh K., et al. Enhanced polyhydroxybutyrate (PHB) production by newly isolated rare actinomycetes *Rhodococcus* sp. strain BSRT1-1 using response surface methodology. *Scientific Reports* 2021; 11 (1): 1–14.

144 Werlang, E.B., Moraes, L.B., Viviane, M. et al. (2021). Polyhydroxybutyrate (PHB) production via bioconversion using *Bacillus pumilus* in liquid phase cultivation of the biomass of *Arthrospira platensis* hydrolysate as a carbon source. *Waste and Biomass Valorization* 12: 3245–3255.

145 Mohamad Fauzi, A.H., Chua, A.S.M., Yoon, L.W. et al. (2019). Enrichment of PHA-accumulators for sustainable PHA production from crude glycerol. *Process Safety and Environmental Protection* 122: 200–208.

146 Urtuvia, V., Maturana, N., Peña, C., and Díaz-Barrera, A. (2020). Accumulation of poly(3-hydroxybutyrate-*co*-3-hydroxyvalerate) by *Azotobacter vinelandii* with different 3HV fraction in shake flasks and bioreactor. *Bioprocess and Biosystems Engineering* 43 (8): 1469–1478.

147 Duvigneau, S., Dürr, R., Behrens, J., and Kienle, A. (2021). Advanced kinetic modeling of bio-co-polymer poly(3-hydroxybutyrate-*co*-3-hydroxyvalerate) production using fructose and propionate as carbon sources. *Processes* 9: 1260.

148 Ghysels, S., Mozumder, M.S.I., De Wever, H. et al. (2018). Targeted poly (3-hydroxybutyrate-*co*-3-hydroxyvalerate) bioplastic production from carbon dioxide. *Bioresource Technology* 249: 858–868.

149 Liu, J., Zhao, Y., Diao, M. et al. (2019). Poly(3-hydroxybutyrate-*co*-3-hydroxyvalerate) production by *Rhodospirillum rubrum* using a two-step culture strategy. *Journal of Chemistry* 2019: 8369179.

150 Bhatia, S.K., Gurav, R., Choi, T.R. et al. (2019). Poly(3-hydroxybutyrate-*co*-3-hydroxyhexanoate) production from engineered *Ralstonia eutropha* using synthetic and anaerobically digested food waste derived volatile fatty acids. *International Journal of Biological Macromolecules* 133: 1–10.

151 Ivankovic, A., Zeljko, K., Talic, S. et al. (2017). Biodegradable packaging in the food industry. *Journal of Food Safety and Food Quality* 68 (2): 23–52.

152 Audic, J.L., Lemiègre, L., and Corre, Y.M. (2014). Thermal and mechanical properties of a polyhydroxyalkanoate plasticized with biobased epoxidized broccoli oil. *Journal of Applied Polymer Science* 131 (6): 1–7.

153 Jin, Y.J., Weng, Y.X., Li, X.X. et al. (2014). Effect of blending with polyamidoamine (PAMAM) dendrimer on the toughness of poly(hydroxybutyrate-*co*-hydroxyvalerate) (PHBV). *Journal of Materials Science* 50 (2): 794–800.

154 Wang, J., Tavakoli, J., and Tang, Y. (2019). Bacterial cellulose production, properties and applications with different culture methods – a review. *Carbohydrate Polymers* 219: 63–76.

155 George, J., Ramana, K.V., Sabapathy, S.N. et al. (2005). Characterization of chemically treated bacterial (*Acetobacter xylinum*) biopolymer: some thermo-mechanical properties. *International Journal of Biological Macromolecules* 37 (4): 189–194.

156 George, J., Ramana, K.V., Bawa, A.S., and Siddaramaiah (2011). Bacterial cellulose nanocrystals exhibiting high thermal stability and their polymer nanocomposites. *International Journal of Biological Macromolecules* 48 (1): 50–57.

157 Halib, N., Amin, M.C.I.M., and Ahmad, I. (2010). Unique stimuli responsive characteristics of electron beam synthesized bacterial cellulose/acrylic acid composite. *Journal of Applied Polymer Science* 116 (5): 2658–2667.

158 Esa, F., Tasirin, S.M., and Rahman, N.A. (2014). Overview of bacterial cellulose production and application. *Agriculture and Agricultural Science Procedia* 2: 113–119.

159 George, J., Ramana, K.V., Sabapathy, S.N., and Bawa, A.S. (2005). Physico-mechanical properties of chemically treated bacterial (*Acetobacter xylinum*) cellulose membrane. *World Journal of Microbiology and Biotechnology* 21 (8–9): 1323–1327.

160 Wippermann, J., Schumann, D., Klemm, D. et al. (2009). Preliminary results of small arterial substitute performed with a new cylindrical biomaterial composed of bacterial cellulose. *European Journal of Vascular and Endovascular Surgery* 37 (5): 592–596.

161 George, J., Sajeevkumar, V.A., Kumar, R. et al. (2008). Enhancement of thermal stability associated with the chemical treatment of bacterial (*Gluconacetobacter xylinus*) cellulose. *Journal of Applied Polymer Science* 108 (3): 1845–1851.

162 Iwamoto, S., Kai, W., Isogai, A., and Iwata, T. (2009). Elastic modulus of single cellulose microfibrils from tunicate measured by atomic force microscopy. *Biomacromolecules* 10 (9): 2571–2576.

163 Moon, R.J., Martini, A., Nairn, J. et al. (2011). Cellulose nanomaterials review: structure, properties and nanocomposites. *Chemical Society Reviews* 40 (7): 3941–3994.

164 de Morais Lima, M., Carneiro, L.C., Bianchini, D. et al. (2017). Structural, thermal, physical, mechanical, and barrier properties of chitosan films with the addition of xanthan gum. *Journal of Food Science* 82 (3): 698–705.

165 Nur Hazirah, M.A.S.P., Isa, M.I.N., and Sarbon, N.M. (2016). Effect of xanthan gum on the physical and mechanical properties of gelatin-carboxymethyl cellulose film blends. *Food Packaging and Shelf Life* 9: 55–63.

166 Singh, R.S. and Kaur, N. (2010). Microbial biopolymers for edible film and coating applications. In: *Biopolymers for Edible Film and Coatings* (ed. N.N. Nawani, M. Khetmalas, P.N. Razdan and A. Pandey), 187–216.

167 Farris, S., Unalan, I.U., Introzzi, L. et al. (2014). Pullulan-based films and coatings for food packaging: present applications, emerging opportunities, and future challenges. *Journal of Applied Polymer Science* 131 (13): 40539.

168 Diab, T., Biliaderis, C.G., Gerasopoulos, D., and Sfakiotakis, E. (2001). Physicochemical properties and application of pullulan edible films and coatings in fruit preservation. *Journal of the Science of Food and Agriculture* 81 (10): 988–1000.

169 Rai, M., Wypij, M., Ingle, A.P. et al. (2021). Emerging trends in pullulan-based antimicrobial systems for various applications. *International Journal of Molecular Sciences* 22 (24): 13596.

170 Zhu, G., Sheng, L., and Tong, Q. (2014). Preparation and characterization of carboxymethyl-gellan and pullulan blend films. *Food Hydrocolloids* 35: 341–347.

171 Liu, Y., Liu, Y., Han, K. et al. (2019). Effect of nano-TiO_2 on the physical, mechanical and optical properties of pullulan film. *Carbohydrate Polymers* 218: 95–102.

172 Yang, L. and Paulson, A.T. (2000). Mechanical and water vapour barrier properties of edible gellan films. *Food Research International* 33 (7): 563–570.

173 Danalache, F., Carvalho, C.Y., Alves, V.D. et al. (2016). Optimisation of gellan gum edible coating for ready-to-eat mango (*Mangifera indica* L.) bars. *International Journal of Biological Macromolecules* 84: 43–53.

174 Moosavi-Nasab, M., Layegh, B., Aminlari, L., and Hashemi, M.B. (2010). Microbial production of levan using date syrup and investigation of its properties. *World Academy of Science, Engineering and Technology* 44 (8): 1248–1254.

175 Dahech, I., Belghith, K.S., Hamden, K. et al. (2011). Antidiabetic activity of levan polysaccharide in alloxan-induced diabetic rats. *International Journal of Biological Macromolecules* 49 (4): 742–746.

176 Srikanth, R., Reddy, C.H.S.S.S., Siddartha, G. et al. (2015). Review on production, characterization and applications of microbial levan. *Carbohydrate Polymers* 120: 102–114.

177 Öner, E.T., Hernández, L., and Combie, J. (2016). Review of levan polysaccharide: from a century of past experiences to future prospects. *Biotechnology Advances* 34 (5): 827–844.

14

Effects of Natural Materials on Food Preservation and Storage

Subhanki Padhi and Winny Routray

National Institute of Technology, Department of Food Process Engineering, Rourkela, Odisha 769008, India

14.1 Introduction

Food accessibility, stability of food, proper utilization of food, and preservation of food can be considered as the strong pillars in achieving food security. These four entities have extensive impact on the general availability of food and can help in solving several associated socioeconomic problems. Global demand for food has been continuously increasing and has been projected to further increase with the rising population. To satisfy the needs of this increasing population, the food production should be increased by 70%. The demand for food cannot be satisfied by only improving the farming operations and must encompass other impactful options including curbing of the global food wastage and their valorization.

14.1.1 Major Objective of Food Preservation and Storage

According to a survey by the United Nation (2013), about 30–50%, i.e. around 2 billion tons of the total global food production is wasted. The main objective behind preservation and storage is to ensure food safety and quality for consumer acceptance. The spoilage of food by the action of microorganisms is a major problem of food industries. The availability of water in the fresh fruits, vegetables, and meat products makes them susceptible for the growth of microorganisms. Apart from microbial growth, the spoilage of food can occur attributed to many factors, including physiological changes, chemical reactions such as oxidation and reduction, and enzymatic changes.

The inception of effective food preservation and storage techniques can help in reducing the food wastage that occurs due to the decomposition of food.

Natural Materials for Food Packaging Application, First Edition.
Edited by Jyotishkumar Parameswaranpillai, Aswathy Jayakumar,
E. K. Radhakrishnan, Suchart Siengchin, and Sabarish Radoor.
© 2023 WILEY-VCH GmbH. Published 2023 by WILEY-VCH GmbH.

The conventional food preservation methods mainly include salting, sun drying, and fermentation, which help in preserving and storing food for a longer period and are also currently extensively practiced worldwide. Other commonly used processing methods for preservation include freezing, which mainly maintains the quality of the commodity at lower temperatures and mechanized drying, which is implemented to remove the excess water content from food products. Furthermore, other advanced technologies employed for food preservation and storage include fermentation, sterilization and pasteurization, smoking, aseptic packaging, vacuum packaging, and modified-atmospheric packaging [1]. Many researchers have also demonstrated the significant contribution of various chemical or synthetic preservatives such as thiocyanates, pyrrolidines, sodium benzoate, sulfites, sulfur dioxide, sorbates, sodium nitrite, imidazole, and formaldehyde, in reducing the microbial spoilage of food products [2]. However, the use of these chemical preservatives has led to many carcinogenic effects on the health conditions of humans and animals [3]. With the continual upsurge in population and the demand for higher amount of food, the need for better preservation and storage technologies that help in retaining the nutritional properties of food and maintain food safety, is eminent [4].

14.1.2 Available Solutions from the Natural Resources and Combination with Technology

With the advancement in research and development, studies have been carried out to utilize natural materials as preservatives. The natural materials are the extracts of different parts of fruits, vegetables, and animals that exhibit antioxidant and antimicrobial properties. The blending of natural antioxidants and antimicrobials into food products helps in enhancing the shelf life, preventing the growth of food-borne pathogens, inhibits oxidation and browning activity, and impedes the nutrient losses.

The use of naturally occurring biochemicals such as antioxidants, antimicrobials, and inhibitors, along with the abovementioned preservation and storage technologies, has been observed to significantly decelerate the spoilage of food. The natural materials and biochemicals can be derived from animals, plants, fungi, algae, and bacteria. The natural materials act as biodegradable and biocompatible materials, having no toxicity, and are cost-effective in nature [5]. The antimicrobial materials and biochemicals contribute toward prevention of microbial spoilage by inhibiting their growth in the food products [6]. Attributed to these properties, natural materials help in preserving and storing food products for a longer period without hampering the quality of the food ingredients. The development of these advanced technologies contributes toward further augmentation of food preservation industries that aim to bestow sustainable and environment-friendly products.

Some of the commonly used biochemicals for food preservation and storage are essential oils, organic acids, phenolic compounds, polysaccharides, vitamins, aromatic compounds, and bacteriocins. The natural materials and biochemicals can be used as individual preservative or in combination with the various processing technologies to preserve the organoleptic properties of the food product. The natural materials should be added in a proper proportion to retain the original taste, flavor, color, aroma and texture of food, and the food should be safe for consumption [6]. This book chapter deals with a brief description on the utilization of various natural materials during food processing operations and as packaging materials along with their corresponding importance in food preservation and storage. The effects of using different natural materials during the various post-harvest operations have also been discussed. The chapter will concentrate on the research conducted in last 15–20 years.

14.2 Biomolecules Utilized for Preservation, Their Properties, and Uses

Sources of the applicable biomolecules to be utilized as preservative and antimicrobials include various plant and animal parts and tissues. Different parts of the plant such as stem, fruits, leaves, and seeds can be used for extracting natural compounds that can be incorporated into food products to improve the safety and quality of food [7]. However, commercial production has also been widely conducted using microorganisms, which have been recognized as a useful source. Some of the extensively explored and applied biomolecules/biomolecules have been discussed in following sections along with their corresponding applications and properties.

14.2.1 Polysaccharides

Generally, the polysaccharides, being highly hydrophilic in nature have a very low water barrier, but they exhibit a selective permeability towards oxygen and carbon dioxide and inhibit the lipid migration. The polysaccharides have been considered as sacrificing agents because they sacrifice their moisture to add moisture to the surface of the food product that was lost first [8]. The polysaccharides such as chitosan, natural gum, starch, carrageenan, and alginate have been used as coating materials.

Chitosan is a natural biodegradable polysaccharide obtained from deacetylation of chitin. The biochemical properties and film-forming capacity of chitosan make it a suitable preservative agent for fruits as a coating material. Guerra et al. [9] reported that chitosan when coated with essential oil of peppermint helped in impeding the fungal infection of grapes during storage. Azevedo et al. [10] studied that the strawberries coated with chitosan and carvacrol effectively helped in suppressing the rate of decay during storage. Campaniello and Bevilacqua [11] studied the effect of chitosan coating on fresh-sliced strawberries. The fresh-cut strawberries were dipped in 1% chitosan solution, modified atmospheric packaging (MAP) was employed using 80% and 5% oxygen concentration, and maintained at different temperatures (4, 8, 12, 15 °C) to study the effect of chitosan coating on the strawberries. It was found that chitosan helped in increasing the stability for a longer period and exhibited higher antimicrobial properties [11].

Many researchers have explored the use of modified and native starches for application as edible coatings. Tavassoli-Kafrani et al. [12] showed that pumpkin coated with cassava starch helped in preserving the carotenoid compounds of pumpkin. Alotaibi and Tahergorabi [13] reported that the combined use of sweet potato starch and essential oil of thyme as a coating for shrimp helped in extending the refrigerated storage of shrimp by decreasing the bacterial count. Hamedi et al. [14] found that chicken slices coated with edible biofilm of alginate and *Ziziphora* essential oil helped in decreasing the microbial growth and improved the quality during storage. The use of shellac, aloe vera gel, and essential oil of lemon as coating of apple slices, helped in sustaining the color of apple and decreasing the ethylene synthesis at the time of storage [15]. Pectin obtained from cell wall of plants shows lower water holding capacity, and thus can be used for preservation and storage of low moisture-containing foods [16].

14.2.2 Essential Oil

The essential oils are secondary metabolites extracted from plant-based natural raw materials, by using different processes, including steam distillation, dry distillation, solvent extraction, hydro-diffusion, supercritical fluid extraction, and solvent-free microwave extraction [17]. Being secondary metabolites of plants, essential oils are generally recognized as safe (GRAS) as per FDA, and they have been often used as replacement for the synthetic additives [18]. The essential oils contain various kinds of natural aromatic, volatile, and bioactive compounds that have applications in different food industries [19]. The presence of antimicrobial, antioxidant, and antifungal properties in the essential oils helps in enhancing the shelf life of the food products and thus, makes them suitable for use in industries. Various kinds of essential oils can be obtained from *Citrus limonum* (lemon), *Lavandula angustifolia* (lavender), *Melaleuca alternifolia* (tea tree), *Thymus vulgaris* (thyme), *Cinnamomum zeylanicum* (cinnamon), *Azadirachta indica* (neem), *Syzygium aromaticum* (clove), *Eucalyptus globulus* (eucalyptus), and *Brassica nigra* (mustard) [20]. The essential oils can be used for food preservation as a part of packaging films and coatings. The Table 14.1 shows antimicrobial activity exhibited by different essential oils.

The antimicrobial activity of essential oils can be explained by either their bactericide or bacteriostatic nature [3]. Bactericide nature refers to the killing of bacterial cells through the application of essential oils, whereas, bacteriostatic nature implies that the application of essential oil first obstructs the growth of bacterial cells and then the microbial cells retrieve their reproduction capacity after a stretch [29]. The role of antioxidant compounds present in the essential oils is to inhibit, modify and stop the oxidation reactions when they are occurring at a lower concentration [30]. The oxidation reactions occurring in food products cause damage to the color of the product and produce off-flavors in the product. Rodriguez-Garcia et al. [31] studied that oregano essential oil exhibited antioxidant properties and it helped in inhibiting the lipid peroxidation process in fat-containing foods and in scavenging free radicals. de Souza et al. [32] also reported that the presence of phenolic compounds, such as carvacrol, eugenol, and thymol, in essential oils helps in exhibiting antioxidant

Table 14.1 Antibacterial and antifungal activity exhibited by different essential oils.

Food product	Microorganisms present	Essential oil	Key ingredient	Properties exhibited	Sources
Iranian white cheese	*Escherichia coli* O157:H7, *Listeria monocytogenes*	Black Cumin	Cuminaldehyde (11.4%)	Shelf-life extension, decrease in bacterial growth	Ehsani et al. [21]
Chicken breast filet	*Listeria monocytogenes*, *Salmonella typhimurium*	Ginger	β-sesquiphellandrene (12.74%), α-zingiberene (24.96%)	Inhibited the microorganism growth in the meat	Noori et al. [22]
Rice	*Aspergillus flavus*	Nutmeg	Elemicin (27.08%), Thujanol (18.55%), Myristicine (21.29%)	Preserves against fungal infection and secretion of aflatoxin B_1	Das et al. [23]
Spices	*Aspergillus flavus*	Basil	γ-terpinene (26.08%), Methyl cinnamate (48.29%)	Restricts the biodeterioration caused by fungal infestation and aflatoxins	Prakash et al. [24]
Bread	*Penicillium expansum*	Lemon Grass	Citral (62.58%)	Prevented the growth of microbes and improved the shelf life	Mani López et al. [25]
Fruits	*Botrytis cinerea*, *Aspergillus niger*, *Penicillium digitatum*	Cinnamon	Cinnamaldehyde (89.15%)	Protection and prevention of spoilage after harvesting	Mousavian et al. [26]
Leafy Vegetables	*Escherichia coli*, *Salmonella choleraesius*	Lemon Grass	Geranial (35%), Neral (50%)	Beneficial against bacterial studies of in vivo and in vitro examination	Ortega-Ramirez et al. [27]
Fish	*Aspergillus ochraceus*, *Aspergillus oryzae*, *Aspergillus fumigatus*, *Aspergillus parasiticus*	Lemon Grass	Myrcene (10.4%), Neral (33%), Geranial (41.3%)	Suppressed the microbial growth in fermented fish	Dégnon et al. [28]

properties. This can be attributed to the capability of phenolic compounds in donating hydrogen atoms to the free radicals and thus changing them to a better firm entity. Table 14.2 shows the changes observed in the properties of food products when essential oil was incorporated in different polymer packaging materials, during different studies worldwide.

14.2.3 Phenolic Compounds

Phenolic compounds are the secondary metabolites of plant extracts. The chemical structure of phenolic compounds comprises at least one aromatic ring with one or more hydroxyl groups. In some food products, phenolic compounds are present naturally, whereas, in some others, these compounds are added as antioxidants to improve the quality of food products. The phenolic compounds can be broadly

Table 14.2 Properties exhibited by some major essential oil after their incorporation in different packaging films.

Essential oil	Polymer film used	Properties exhibited	Sources
Eucalyptus	Poly(butylene adipate-*co*-terephthalate) and polylactic acid	Increased UV blockage by 40%; Decreased the growth of *Staphylococcus aureus* and *Escherichia coli*; Impedes *E. coli* in biofilm	Sharma et al. [33]
Lavender	Furcellaran, gelatin, and starch	Exhibited antimicrobial and antioxidant activity	Jamróz et al. [34]
Clove	Poly(butylene adipate-*co*-terephthalate) and polylactic acid	Shows UV blockage property of 80%; *Staphylococcus aureus* is completely killed	Sharma et al. [35]
	β-cyclodextrin	Elastic property decreased; Water is absorbed from relative humidity of 60%	Maestrello et al. [36]
Rosemary, Mint	Pectin, starch, and chitosan polymer	Reduced tensile strength and water barrier properties; Improved flexibility; Zone of inhibitions against *Bacillus subtilis*, *E. coli*, and *L. monocytogenes* increased at least by 40%	Akhter et al. [37]
Pine	Poly(butylene adipate-*co*-terephthalate) and polylactic acid	Lower Young's modulus and greater elongation at break	Hernández-López et al. [38]
Laurel	Polyethylene (PE) films coated with chitosan	Exhibits antimicrobial activity; Storage time of pork is increased	Wu et al. [39]

classified as phenolic acids, flavonoids, lignans, and hydroxycinnamic acids [2]. This classification is based on the number of phenolic rings, chemical structure, and the elements that structurally link the rings. The detailed classification of phenolic compounds is presented in Figure 14.1. The lipophilic nature of phenolic compounds helps in exhibiting their antimicrobial activity [6]. Researchers have studied that the lipophilic compounds impair the structure and functional properties of microbial cells by destroying their osmotic balance and membrane permeability [30]. This activity further damages the functionality of the cells and several associated biomolecules, including amino acids, nucleic acids, ATP, and ions, exude out from the cell, which lowers the pH within the bacterial cells [29]. The degree of toxicity of phenolic compounds to the microorganisms is directly related to the hydroxyl groups that are present in the phenolic rings [40]. The higher the hydroxylation, the higher is the level of toxicity to the microorganisms.

Furthermore, flavonoids, flavonols, and flavones, [40], can be distinguished from each other by their phenolic rings. These compounds are very efficacious against microorganisms due to their capacity of binding to and inactivating the proteins and bacterial cell walls [41].

14.2.4 Aromatic Compounds

Aromatic compounds are the naturally occurring secondary metabolites, widely extracted from plant-based commodities and samples. They help in increasing the shelf life of grains, legumes and fruits [42]. Aromatic compounds are the derivatives of simple phenols and polyphenol carboxylic acids, exhibiting antioxidant, antibacterial, and antifungal properties. Hakkim et al. [43] observed that different naturally occurring phenolic compounds such as caffeic acid, rosmarinic acid, cinnamic acid, coumaric acid, and hydroxyphenyl lactic acid showed antibacterial and antifungal activity against pathogens including *Staphylococcus aureus, Escherichia coli, Pseudomonas*

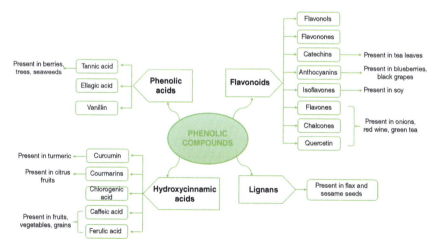

Figure 14.1 Taxonomy of phenolic compounds.

vulgaris, *Listeria monocytogenes*, *Salmonella Typhi*, and *Bacillus licheniformis* in freshly cut slices of apple. Rosemary powder and predominantly rosmarinic acid have been observed to exhibit antioxidant and antimicrobial activities, attributed to which they are utilized as preservatives for increasing the shelf life of different meat-based products [44–46]. In a different study, it was observed that propolis exhibits antimycotic as well as antimicrobial activities. Hence, it acts as a natural preservative and helps in preserving the aroma and color of food products [5]. Propolis is used as a preservative in different food products such as fermented Turkish sausage [47], fresh fish burgers [48], and apple juice [49], and has been observed to extend the shelf life significantly along with the maintenance of quality. There are many other sources, which can be further investigated for their efficacy in maintenance of quality and preservation of various food commodities.

14.2.5 Proteins

The proteins obtained from soybeans, wheat, milk, corn, and peanuts can be utilized as coating materials for different food products. Casein, gluten, zein, whey, and collagen proteins have been utilized for making essential oil-based edible coating. During a study, zein films were made in combination with *Zataria multiflora* Boiss. essential oil and monolaurin; wherein, their corresponding effects on microorganisms, including *L. monocytogenes* and *E. coli* O157:H7 present in beef, were evaluated [50]. It was found that this film helped in significantly reducing the growth of both the microorganisms during the storage time. In a different study, whey protein was combined with oregano essential oil to form an active edible coating, and the corresponding effect on the meat products was studied by Catarino et al. [51]. The whey protein and oregano essential oil film led to inhibition of the lipid peroxidation, decrease of the microbial load, retention of the color of the meat and thus leading to an extended shelf life.

14.2.6 Bacteriocins

Bacteriocins are the natural antimicrobial compounds that are obtained from other microorganisms lactic acid bacteria (LAB). These consist of active ribosomal proteins and peptides developed from gram-negative and gram-positive LAB [52]. Bacteriocins exhibit hydrophilic properties and help in providing ancillary hurdles to unwanted growth of fungi and bacteria [41]. Attributed to the non-immunogenic and non-toxin properties of bacteriocins, these compounds are used as natural food preservatives to hinder the growth of molds, yeast, and bacteria [53]. The most commonly used antimicrobial bacteriocins are pediocin and nisin [53]. For bakery, cooked meat, cheese, yoghurt, buttermilk, processed fruits, and vegetables, LAB are utilized for bacteriocin fortification, which further enhances the stability, quality, and consistency of the abovementioned fermented food products [54].

Nisin is synthesized by strains of *Lactococcus lactis* and exhibits natural antimicrobial properties [55]. It helps in boosting the shelf life of meat-based products, vegetable and fruit juices, and decreases the weight loss in Ricotta cheese during the storage.

It also helps in impeding the growth of bacteria such as *Staphylococcus aureus*, *Bacillus cereus*, *L. monocytogenes*, and *Clostridium botulinum* in the cases of canned foods, cheese, ice cream, beverages, seafood, eggs, and meat-based products [5]. The incorporation of nisin has been observed to be effective for impeding the growth of *L. monocytogenes* in Babybel cheese [56], cold-smoked salmon [57], pork bologna [58], and beef [59].

Pediocins are also natural food preservatives, which are mainly synthesized by *Pediococcus acidilactici* [60]. They help in damaging only the vegetative cells of *L. monocytogenes* and not the sporulating cells [41]. They exhibit antimicrobial activity against microbes and help in boosting the shelf life of meat-based products, dairy-based products, and wines [61].

14.2.7 Other Animal-Based Antimicrobials

The different animal-based natural antimicrobials that have been used as preservatives include various biochemicals and enzymes such as chitosan, lactoferrin, lysozyme, peptides, and lactoperoxidase. Lactoferrin is a glycoprotein, which is obtained from bovine milk; synergic effect has been exhibited by nisin and lactoferrin towards *E. coli* O157:H7 and *L. monocytogenes* [41]. Lactoferrin helps in controlling various food-borne microbes (*Pseudomonas fluorescens*, *Carnobacterium viridans*, *L. monocytogenes* and *E. coli*) and thus increases the shelf life of food products such as beef, meatballs, pork, sausages, and milk [62, 63].

Lysozyme is a hydrolytic enzyme and is obtained from milk and eggs. Attributed to its antimicrobial activity and chemical stability, it can be used as a natural preservative in dairy-based products, fresh fruits and vegetables, processed and raw meat-based products, seafoods, and sauces [41, 57]. Lactoperoxidase enzyme is obtained from raw bovine milk and it shows antibacterial properties against gram-positive as well as gram-negative bacteria [41].

14.3 Different Extraction Processes Employed for Natural Materials

Extraction and purification processes are decided based on the characteristic properties of source materials and the targeted biomolecules/biochemicals. Often these extraction methods are preceded by various sample preparation steps to ease the extraction of these biochemicals and synchronize with the extraction methods [64]. These sample preparation steps can include a single step or a combination of various procedures such as size reduction, drying, mixing, and centrifugation. Solvent-based extraction methods and distillation are still the most common methods of application, which are further enhanced in terms of yield, efficiency, and functionality using different advanced treatment and extraction methods, which often include ultrasonication, high pressure and subcritical extraction, microwave and radio-frequency application, supercritical fluid extraction, pulsed electric field application, and cold plasma application. Purification and separation steps are also important for the final bioactivity of the extracts, which has been achieved often by traditional and advanced filtration and chromatographic methods. The detailed discussion about the individual effects of the various parameters of these methods is out of the scope of this chapter. However, detailed comprehensive reports on the extraction and purification methods for plant and animal-based natural compounds can be found in different research papers. Ferreira et al. [65] studied the various extraction and application of different plant and animal-based polysaccharides. Elieh-Ali-Komi and Hamblin [66] and Hintz et al. [40] studied the extraction of proteins from natural materials. Aziz et al. [17] also studied the different extraction methods for extracting polyphenolic compounds from natural materials. Table 14.3 has summarized various methods of extraction of natural compounds and their applications in food processing.

Table 14.3 Extraction processes and application of various plant and animal-based natural compounds.

Source (plant/animal based)		Extraction process	Uses	Reference
Polysaccharides	Plant based (Pectin, Starch, Cellulose)	Ultrasonication; Homogenization along with steam explosion, acid hydrolysis, and by use of enzymes	As antimicrobial and antioxidant materials in food packaging system	Hintz et al. [40]
	Animal based (Chitin, Chitosan)	Alkaline treatment and deproteinization; Hydrochloric acid (dilute) treatment with demineralization	As polymer composite material to exhibit antimicrobial property	Campaniello et al. [11]
Proteins	Plant based (Gluten, Zein, Gliadin, Polypeptides)	Alkaline extraction; Protein precipitation; Enzyme-aided extraction; Ultrasound-aided extraction	Mixed with different packaging films to enhance their barrier and mechanical properties	Hintz et al. [40]
	Animal based (Collagen)	Chemical and Enzymatic hydrolysis	As polymer composites and enhances mechanical and water barrier properties	Elieh-Ali-Komi and Hamblin [66]
Polyphenols (Plant-based extracts)		Hydro-diffusion; Steam distillation; Dry distillation; Solvent extraction, Supercritical fluid extraction; Solvent-free microwave extraction	Exhibits antimicrobial and antioxidant activity when incorporated into food products	Gutiérrez-del-Río et al. [2]

14.4 Effects of Natural Materials on Different Product Quality and Storage

The incorporation of different biopolymers and biochemicals affects the physiochemical properties of the final product and further processing employed for their development. Products are often developed through the application of different processing technologies and traditional methods; wherein specific modifications and control mechanisms are required for optimizing the properties of final product and the requisite processing parameters. Few studies regarding investigation of drying process and packaging employed using the fortified matrices have been discussed in following subsections to demonstrate the effects of the biopolymer and biochemical additives on processing methods, packaging characteristics and storage, and corresponding properties of the developed products.

14.4.1 Drying Methods and Corresponding Properties

Drying is a commonly employed procedure for sample preparation, treatment as well as a processing step for final product development. During a study by Shi, Fang [67], the impact of adding whey protein isolate and maltodextrin to honey during spray drying was studied. Initially it was observed that there was no powder obtained when only honey was spray dried. When honey was mixed with maltodextrin and whey protein isolate in the ratio of 60:39.5:0.5, the honey powder obtained was around 57.4%. The addition of whey protein isolate helped in the movement of protein molecules toward the droplet–air interface and the protein molecules exhibited magnificent film-forming capability after drying. The temperature of the inlet and outlet of the spray dryer was kept at 150 ± 1 and $85\pm1\,°C$, respectively. In a different study, Rodriguez et al. [68] observed the impact of natural antioxidants like rosemary, chamomile extracts, and essential oils on the oxidative stability of chia seed oil after freeze-drying. The natural antioxidants helped in protecting the freeze-dried chia seed oil from lipid oxidation. Furthermore, the effect of zein coating on the microencapsulation of flax seed oil using spray and freeze drying was evaluated by Quispe-Condori et al. [69]; wherein, it was observed that on addition of zein protein, the bulk density of the obtained powder declined.

14.4.2 Enhancement of Packaging Characteristics

The combination of natural antimicrobial products with packaging material helps in preventing the contamination of foods by microbes and increasing the shelf life of food products. Bacteriocins when used as antimicrobial in packaging films helped in extending the shelf life and enhancing the quality and safety of food products [70, 71]. The antioxidant preservatives contribute towards inhibition of the rate of oxidation and impediment of the rancidity of meat products. Plant extracts have been widely explored along with other components for obtaining desirable properties in the final packaging materials, which have been further enhanced through synergistic interactions. Also, it has been observed rather than combining specific compounds, extracts have been used, which are generally a mixture of different compounds and could lead to better effects.

Peng et al. [72] found that the antioxidant action of the chitosan film increased, and the water vapor permeability decreased after addition of black tea and green tea extracts to the chitosan film. Also, the chitosan film incorporated with green tea extract showed higher 2,2-diphenyl-1-picrylhydrazyl (DPPH) activity than the black tea extract. Li, Miao [73] studied the effect of incorporation of gingko leaf extract, grape seed extract, green tea extract, and ginger extract in gelatin-based film as antioxidant agents. They observed that the DPPH activity was highest in the film incorporated with gingko leaf extract and the moisture scavenging activity was highest in the film incorporated with green tea extract. Bonilla and Sobral [74] observed the effect of incorporation of rosemary, cinnamon, and guarana into edible films of gelatin and chitosan. They reported that the incorporation helped in reducing the permeability of water vapor into the film, improved the elasticity behavior of films, and inhibited the growth of microbes like *E. coli* and *Staphylococcus aureus*. The presence of ascorbic acid, citric acid, hesperidins, and naringin in grape seed extract makes it a suitable antimicrobial agent for incorporation into edible films, which has been reported to be effectively preventing the growth of *L. monocytogenes* when infused into edible soy protein films [75]. The incorporation of grape seed extract into carrageenan-based film exhibited higher antimicrobial activity against microbial cells and it can be used as an active packaging material [76]. The infusion of cinnamon essential oil into edible chitosan-based film helped in improving the antimicrobial and mechanical properties of the film [77].

14.4.3 Maintenance of Physiochemical Properties of Raw and Processed Products

To make the fruits, vegetables, dairy products, and meat-based products available for consumers throughout the year, efficient and effective preservation and storage techniques are needed. With the rising consciousness amongst the consumers regarding the health risks due to chemical preservatives, there has been an increasing demand for the natural preservatives [78]. Many researchers have found that different naturally obtained preservatives are helpful in enhancing the shelf life of food products and impeding the microbial growth as discussed in previous sections on different biochemicals. However, apart from addition of shelf life through prevention of microbial spoilage, physicochemical properties of the processed products are maintained and modified during the storage, which maintains the acceptability and nutritional properties of the products.

The change in the color of meat at the time of storage from red to brown is attributed to the lipid oxidation process. Addition of natural antioxidant and antimicrobials can help in increasing shelf life and inhibiting pathogen growth. Furthermore, different plant extracts such as avocado seed extract, green tea extract, grape seed extract, lychee seed extract, olive oil waste extract, apple pomace powder, gingko biloba leaf extract, and broccoli powder extract have been used as natural antioxidants to prevent the lipid oxidation in case of porcine patties, minced beef patties, lamb patties, ground beef patties, raw meat paste, meatballs, and goat meat nuggets, and to boost the shelf life [79–83]. Yin and Cheng [84] studied that the addition organo-sulfur compounds extracted from garlic (diallyl disulfide, diallyl sulfide, n-acetyl cysteine, s-acetyl cysteine) can help in preventing the lipid oxidation and thus extend the storage time.

Fish being highly perishable in nature is usually preserved by freezing it. Freezing is not enough in maintaining the quality of fish as it does not help in preventing discoloration, oxidation of lipids, growth of microbes, and rancidity. Bacteriocin such as nisin, pediocin, and lacticin are used as antimicrobial preservatives for fish-based products. Also, the use of organic acids has proved in preserving the quality of fish [85].

14.5 Conclusion

Food safety has become a priority in todays' world along with better technologies for production of food. Nowadays, with the increasing demand for food products with clean-label ingredients, natural preservatives are being increasingly utilized for the preservation of the properties of raw as well as processed food products. Overall, it can be concluded that some of the important properties that natural preservatives should exhibit include nontoxicologic and biodegradable properties, chemical and physical stability, enhancement of shelf life, and antioxidant as well as antimicrobial action. It should not affect the color, aroma, and texture of the original product, provide consistency in action, and biocompatibility with the other different materials present in the product. The development of biodegradable active packaging materials infused with natural materials has been contributing towards exhibiting antimicrobial and antioxidant properties and has further helped in boosting the mechanical, thermal, and barrier properties of the film. The natural antimicrobial packaging system can be considered innovative, efficient, environment friendly, and effective for different food packaging. The consumer acceptance, industrial application, and cost of these natural preservative-incorporated food products are some of the determining factors for the successful application of these products. Incorporation of natural ingredients has also helped in maintenance of the freshness and consistency of the product without having any toxic effect on human health. The combined use of improved food processing and preservation techniques, and incorporation of natural agents as antioxidant and antimicrobial compounds into food products can prove to be the best preservation and storage technology for different types of food products. Hence, further investigation should be focused on optimization of these ingredient formulations and the processing parameters for maintaining the highest concentrations of these ingredients, which would lead to extension of shelf life with highest quality.

References

1 Olurankinse, C. (2014). Strategies for sustainable food processing and preservation. *IOSR Journal of Environmental Science, Toxicology and Food Technology* 8 (6): 31–36.

2 Gutiérrez-del-Río, I., Fernández, J., and Lombó, F. (2018). Plant nutraceuticals as antimicrobial agents in food preservation: terpenoids, polyphenols and thiols. *International Journal of Antimicrobial Agents* 52 (3): 309–315.

3 Falleh, H., Jemaa, M.B., Saada, M., and Ksouri, R. (2020). Essential oils: a promising eco-friendly food preservative. *Food Chemistry* 330: 127268.

4 Martindale, W. and Schiebel, W. (2017). The impact of food preservation on food waste. *British Food Journal* 119 (12): 2510–2518.

5 Mogoşanu, G.D., Grumezescu, A.M., Bejenaru, C., and Bejenaru, L.E. (2017). Natural products used for food preservation. *Food Preservation*: : Elsevier 365–411.

6 Martínez-Graciá, C., González-Bermúdez, C.A., Cabellero-Valcárcel, A.M. et al. (2015). Use of herbs and spices for food preservation: advantages and limitations. *Current Opinion in Food Science* 6: 38–43.

7 Mir, S.A., Dar, B., Wani, A.A., and Shah, M.A. (2018). Effect of plant extracts on the techno-functional properties of biodegradable packaging films. *Trends in Food Science and Technology* 80: 141–154.

8 Dehghani, S., Hosseini, S.V., and Regenstein, J.M. (2018). Edible films and coatings in seafood preservation: a review. *Food Chemistry* 240: 505–513.

9 Guerra, I.C.D., de Oliveira, P.D.L., Santos, M.M.F. et al. (2016). The effects of composite coatings containing chitosan and *Mentha* (*piperita* L. or *x villosa* Huds) essential oil on postharvest mold occurrence and quality of table grape cv. Isabella. *Innovative Food Science & Emerging Technologies* 34: 112–121.

10 Azevedo, A.N., Buarque, P.R., Cruz, E.M.O. et al. (2014). Response surface methodology for optimisation of edible chitosan coating formulations incorporating essential oil against several foodborne pathogenic bacteria. *Food Control* 43: 1–9.

11 Campaniello, D., Bevilacqua, A., Sinigaglia, M., and Corbo, M.R. (2008). Chitosan: antimicrobial activity and potential applications for preserving minimally processed strawberries. *Food Microbiology* 25 (8): 992–1000.

12 Tavassoli-Kafrani, E., Shekarchizadeh, H., and Masoudpour-Behabadi, M. (2016). Development of edible films and coatings from alginates and carrageenans. *Carbohydrate Polymers* 137: 360–374.

13 Alotaibi, S. and Tahergorabi, R. (2018). Development of a sweet potato starch-based coating and its effect on quality attributes of shrimp during refrigerated storage. *LWT* 88: 203–209.

14 Hamedi, H., Kargozari, M., Shotorbani, P.M. et al. (2017). A novel bioactive edible coating based on sodium alginate and galbanum gum incorporated with essential oil of *Ziziphora persica*: the antioxidant and antimicrobial activity, and application in food model. *Food Hydrocolloids* 72: 35–46.

15 Chauhan, O., Raju, P., Singh, A., and Bawa, A. (2011). Shellac and aloe-gel-based surface coatings for maintaining keeping quality of apple slices. *Food Chemistry* 126 (3): 961–966.

16 Ju, J., Xie, Y., Guo, Y. et al. (2019). Application of edible coating with essential oil in food preservation. *Critical Reviews in Food Science and Nutrition* 59 (15): 2467–2480.

17 Aziz, Z.A., Ahmad, A., Setapar, S.H.M. et al. (2018). Essential oils: extraction techniques, pharmaceutical and therapeutic potential-a review. *Current Drug Metabolism* 19 (13): 1100–1110.

18 Ju, J., Xu, X., Xie, Y. et al. (2018). Inhibitory effects of cinnamon and clove essential oils on mold growth on baked foods. *Food Chemistry* 240: 850–855.

19 Sharma, S., Barkauskaite, S., Jaiswal, A.K., and Jaiswal, S. (2020). Essential oils as additives in active food packaging. *Food Chemistry* 343: 128403.

20 Bhavaniramya, S., Vishnupriya, S., Al-Aboody, M.S. et al. (2019). Role of essential oils in food safety: antimicrobial and antioxidant applications. *Grain & Oil Science and Technology* 2 (2): 49–55.

21 Ehsani, A., Hashemi, M., Naghibi, S.S. et al. (2016). Properties of *Bunium persicum* essential oil and its application in Iranian white cheese against *Listeria monocytogenes* and *Escherichia coli* O157: H7. *Journal of Food Safety* 36 (4): 563–570.

22 Noori, S., Zeynali, F., and Almasi, H. (2018). Antimicrobial and antioxidant efficiency of nanoemulsion-based edible coating containing ginger (*Zingiber officinale*) essential oil and its effect on safety and quality attributes of chicken breast fillets. *Food Control* 84: 312–320.

23 Das, S., Kumar Singh, V., Kumar Dwivedy, A. et al. (2020). Assessment of chemically characterised *Myristica fragrans* essential oil against fungi contaminating stored scented rice and its mode of action as novel aflatoxin inhibitor. *Natural Product Research* 34 (11): 1611–1615.

24 Prakash, B., Shukla, R., Singh, P. et al. (2011). Efficacy of chemically characterized *Ocimum gratissimum* L. essential oil as an antioxidant and a safe plant based antimicrobial against fungal and aflatoxin B1 contamination of spices. *Foodservice Research International* 44 (1): 385–390.

25 Mani López, E., Valle Vargas, G.P., Palou, E., and López Malo, A. (2018). *Penicillium expansum* inhibition on bread by lemongrass essential oil in vapor phase. *Journal of Food Protection* 81 (3): 467–471.

26 Mousavian, M., Bazgir, E., and Moradpour, A. (2018). Cinnamon bark essential oil compounds and its antifungal effects against fungal rotting of fruits. *Journal of Crop Improvement* 19 (4): 907–920.

27 Ortega-Ramirez, L.A., Silva-Espinoza, B.A., Vargas-Arispuro, I. et al. (2017). Combination of *Cymbopogon citratus* and *Allium cepa* essential oils increased antibacterial activity in leafy vegetables. *Journal of the Science of Food and Agriculture* 97 (7): 2166–2173.

28 Dègnon, R., Allagbé, A., Adjou, E., and Dahouenon-Ahoussi, E. (2019). Antifungal activities of *Cymbopogon citratus* essential oil against *Aspergillus* species isolated from fermented fish products of Southern Benin. *Journal of Food Quality and Hazards Control* 6 (2): 53–57.

29 Calo, J.R., Crandall, P.G., O'Bryan, C.A., and Ricke, S.C. (2015). Essential oils as antimicrobials in food systems-a review. *Food Control* 54: 111–119.

30 Prakash, B., Kedia, A., Mishra, P.K., and Dubey, N. (2015). Plant essential oils as food preservatives to control moulds, mycotoxin contamination and oxidative deterioration of agri-food commodities–potentials and challenges. *Food Control* 47: 381–391.

31 Rodriguez-Garcia, I., Silva-Espinoza, B., Ortega-Ramirez, L. et al. (2016). Oregano essential oil as an antimicrobial and antioxidant additive in food products. *Critical Reviews in Food Science and Nutrition* 56 (10): 1717–1727.

32 de Souza, W.F.M., Mariano, X.M., Isnard, J.L. et al. (2019). Evaluation of the volatile composition, toxicological and antioxidant potentials of the essential oils and teas of commercial Chilean boldo samples. *Food Research International* 124: 27–33.

33 Sharma, S., Barkauskaite, S., Jaiswal, S. et al. (2020). Development of essential oil incorporated active film based on biodegradable blends of poly(lactide)/poly(butylene adipate-co-terephthalate) for food packaging application. *Journal of Packaging Technology and Research* 4 (3): 235–245.

34 Jamróz, E., Juszczak, L., and Kucharek, M. (2018). Investigation of the physical properties, antioxidant and antimicrobial activity of ternary potato starch-furcellaran-gelatin films incorporated with lavender essential oil. *International Journal of Biological Macromolecules* 114: 1094–1101.

35 Sharma, S., Barkauskaite, S., Duffy, B. et al. (2020). Characterization and antimicrobial activity of biodegradable active packaging enriched with clove and thyme essential oil for food packaging application. *Food* 9 (8): 1117.

36 Maestrello, C., Tonon, L., Madrona, G. et al. (2017). Production and characterization of biodegradable films incorporated with clove essential oil/ß-cyclodextrin microcapsules. *Chemical Engineering Transactions* 57: 1393–1398.

37 Akhter, R., Masoodi, F., Wani, T.A., and Rather, S.A. (2019). Functional characterization of biopolymer based composite film: incorporation of natural essential oils and antimicrobial agents. *International Journal of Biological Macromolecules* 137: 1245–1255.

38 Hernández-López, M., Correa-Pacheco, Z.N., Bautista-Baños, S. et al. (2019). Bio-based composite fibers from pine essential oil and PLA/PBAT polymer blend. Morphological, physicochemical, thermal and mechanical characterization. *Materials Chemistry and Physics* 234: 345–353.

39 Wu, Z., Zhou, W., Pang, C. et al. (2019). Multifunctional chitosan-based coating with liposomes containing laurel essential oils and nanosilver for pork preservation. *Food Shemistry* 295: 16–25.

40 Hintz, T., Matthews, K.K., and Di, R. (2015). The use of plant antimicrobial compounds for food preservation. *BioMed Research International* 2015.

41 Tiwari, B.K., Valdramidis, V.P., O'Donnell, C.P. et al. (2009). Application of natural antimicrobials for food preservation. *Journal of Agricultural and Food Chemistry* 57 (14): 5987–6000.

42 Sarkar, D. and Shetty, K. (2014). Metabolic stimulation of plant phenolics for food preservation and health. *Annual Review of Food Science and Technology* 5: 395–413.

43 Hakkim, F.L., Essa, M.M., Arivazhagan, G. et al. (2012). Evaluation of food protective property of five natural products using fresh-cut apple slice model. *Pakistan Journal of Biological Sciences: PJBS* 15 (1): 10–18.

44 Piskernik, S., Klančnik, A., Riedel, C.T. et al. (2011). Reduction of *Campylobacter jejuni* by natural antimicrobials in chicken meat-related conditions. *Food Control* 22 (5): 718–724.

45 Lara, M., Gutierrez, J., Timón, M., and Andrés, A. (2011). Evaluation of two natural extracts (*Rosmarinus officinalis* L. and *Melissa officinalis* L.) as antioxidants in cooked pork patties packed in MAP. *Meat Science* 88 (3): 481–488.

46 Gök, V., Obuz, E., Şahin, M.E., and Serteser, A. (2011). The effects of some natural antioxidants on the color, chemical and microbiological properties of sucuk (Turkish dry-fermented sausage) during ripening and storage periods. *Journal of Food Processing and Preservation* 35 (5): 677–690.

47 Ozturk, I. (2015). Antifungal activity of propolis, thyme essential oil and hydrosol on natural mycobiota of sucuk, a Turkish fermented sausage: monitoring of their effects on microbiological, color and aroma properties. *Journal of Food Processing and Preservation* 39 (6): 1148–1158.

48 Spinelli, S., Conte, A., Lecce, L. et al. (2015). Microencapsulated propolis to enhance the antioxidant properties of fresh fish burgers. *Journal of Food Process Engineering* 38 (6): 527–535.

49 Silici, S. and Karaman, K. (2014). Inhibitory effect of propolis on patulin production of *Penicillium expansum* in apple juice. *Journal of Food Processing & Preservation* 38 (3): 1129–1134.

50 Moradi, M., Tajik, H., Rohani, S.M.R., and Mahmoudian, A. (2016). Antioxidant and antimicrobial effects of zein edible film impregnated with *Zataria multiflora* Boiss. essential oil and monolaurin. *LWT - Food Science and Technology* 72: 37–43.

51 Catarino, M.D., Alves-Silva, J.M., Fernandes, R.P. et al. (2017). Development and performance of whey protein active coatings with *Origanum virens* essential oils in the quality and shelf life improvement of processed meat products. *Food Control* 80: 273–280.

52 Nishie, M., Nagao, J.-I., and Sonomoto, K. (2012). Antibacterial peptides "bacteriocins": an overview of their diverse characteristics and applications. *Biocontrol Science* 17 (1): 1–16.

53 Yang, S.-C., Lin, C.-H., Sung, C.T., and Fang, J.-Y. (2014). Antibacterial activities of bacteriocins: application in foods and pharmaceuticals. *Frontiers in Microbiology* 5: 241.

54 Seetaramaiah, K., Smith, A.A., Murali, R., and Manavalan, R. (2011). Preservatives in food products-review. *International Journal of Pharmaceutical and Biological Science Archive* 2 (2): 583–599.

55 Perin, L.M. and Nero, L.A. (2014). Antagonistic lactic acid bacteria isolated from goat milk and identification of a novel nisin variant *Lactococcus lactis*. *BMC Microbiology* 14 (1): 1–9.

56 Cao-Hoang, L., Grégoire, L., Chaine, A., and Waché, Y. (2010). Importance and efficiency of in-depth antimicrobial activity for the control of listeria development with nisin-incorporated sodium caseinate films. *Food Control* 21 (9): 1227–1233.

57 Konieczny, P. and Kijowski, J. (2005). Animal origin food preservation and its safety issues. *Polish Journal of Food and Nutrition Sciences* 14 (1): 21.

58 Samelis, J., Bedie, G., Sofos, J. et al. (2005). Combinations of nisin with organic acids or salts to control *Listeria* monocytogenes on sliced pork bologna stored at 4°C in vacuum packages. *LWT - Food Science and Technology* 38 (1): 21–28.

59 Khan, A., Vu, K.D., Riedl, B., and Lacroix, M. (2015). Optimization of the antimicrobial activity of nisin, Na-EDTA and pH against gram-negative and gram-positive bacteria. *LWT - Food Science and Technology* 61 (1): 124–129.

60 Papagianni, M. and Anastasiadou, S. (2009). Pediocins: the bacteriocins of *Pediococci*. Sources, production, properties and applications. *Microbial Cell Factories* 8 (1): 1–16.

61 Wang, J., Li, L., Zhao, X., and Zhou, Z. (2015). Partial characteristics and antimicrobial mode of pediocin produced by *Pediococcus acidilactici* PA003. *Annals of Microbiology* 65 (3): 1753–1762.

62 Colak, H., Hampikyan, H., Bingol, E.B., and Aksu, H. (2008). The effect of nisin and bovine lactoferrin on the microbiological quality of turkish-style meatball (TEKIRDAĞ KÖFTE). *Journal of Food Safety* 28 (3): 355–375.

63 Aly, E., Ros, G., and Frontela, C. (2013). Structure and functions of lactoferrin as ingredient in infant formulas. *Journal of Food Research* 2 (4): 25.

64 Munteanu, S.B. and Vasile, C. (2020). Vegetable additives in food packaging polymeric materials. *Polymers* 12 (1): 28.

65 Ferreira, A.R., Alves, V.D., and Coelhoso, I.M. (2016). Polysaccharide-based membranes in food packaging applications. *Membranes* 6 (2): 22.

66 Elieh-Ali-Komi, D. and Hamblin, M.R. (2016). Chitin and chitosan: production and application of versatile biomedical nanomaterials. *International Journal of Advanced Research* 4 (3): 411.

67 Shi, Q., Fang, Z., and Bhandari, B. (2013). Effect of addition of whey protein isolate on spray-drying behavior of honey with maltodextrin as a carrier material. *Drying Technology* 31 (13-14): 1681–1692.

68 Rodriguez, E.S., Julio, L.M., Henning, C. et al. (2019). Effect of natural antioxidants on the physicochemical properties and stability of freeze-dried microencapsulated chia seed oil. *Journal of the Science of Food and Agriculture* 99 (4): 1682–1690.

69 Quispe-Condori, S., Saldaña, M.D., and Temelli, F. (2011). Microencapsulation of flax oil with zein using spray and freeze drying. *LWT- Food Science and Technology* 44 (9): 1880–1887.

70 Beena Divya, J., Kulangara Varsha, K., Madhavan Nampoothiri, K. et al. (2012). Probiotic fermented foods for health benefits. *Engineering in Life Sciences* 12 (4): 377–390.

71 Beshkova, D. and Frengova, G. (2012). Bacteriocins from lactic acid bacteria: microorganisms of potential biotechnological importance for the dairy industry. *Engineering in Life Sciences* 12 (4): 419–432.

72 Peng, Y., Wu, Y., and Li, Y. (2013). Development of tea extracts and chitosan composite films for active packaging materials. *International Journal of Biological Macromolecules* 59: 282–289.

73 Li, J.-H., Miao, J., Wu, J.-L. et al. (2014). Preparation and characterization of active gelatin-based films incorporated with natural antioxidants. *Food Hydrocolloids* 37: 166–173.

74 Bonilla, J. and Sobral, P.J. (2016). Investigation of the physicochemical, antimicrobial and antioxidant properties of gelatin-chitosan edible film mixed with plant ethanolic extracts. *Food Bioscience* 16: 17–25.

75 Theivendran, S., Hettiarachchy, N.S., and Johnson, M.G. (2006). Inhibition of *Listeria monocytogenes* by nisin combined with grape seed extract or green tea extract in soy protein film coated on turkey frankfurters. *Journal of Food Science* 71 (2): M39–M44.

76 Kanmani, P. and Rhim, J.-W. (2014). Development and characterization of carrageenan/grapefruit seed extract composite films for active packaging. *International Journal of Biological Macromolecules* 68: 258–266.

77 Guo, Y., Chen, X., Yang, F. et al. (2019). Preparation and characterization of chitosan-based ternary blend edible films with efficient antimicrobial activities for food packaging applications. *Journal of Food Science* 84 (6): 1411–1419.

78 Dwivedi, S., Prajapati, P., Vyas, N. et al. (2017). A review on food preservation: methods, harmful effects and better alternatives. *Asian Journal of Pharmacy and Pharmacology* 3: 193–199.

79 Kim, S.-J., Min, S.C., Shin, H.-J. et al. (2013). Evaluation of the antioxidant activities and nutritional properties of ten edible plant extracts and their application to fresh ground beef. *Meat Science* 93 (3): 715–722.

80 Banerjee, R., Verma, A.K., Das, A.K. et al. (2012). Antioxidant effects of broccoli powder extract in goat meat nuggets. *Meat Science* 91 (2): 179–184.

81 Rodríguez-Carpena, J.G., Morcuende, D., and Estévez, M. (2011). Avocado by-products as inhibitors of color deterioration and lipid and protein oxidation in raw porcine patties subjected to chilled storage. *Meat Science* 89 (2): 166–173.

82 Muíño, I., Díaz, M.T., Apeleo, E. et al. (2017). Valorisation of an extract from olive oil waste as a natural antioxidant for reducing meat waste resulting from oxidative processes. *Journal of Cleaner Production* 140: 924–932.

83 Kobus-Cisowska, J., Flaczyk, E., Rudzińska, M., and Kmiecik, D. (2014). Antioxidant properties of extracts from *Ginkgo biloba* leaves in meatballs. *Meat Science* 97 (2): 174–180.

84 Yin, M.-C. and Cheng, W.-S. (2003). Antioxidant and antimicrobial effects of four garlic-derived organosulfur compounds in ground beef. *Meat Science* 63 (1): 23–28.

85 Mei, J., Ma, X., and Xie, J. (2019). Review on natural preservatives for extending fish shelf life. *Food* 8 (10): 490.

15

Marketing, Environmental, and Future Perspectives of Natural Materials in Packaging

Prakash Binu, Sasi Arun Sasi, Velamparambil Gopalakrishnan Gopikrishna, Abdul Shukkur, Balu Balachandran, and Mahesh Mohan

Mahatma Gandhi University, School of Environmental Sciences, P D Hills, Kottayam 686560, Kerala

15.1 Introduction

Packaging appeared in its primitive form when humanity began to store food and other goods for the next day. Package design and construction play an unique role in determining the shelf life of food ingredients [1]. Packaging was supposed can play a role in extending the shelf life of food Providing improved barrier properties. On account of the impact of nondegradable plastic packaging on the environment and human health. Many companies are fighting to drastically reduce plastic production, increase recycling, and promote more sustainable food packaging. Packaging is an integral part of addressing the major sustainable food consumption challenges in international affairs, which is clearly to minimize the environmental footprint of packaged foods. Packaging industry is undergone pressure mainly due to elevating environmental concerns, it is a continuous source of large quantity of plastic waste, so please review extensive research needs to be done on renewable alternatives. The use of fossil-based plastics is limited due to their nonrecyclable or nonbiodegradable nature [2]. Product shelf life is determined by three factors: product characteristics, individual pack storage, and distribution conditions [3]. The main task of food packaging must protect food from external influences and damage, contain food, and provide the consumer with ingredient and nutritional information [4]. The use of nonbiodegradable materials in a variety of packaging applications has created contamination problems. Natural packaging materials offer great alternatives to many of the plastics on the market [5].

Moreover, packaging is an important factor in maintaining food quality, primarily by controlling gas exchange and steam exchange with the outside air, and helps maintain food quality during storage. Biopolymers made from renewable raw materials

Natural Materials for Food Packaging Application, First Edition.
Edited by Jyotishkumar Parameswaranpillai, Aswathy Jayakumar,
E. K. Radhakrishnan, Suchart Siengchin, and Sabarish Radoor.
© 2023 WILEY-VCH GmbH. Published 2023 by WILEY-VCH GmbH.

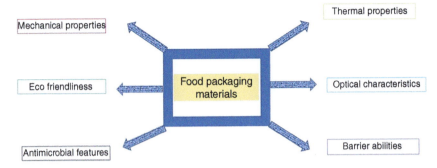

Figure 15.1 Food packaging materials.

must be biodegradable, especially compostable so that they can function as fertilizers and soil conditioners. In the last few decades, research on the development of new biodegradable packaging materials has increased. Large surface area for dramatic improvement in nanopolymers (NP). Large surface area for dramatic improvement in nanopolymers especially its, mechanical performance such as flexibility is reduced gas permeability, temperature and humidity stability are more effective [6]. In fact, it is important not only to use eco-friendly ingredients, but also to focus on friendly packaging to mitigate pollution. Sustainable packaging is generally a innovative concept and has received a lot of attention. Han et al. [7] stated that, the focus of food packaging is to wrap food in a cost-effective method that meets industry and consumer demands, minimizes environmental impact, and ensures food safety. The choice of packaging material actually comprises their environmental impact, end-of-life treatment, areas of end use, and reusability during the period effectively fulfills in the core packaging function (Figure 15.1). This book chapter illustrates the diverse aspects of marketing, environmental, and future perspectives of natural materials in packaging.

15.2 Biodegradable Food Packaging

Various biodegradable polymers are promising packaging materials, some of which are already in use [8]. Green Product innovation is a multi-faceted process, with three main types (materials, energy, and pollution) focused on the environment at the forefront [9]. The food packaging industry is currently looking for lightweight biodegradable packaging to reduce material consumption, waste, and shipping costs. There is almost no recycling traditional plastics obtained synthetically from petroleum have low biodegradability and are considered environmentally harmful waste [10]. Most of the materials utilized in the packaging industry today, especially are made from fossil fuels and are virtually nonbiodegradable. Discussion for developing biodegradable polymers from solving waste disposal problems garbage that replaces renewable raw materials [11].

"Green" is widely used and not only evolved commonly as a logo as well as label, nevertheless normally refers to composite materials, products, and technologies that have low influence on the environment and human life due to safety regulations [12]. The food packaging industry is currently looking for lightweight, biodegradable packaging reduce material usage, waste, and transportation costs [13]. Package manufacturing is one of the most constant causes of material-related environmental pollution [14]. A latest kind of packaging that integrate food packaging materials and antibacterial substances to diminish microbial surface contamination of foods, increase the safety of microbial products and extend shelf life is appearing attention in the packaging industry [15]. The green packaging material can be biocomposite material, bionanocomposite, or nanopaper. The food packaging process enterprise is at present in pursuing biodegradable packaging, which is lightweight as decreasing substance use in addition to transportation costs [16]. Bamboo is biodegradable, durable, and heat resistant, so it has many desirable properties in food packaging.

Recycling is used to reprocess materials into new products, as opposed to reusing returned products in their original form [17]. The form of unmodified cellulose is very easily biodegradable [18]. Despite all these factors, starch remains the most promising polysaccharide for food packaging because it is easier to process than cellulose. Indeed, starch is very cheap and highly biodegradable [19]. Physical blending of various morphologies and physical properties, such as functioning packaging technology and natural fiber reinforcement, are investigated to upgrade the characteristics of biodegradable polymers. Biodegradable nanocomposite food packaging category contains new biodegradable materials it can usually be created from polylactide or poly(lactic acid) or PLA and montmorillonite (MMT). MMT is obtained from relatively inexpensive and widely available natural clay from volcanic ash/rock. Paper and board materials have excellent mechanical properties as packaging materials, but gas and water vapor permeability for many food applications is often very high [20]. PLA is a biodegradable material currently being developed for use in food packaging [21].

Because traditional packaging materials are natural ingredients, they may be abundant as compared with others with no impact on manufactured materials, costs, or environment concerning these types of materials in any respect processing or technology. The growth of sustainable or green packages has the possible to diminished the environmental impact of food packaging with either edible or biodegradable materials, plant extracts, and nanomaterials [22]. Besides, biodegradable films can be made from chitin in the crust and chitosan from the exoskeleton of insects. Chitin is a biopolymer with a chemical shape just like cellulose [23]. A biodegradable laminate of chitosan cellulose and polycaprolactone film has been developed and used nowadays [24]. Moreover, collagen-based bioplastics are synthesized using a discharge procedure and have a variety of uses [25]. Soy protein is used in the making of bioplastic films for various packaging applications. Soy protein-based films are smoother and more than that it is transparent, more flexible, and cheaper than other protein-based bioplastics. Isolated soy protein as a raw material is high level to further sources mainly because of its excellent film-forming properties, low cost, and excellent barrier property against oxygen [26]. Whey protein has exceptional functional properties and film-forming ability. Biodegradability is not only a functional requirement, but also an important environment attribute.

15.3 Different Bio-Based Packaging Materials

Biodegradable polymers are derived from renewable agricultural sources, animal sources, waste from the marine food processing industry, or microbial sources. According to Cheung et al. [27], synthetic polymers can also be partially degraded by blending with biopolymers containing biodegradable components. Bionanocomposite materials as biosensors have the capability to react to alters in the environment such as temperature and humidity. Bio-packaging materials can contain other natural extracts/ingredients such as lignin and wax that act as preservatives and prevent the initial deterioration of food [28]. Bio-based packaging can create new atmospheric conditions that improve quality of certain foods especially vegetables and enhanced shelf life. Biopolymers are consisting of monomeric units that covalently bond to form chain molecules. The prefix bio means that the biopolymer is biodegradable. Therefore, biopolymers have the ability to be broken down or broken down by the action of naturally take place in organisms [29].

15.3.1 Bioplastics

Bioplastic tissue and bioplastics can be defined as plastics based on renewable raw materials (bio) or biodegradable and/or compostable plastic [30]. Much research has been done on the potential use of diverse bio-based materials in the development of sustainable packaging materials. The main source of bioplastic starch is corn, but there are also potato, wheat, rice, barley, and oat starches. In addition, participants will work with the food supply chain to find affordable, innovative and environmentally friendly, and imaginative food packaging alternatives to protect and monitor the quality of packaged food [31]. The purpose of bioplastics is to mimic the biomass life cycle, including the conservation of fossil resources, water, and carbon dioxide (CO_2) production [32]. In particular, the properties of bioplastics have been improved using a variety of methods. Improved gas and water barrier properties. Some of the strategies include coating, mixing, adding nanoparticles, adding cellulose, and chemically and physically modifying. Figure 15.2 discussed significance of bioplastics.

For the past ten years, the food, packaging, and distribution industry have more development and progress in the application of food bioplastics in food packaging. Moreover, there is great potential for incorporating nanoclay into bio-based food packaging [33]. Biodegradable plastics have minimal impact on landfills and generally eliminate the oxygen and moisture needed for biodegradation, so landfills.

15.3.2 Biopolymers

Over the past decade, environmental issues have become more and more important to both the food industry and consumers, consisting materials from nonrenewable sources leading to bio-based packaging materials. Over the last decade, environmental aspects are becoming increasingly important for both the food industry and consumers, such as leading to materials from nonrenewable sources and bio-based packaging materials [34]. As discussed in Figure 15.3 biopolymers have a classification of natural, synthetic, and microbial. Therefore, biomacromolecules can be degraded or degraded by the action of natural organisms [35]. The most common

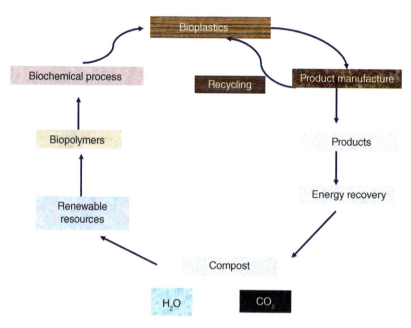

Figure 15.2 Significance of bioplastics.

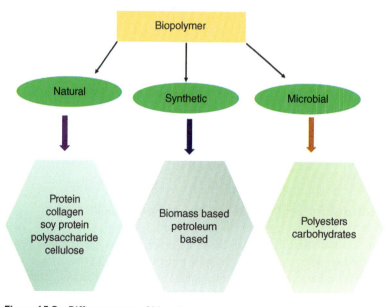

Figure 15.3 Different types of biopolymers.

type of biopolymer for food packaging applications are cellulose, chitosan, starch, and agar obtained from both carbohydrates and gelatin. Several studies have suggested that among biopolymers, starches and derivatives are the most widely studied types for the production of bionanocomposite materials for various aspects of food packaging applications [36–38].

In the food packaging sector, starch-based ingredients won much attention for its biodegradability, edible, and general availability. A ubiquitous biomaterial, starch is one of the richest and cheapest sources of water-soluble polysaccharides due to its unique biodegradability property [24]. The prospective utilization of lignin-based biomass as an inexpensive reinforcement used in polyurethane preparation as an alternative to fossil-based polymers [39]. Lignin, actually used as a biodegradable solution of phenol in petrochemical products phenol formaldehyde (PF) resin [40]. Natural clays are inexpensive materials that naturally occur as aluminum silicate primarily composed of fine-grained minerals and various clays [41].

15.4 Nano Food Packaging

Food nanotechnology is an emerging field of increasing interest, opening up latest opportunities for the food industry throughout the universe. Application of nanoparticles to food packaging including nanoencapsulation, nanocomposites, nanoemulsions, and edible nanocoatings [42]. The process of creating a nanofood packages include heat/mass transfer, nanoscale reaction engineering, nanobiotechnology, and molecular synthesis. Bionanocomposites can be potential food packaging along with food packaging foods in some way by releasing beneficial compounds such as antibacterial agents and antioxidants, or by eliminating unwanted elements such as oxygen and water vapor [43, 44]. Bionanocomposites can also be used in good food packaging to recognize packaged food characteristics such as microbial contamination [45]. Most types of nano-sized that is, less than 100 nm fillers biopolymer. Nanofillers usually studied in food packaging implications can be divided into various types, such as nanoparticles and nanofibrils [46]. Bionanocomposites are a promising alternative to traditional plastics, principally those made from petroleum, for food packaging. According to Sharma and Dhanjal [47], gelatin-based nanocomposite films will benefit from competing and eliminating bacterial invaders that may extend shelf life and improve food quality. Polymers made by microorganisms or genetically modified bacteria among them the most well-known type of biopolymer is polyhydroxyalkanoate [48].

15.5 Natural Antimicrobial Agents in Food Packaging

Natural antibacterial agents are one of the most important sources of antibacterial packaging. They are natural and not harmful to health [49]. Antibacterial and active packaging, on the other hand, are currently considered as very effective concerning biology and control the activity of bacteria and fungi, also natural polymers are used in several studies as they provide good food protection. Antibacterial packaging is a new development containing antibacterial agents a polymer film that suppresses the activity of target microorganisms [50]. Antimicrobial packages are divided into two main groups known as biodegradable packages and nonbiodegradable packages. In general, antimicrobial biodegradable films have been made by natural polymers with the inherent antimicrobial reactivity or by adding antimicrobial agents to the natural polymers [51]. Examples of renewable biopolymers are polysaccharides, proteins, lipids, gums, and their complexes. Many studies [50, 52, 53] have shown that antimicrobial packaging can effectively suppress bacteria when the appropriate amount of antimicrobial active ingredient is incorporated into the polymer film. There are many reports on the effectiveness of antimicrobial in plant extracts and food packaging components. Antibacterial packaging is one of the more innovative and good types of active packaging developed in the last decade [54]. Studies are using this protocol to investigate the antibacterial activity of edible films fortified with various botanical essential oils [55–58].

15.6 Edible Films in Food Packaging

Interest in the development of antibacterial packaging containing natural antibacterial agents is increasing. The antibacterial properties of some plant extracts and essential oils from spices/herbs have long been studied and used to preserve foods [59]. Several studies [34, 60] reported that the use of nanocomposites promises to expand the use of edible and biodegradable films. Edible packaging materials or coatings are nontoxic, adhere to the surface of food, are tasteless, must have a pleasant taste, have excellent barrier properties, and prevent dehydration of food [61]. Figure 15.4 depicts classification of different edible polymers. Edible

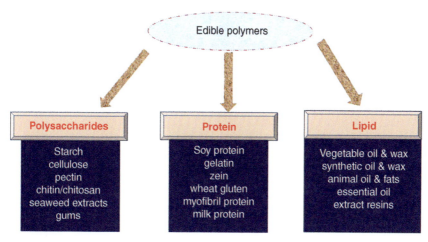

Figure 15.4 Edible polymers.

coatings are soluble formulations that are applied thinly to the surface of the food. Edible film layers are made directly on the surface of food or between different things layers of components to prevent the transfer of moisture and oxygen ([62, 63]). Edible film sources of vegetables and animals such as corn protein and whey protein, collagen, and gelatin [64]. Chitosan, one of the most abundant natural macromolecules, is the flagship product of such innovative edible antibacterial packaging ([65]).

Ortiz et al. [66] reported microfibrillated cellulose mixed with soy protein and glycerin a plasticizer for obtaining biodegradable packaging. Edible films and coatings made from milk protein contain reduced oxygen very important permeability to avoid rancidification and lipid oxidation. Proteins are gaining new notice as degradable, reproducible polymer films. Moreover, traditionally proteins are utilized as both adhesives and edible films [67]. Besides, protein films show excellent gas barrier properties and are not completely hydrophobic, but many are water resistant [68]. Edible polymers have emerged as an alternative to synthetic plastics for food applications and have received a great deal of attention in recent years for being superior to synthetic polymers [69]. Furthermore, edible polymers also act as carriers for antibacterial agents and antioxidants. The antibacterial activity of garlic oil is oil, *Escherichia coli*, tested by mixing bacterial inoculations of *Cereus*, *Staphylococcus aureus*, and *Staphylococcus typhimurium*. It was found that garlic oil suppresses the growth of all bacteria and can be linked into edible packages [70]. Essential oils and other bioactive compounds are involved in the chemical structure of membrane formation by interaction with plasticizers and polymers by reducing the spread of bacterial activity [71]. The green tea extract is incorporated into the chitosan film and exhibit excellent antioxidant ability by enhancing the concentration of polymer film of green tea extract. Figure 15.5 shows various bioactive or biological agents for smart food packaging.

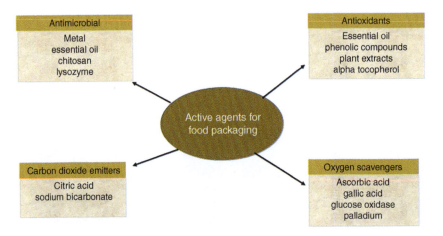

Figure 15.5 Bioactive agents for smart food packaging.

15.7 Environment and Food Packaging

The environmental impact is becoming more and more significant area to consumers [72]. Environmental and food safety are important areas that influence the development of biodegradable packages. A green packaging strategy to decrease the negative effect of packaging on the environment. Environmental awareness in the manufacturing process of new products such as packaging. In addition, new political, economic, and design consciousness has led to the development of new products that are considered more ethical and environmentally friendly. Figure 15.6 explained the eco-friendly packaging factors. Many different packaging materials can provide a chemical barrier. Glass and metal provide a nearly perfect barrier to chemicals and other environmental influences [73]. The environmental impact of packaging process is a major issue worldwide and is recognized as an increasingly critical issue [74]. Packaging is one of these industries as it has a significant impact on the cost and eco-efficiency of the entire supply chain from the procurement of packaging materials [75]. Jayaramudu et al. (2013) stated that, to fulfill the growing need for sustainability and environmental safety more and more research is directed to the growth of food packaging materials quickly and entirely mineralize in the ecosystem (Figure 15.7).

Today, especially our natural environment is toxically affected by the accumulation of packaging waste. Packaging recycling connects the fields of packaging science and the environment to recycle packaging, it is necessary to understand the nature of the packaging material and its environmental impact [76]. Packaging waste treatment plans and strategies are taken seriously in many countries around the world, with a strong tendency toward eco-design in all areas [77, 78]. As awareness of the negative environmental and social impacts of packaging grows, so does the interest of various parties in the supply chain. Package design/redesign was approached with the goal of reducing the potential cost and environmental impact of supply chain functionality [79]. Figure 15.8 showed the environmental suitability of management. When evaluating the environmental implicationsof food packaging, the positive advantage of diminishing food waste throughout the supply chain should be taken into account [80].

Figure 15.6 Eco-friendly packaging.

Figure 15.7 Environmental impacts of food packaging. Source: HOT BARGAINS; Paulo de Oliveira/AGE Fotostock; fotofabrika / Adobe Stock.

Figure 15.8 Environmental suitability of packaging.

The negative environmental impacts linked with the life cycle of various plastics and plastic packaging are: this is because the amount of waste has increased and the circular economy system is not functioning.

15.8 Sustainable Packaging

Sustainability is a complex term and is very easy to interpret. The concept of sustainability is one of the most discussed in the packaging sector these days. For many product manufacturers, the inclusion of sustainability principles in business

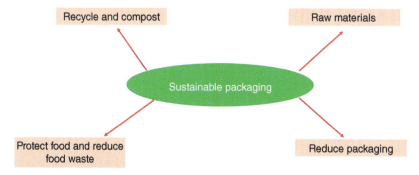

Figure 15.9 Sustainable packaging.

practices can usually be visualized by others (society) through packaging. The goal of sustainable food packaging (As discussed in Figure 15.9) integrating functional and/or innovative materials into packaging to improve economic and environmental health ([81]). Sustainable packaging is still a complex concept, with research documents for evaluating package design and materials used [82]. Food packaging is generally considered an unnecessary expense considered an additional economic and environmental costs rather than being properly valued to reduce food loss. Various reports [83–85] showed that uses of recyclable materials give the impression that the packaging is environmentally friendly.

Currently, sustainable packaging is an integral part of the food supply system key issues in sustainable food consumption policy and production due to its environmental impact. Further improvement of recyclability, reusability, composting, and energy recovery are other tools it is being considered by the industry to reduce waste generation. Information on packaging sustainability and consumer perception were determined separately [86]. Information density about food packaging sustainability is important more significant than the information quantity of consumer perception of food packaging. Therefore, sustainable packaging is a complex idea and needs to be implemented in the system approach and critical thinking [87].

Bio-based packaging materials were defined as "materials" by Robertson [88]. It is mainly obtained from renewable resources every year. The packaging sector uses a variety of natural and synthetic biodegradable polymers. Polymers are made from starch and cellulose films, fermented organic materials, and edible films.

15.9 Marketing of Natural Materials in Packaging

The packaging industry is playing a more dynamic role in the global economy, and its market value has risen sharply in recent years. Besides, increased environmental awareness is clearly reflected in marketing. In addition, packaging helps improve food acceptance at the market level. Food packaging with more desirable features is continuously sought in reaction to the consumer issues for food safety, marketplace globalization, and developing environmental awareness [89]. The packaging industry is the world's largest consumer of plastic and a major source of plastic waste that enters the environment at an alarming rate. The packaging is the face of the product and often the only thing consumers see before purchasing. Resulting in, unique or innovative packages can be sold in an emerging environment [90]. The true value of packaging is that it has become an integral part of the product. Corporate functions separate ingredients, but consumers do not distinguish between products, packaging, and stock [89]. Companies are operating on a brand-new line of merchandize for ethical, renewable, and sustainable packaging. Of course, the need to invest in a new filling line to enable more environmentally or eco-friendly, recyclable, reusable, and sustainable packaging [91]. There are basically four main packaging elements that can affect consumer's purchase decisions that can be classified into two different categories, visual and informative elements [92]. The visual element consists of the graphic and the size/shape of the package. Today, product packaging plays an increasingly chief role in the competition in the market. Also, the packaging is based on the product consumption rate in a particular area. Food packaging ensures food safety and guarantees food quality throughout the supply chain. Biodegradable packaging material is a natural and easily degradable protective layer. Although it is often used as food packaging. But growing consumer awareness of this package has brought about a huge surge in the market as a whole [93]. The cost of these fully degradable materials from renewable natural resources is relatively higher than nondegradable polymer plastics [94].

Consumers assess the sustainability of packaging based primarily on the circular economy implemented [95]. In terms of expected recyclability and natural-looking packaging materials, paper/cardboard is associated with positive emotions, and attributes such as homely and fresh products convey a sense of wellness. Emphasis is based on natural, abundant, and inexpensive polymers, but these need to be modified and carefully selected. Besides, additives provide the properties required by the packaging industry. Instead of that, consumers are more interested to buy unwrapped and traditional food packaging such as glass, paper, cardboard, any recycled packaging, and plastic packaging products with reduced packaging materials [96]. Consumers will be willing to pay more for green products.

15.10 Future Perspectives of Natural Materials in Packaging

The packaging sector clearly require more well established research and manufacture of new materials as an alternative to both plastic and fossil-based materials. The changing attitudes of social and state law provide another stimulus for the development of biodegradable products. Hopefully, nanotechnology has tremendous potential to benefit society through its application in food packaging. It can make the product cheaper and the production more efficient. Research and development of bio-based nanocomposites for food application areas such as packaging and other food contact surfaces is expected to increase in the next decade. Research in the packaging industry today has two main directions. First study the approach is to modify current packaging materials to impart antibacterial and antioxidant activity, as well as better mechanical and barrier properties. Main challenge that is faced in food packaging industry is the manufacturing of bio-based primary products. The packaging should match the shelf life of the packaging with the shelf life of the product. Another crucial issue is opinion of public because it is widely recognized that consumers should be more environmentally friendly. The future of biopolymers depends not only on their competitiveness but also on their affordability.

15.11 Conclusion

Food packaging is a very crucial and significant element of food safety. Food safety is one of the most important factors affecting public health and social welfare. Application ofnatural material in the area of food packaging technology providing new biologically degradable and environmentally friendly alternatives to petrochemicals base plastic. This can resolve the problem of waste accumulation through nonbiodegradable petrochemical-based plastics nature. Moreover, films made from natural macromolecules provide excellent moisture resistance for fruits and vegetables and do not harm food or human health.

References

1 Sulieman, A.M.E., Hussein, S.W., and Elkhalifa, E.A. (2007). The effect of packaging materials on the quality and shelf-life of *Jibna-beida* (white cheese). *Gezira Journal of Engineering and Applied Sciences* 2 (2).
2 Saalah, S., Saallah, S., Rajin, M., and Yaser, A.Z. (2020). Management of biodegradable plastic waste: a review. *Advances in Waste Processing Technology* 127–143.
3 Mills, J., Donnison, A., and Brightwell, G. (2014). Factors affecting microbial spoilage and shelf-life of chilled vacuum-packed lamb transported to distant markets: a review. *Meat Science* 98 (1): 71–80.
4 Coles, R., McDowell, D., and Kirwan, M.J. (ed.) (2003). *Food Packaging Technology*, vol. 5. CRC press.
5 Muthuraj, R., Misra, M., and Mohanty, A.K. (2018). Biodegradable compatibilized polymer blends for packaging applications: a literature review. *Journal of Applied Polymer Science* 135 (24): 45726.
6 Othman, S.H. (2014). Bio-nanocomposite materials for food packaging applications: types of biopolymers and nano-sized filler. *Agriculture and Agricultural Science Procedia* 2: 296–303.
7 Han, J.W., Ruiz-Garcia, L., Qian, J.P., and Yang, X.T. (2018). Food packaging: a comprehensive review and future trends. *Comprehensive Reviews in Food Science and Food Safety* 17 (4): 860–877.
8 Gross, R.A. and Kalra, B. (2002). Biodegradable polymers for the environment. *Science* 297 (5582): 803–807.

9 Scheidel, A., Temper, L., Demaria, F., and Martínez-Alier, J. (2018). Ecological distribution conflicts as forces for sustainability: an overview and conceptual framework. *Sustainability Science* 13 (3): 585–598.

10 Popa, M., Mitelut, A., Niculita, P. et al. (2011). Biodegradable materials for food packaging applications. *Journal of Environmental Protection and Ecology* 12 (4): 1825–1834.

11 Luyt, A.S. and Malik, S.S. (2019). Can biodegradable plastics solve plastic solid waste accumulation? In: *Plastics to Energy*, 403–423. William Andrew Publishing.

12 Banerjee, A. and Solomon, B.D. (2003). Eco-labeling for energy efficiency and sustainability: a meta-evaluation of US programs. *Energy Policy* 31 (2): 109–123.

13 Ashok, A., Rejeesh, C., and Renjith, R. (2016). Biodegradable polymers for sustainable packaging applications: a review. *International Journal of Bionics and Biomaterials* 1 (11).

14 Otto, S., Strenger, M., Maier-Nöth, A., and Schmid, M. (2021). Food packaging and sustainability–consumer perception vs. correlated scientific facts: a review. *Journal of Cleaner Production* 126733.

15 Irkin, R. and Esmer, O.K. (2015). Novel food packaging systems with natural antimicrobial agents. *Journal of Food Science and Technology* 52 (10): 6095–6111.

16 Li, W., Coffin, D.R., Jin, T.Z. et al. (2012). Biodegradable composites from polyester and sugar beet pulp with antimicrobial coating for food packaging. *Journal of Applied Polymer Science* 126 (S1): E362–E373.

17 King, A.M., Burgess, S.C., Ijomah, W., and McMahon, C.A. (2006). Reducing waste: repair, recondition, remanufacture or recycle? *Sustainable Development* 14 (4): 257–267.

18 Ferreira, F.V., Pinheiro, I.F., Gouveia, R.F. et al. (2018). Functionalized cellulose nanocrystals as reinforcement in biodegradable polymer nanocomposites. *Polymer Composites* 39: E9–E29.

19 Avella, M., De Vlieger, J.J., Errico, M.E. et al. (2005). Biodegradable starch/clay nanocomposite films for food packaging applications. *Food Chemistry* 93 (3): 467–474.

20 Raheem, D. (2013). Application of plastics and paper as food packaging materials-an overview. *Emirates Journal of Food and Agriculture* 177–188.

21 Tawakkal, I.S., Cran, M.J., Miltz, J., and Bigger, S.W. (2014). A review of poly (lactic acid)-based materials for antimicrobial packaging. *Journal of Food Science* 79 (8): R1477–R1490.

22 Molina-Besch, K. and Pålsson, H. (2016). A supply chain perspective on green packaging development-theory versus practice. *Packaging Technology and Science* 29 (1): 45–63.

23 Zargar, V., Asghari, M., and Dashti, A. (2015). A review on chitin and chitosan polymers: structure, chemistry, solubility, derivatives, and applications. *ChemBioEng Reviews* 2 (3): 204–226.

24 Khalil, H.A., Banerjee, A., Saurabh, C.K. et al. (2018). Biodegradable films for fruits and vegetables packaging application: preparation and properties. *Food Engineering Reviews* 10 (3): 139–153.

25 Tabasum, S., Younas, M., Zaeem, M.A. et al. (2019). A review on blending of corn starch with natural and synthetic polymers, and inorganic nanoparticles with mathematical modeling. *International Journal of Biological Macromolecules* 122: 969–996.

26 Singh, P., Kumar, R., Sabapathy, S.N., and Bawa, A.S. (2008). Functional and edible uses of soy protein products. *Comprehensive Reviews in Food Science and Food Safety* 7 (1): 14–28.

27 Cheung, Y.W., Dobreski, D.V., Turner, R., Wheeler, M. and Handa, Y.P., Pactiv LLC, 2011. *Polymer blends of biodegradable or bio-based and synthetic polymers and foams thereof*. US Patent 7,977,397.

28 Ballesteros, L.F., Michelin, M., Vicente, A.A. et al. (2018). Use of lignocellulosic materials in bio-based packaging. In: *Lignocellulosic Materials and Their Use in Bio-Vased Packaging*, 65–85. Cham: Springer.

29 Liu, Y., Lehn, J.M., and Hirsch, A.K. (2017). Molecular biodynamers: dynamic covalent analogues of biopolymers. *Accounts of Chemical Research* 50 (2): 376–386.

30 Jabeen, N., Majid, I., and Nayik, G.A. (2015). Bioplastics and food packaging: a review. *Cogent Food & Agriculture* 1 (1): 1117749.

31 Gold, S., Hahn, R., and Seuring, S. (2013). Sustainable supply chain management in "Base of the Pyramid" food projects—a path to triple bottom line approaches for multinationals? *International Business Review* 22 (5): 784–799.

32 Kumar, S. and Thakur, K.S. (2017). Bioplastics-classification, production and their potential food applications. *Journal of Hill Agriculture* 8 (2): 118–129.

33 Peelman, N., Ragaert, P., De Meulenaer, B. et al. (2013). Application of bioplastics for food packaging. *Trends in Food Science & Technology* 32 (2): 128–141.

34 Ray, S.S. and Bousmina, M. (2005). Biodegradable polymers and their layered silicate nanocomposites: in greening the 21st century materials world. *Progress in Materials Science* 50 (8): 962–1079.

35 Kolybaba, M., Tabil, L.G., Panigrahi, S. et al. (2006). Biodegradable polymers: past, present, and future. In: *ASABE/CSBE North Central Intersectional Meeting*, 1. American Society of Agricultural and Biological Engineers.

36 Heydari, A., Alemzadeh, I., and Vossoughi, M. (2013). Functional properties of biodegradable corn starch nanocomposites for food packaging applications. *Materials & Design* 50: 954–961.

37 Pan, H., Xu, D., Liu, Q. et al. (2014). Preparation and characterization of corn starch-nanodiamond composite films. In: *Applied Mechanics and Materials*, vol. 469, 156–161. Trans Tech Publications Ltd.

38 Tang, X., Alavi, S., and Herald, T.J. (2008). Barrier and mechanical properties of starch-clay nanocomposite films. *Cereal Chemistry* 85 (3): 433–439.

39 Mimini, V., Kabrelian, V., Fackler, K. et al. (2019). Lignin-based foams as insulation materials: a review. *Holzforschung* 73 (1): 117–130.

40 El Mansouri, N.E., Yuan, Q., and Huang, F. (2018). Preparation and characterization of phenol-formaldehyde resins modified with alkaline rice straw lignin. *BioResources* 13 (4): 8061–8075.

41 Bergaya, F. and Lagaly, G. (2006). General introduction: clays, clay minerals, and clay science. *Developments in Clay Science* 1: 1–18.

42 Sarkar, S. and Dey, J.K. (2021). Nanotechnology: an insight for food processing. *Biotica Research Today* 3 (5): 364–366.

43 Sekhon, B.S. (2010). Food nanotechnology–an overview. *Nanotechnology, Science and Applications* 3: 1.

44 Nerus, T. (2020). Emerging of bio-nano composite gelatine-based film as biodegradable food packaging: a review. *Food Research* 4 (4): 944–956.

45 Sorrentino, A., Gorrasi, G., and Vittoria, V. (2007). Potential perspectives of bio-nanocomposites for food packaging applications. *Trends in Food Science & Technology* 18 (2): 84–95.

46 Shankar, S. and Rhim, J.W. (2018). Bionanocomposite films for food packaging applications. *Reference Module in Food Science* 1: 1–10.

47 Sharma, D. and Dhanjal, D.S. (2016). Bio-nanotechnology for active food packaging. *Journal of Applied Pharmaceutical Science* 6 (09): 220–226.

48 Philip, S.E., Odell, M., Keshavarz, T., and Roy, I. (2007). Polyhydroxy-alkanoates-the biodegradable polymers. In *International Conference on Biodegradable Polymers: Their Production, Characterisation and Application*.

49 Saeed, F., Afzaal, M., Tufail, T., and Ahmad, A. (2019). Use of natural antimicrobial agents: a safe preservation approach. *Active Antimicrobial Food Packaging* 18 (0).

50 Sung, S.Y., Sin, L.T., Tee, T.T. et al. (2013). Antimicrobial agents for food packaging applications. *Trends in Food Science & Technology* 33 (2): 110–123.

51 Cha, D.S. and Chinnan, M.S. (2004). Biopolymer-based antimicrobial packaging: a review. *Critical Reviews in Food Science and Nutrition* 44 (4): 223–237.

52 Han, J.H. (2003). Antimicrobial food packaging. *Novel Food Packaging Techniques* 8: 50–70.

53 Huang, T., Qian, Y., Wei, J., and Zhou, C. (2019). Polymeric antimicrobial food packaging and its applications. *Polymers* 11 (3): 560.

54 Kapetanakou, A.E. and Skandamis, P.N. (2016). Applications of active packaging for increasing microbial stability in foods: natural volatile antimicrobial compounds. *Current Opinion in Food Science* 12: 1–12.

55 Avila-Sosa, R., Palou, E., Munguía, M.T.J. et al. (2012). Antifungal activity by vapor contact of essential oils added to amaranth, chitosan, or starch edible films. *International Journal of Food Microbiology* 153 (1–2): 66–72.

56 Hashemi, S.M.B. and Khaneghah, A.M. (2017). Characterization of novel basil-seed gum active edible films and coatings containing oregano essential oil. *Progress in Organic Coatings* 110: 35–41.

57 Otoni, C.G., Avena-Bustillos, R.J., Olsen, C.W. et al. (2016). Mechanical and water barrier properties of isolated soy protein composite edible films as affected by carvacrol and cinnamaldehyde micro and nanoemulsions. *Food Hydrocolloids* 57: 72–79.

58 Utama-Ang, N., Sida, S., Wanachantararak, P., and Kawee-Ai, A. (2021). Development of edible Thai rice film fortified with ginger extract using microwave-assisted extraction for oral antimicrobial properties. *Scientific Reports* 11 (1): 1–10.

59 Ochoa, T.A., Almendárez, B.E.G., Reyes, A.A. et al. (2017). Design and characterization of corn starch edible films including beeswax and natural antimicrobials. *Food and Bioprocess Technology* 10 (1): 103–114.

60 Lagaron, J.M., Cabedo, L., Cava, D. et al. (2005). Improving packaged food quality and safety. Part 2: Nanocomposites. *Food Additives and Contaminants* 22 (10): 994–998.

61 Cutter, C.N. (2006). Opportunities for bio-based packaging technologies to improve the quality and safety of fresh and further processed muscle foods. *Meat Science* 74 (1): 131–142.

62 Said, N.S. and Sarbon, N.M. (2019). Protein-based active film as antimicrobial food packaging: a. *Active Antimicrobial Food Packaging* 53.

63 Pirozzi, A., Pataro, G., Donsì, F., and Ferrari, G. (2021). Edible coating and pulsed light to increase the shelf life of food products. *Food Engineering Reviews* 13: 544–569.

64 Lacroix, M. and Vu, K.D. (2014). Edible coating and film materials: proteins. In: *Innovations in Food Packaging*, 277–304. Academic Press.

65 Kraśniewska, K., Pobiega, K., and Gniewosz, M. (2019). Pullulan–biopolymer with potential for use as food packaging. *International Journal of Food Engineering* 15 (9).

66 Ortiz, C.M., Salgado, P.R., Dufresne, A., and Mauri, A.N. (2018). Microfibrillated cellulose addition improved the physicochemical and bioactive properties of biodegradable films based on soy protein and clove essential oil. *Food Hydrocolloids* 79: 416–427.

67 Song, F., Tang, D.L., Wang, X.L., and Wang, Y.Z. (2011). Biodegradable soy protein isolate-based materials: a review. *Biomacromolecules* 12 (10): 3369–3380.

68 Rhim, J.W., Mohanty, K.A., Singh, S.P., and Ng, P.K. (2006). Preparation and properties of biodegradable multilayer films based on soy protein isolate and poly (lactide). *Industrial & Engineering Chemistry Research* 45 (9): 3059–3066.

69 Kumar, S., Mukherjee, A., and Dutta, J. (2020). Chitosan based nanocomposite films and coatings: emerging antimicrobial food packaging alternatives. *Trends in Food Science & Technology* 97: 196–209.

70 Sung, S.Y., Sin, L.T., Tee, T.T. et al. (2014). Control of bacteria growth on ready-to-eat beef loaves by antimicrobial plastic packaging incorporated with garlic oil. *Food Control* 39: 214–221.

71 Falguera, V., Quintero, J.P., Jiménez, A. et al. (2011). Edible films and coatings: structures, active functions and trends in their use. *Trends in Food Science & Technology* 22 (6): 292–303.

72 Lazzarini, G.A., Zimmermann, J., Visschers, V.H., and Siegrist, M. (2016). Does environmental friendliness equal healthiness? Swiss consumers' perception of protein products. *Appetite* 105: 663–673.

73 Pan, Y., Farmahini-Farahani, M., O'Hearn, P. et al. (2016). An overview of bio-based polymers for packaging materials. *Journal of Bioresources and Bioproducts* 1 (3): 106–113.

74 Reisch, L., Eberle, U., and Lorek, S. (2013). Sustainable food consumption: an overview of contemporary issues and policies. *Sustainability: Science, Practice and Policy* 9 (2): 7–25.

75 Zailani, S., Shaharudin, M.R., Govindasamy, V. et al. (2015). The eco-efficiency practices of the sustainable packaging and its effect towards sustainable supply chain performance. In: *2015 International Symposium on Technology Management and Emerging Technologies (ISTMET)*, 448–453. IEEE.

76 Ross, S. and Evans, D. (2003). The environmental effect of reusing and recycling a plastic-based packaging system. *Journal of Cleaner Production* 11 (5): 561–571.

77 Song, Q., Li, J., and Zeng, X. (2015). Minimizing the increasing solid waste through zero waste strategy. *Journal of Cleaner Production* 104: 199–210.

78 Williams, H. and Wikström, F. (2011). Environmental impact of packaging and food losses in a life cycle perspective: a comparative analysis of five food items. *Journal of Cleaner Production* 19 (1): 43–48.

79 Verghese, K. and Lewis, H. (2007). Environmental innovation in industrial packaging: a supply chain approach. *International Journal of Production Research* 45 (18–19): 4381–4401.

80 Dilkes-Hoffman, L.S., Lane, J.L., Grant, T. et al. (2018). Environmental impact of biodegradable food packaging when considering food waste. *Journal of Cleaner Production* 180: 325–334.

81 Hottle, T.A., Bilec, M.M., and Landis, A.E. (2013). Sustainability assessments of bio-based polymers. *Polymer Degradation and Stability* 98 (9): 1898–1907.

82 Ma, X. and Moultrie, J. (2017). What stops designers from designing sustainable packaging?—a review of eco-design tools with regard to packaging design. In: *International Conference on Sustainable Design and Manufacturing*, 127–139. Cham: Springer.

83 Hoek, A.C., Pearson, D., James, S.W. et al. (2017). Shrinking the food-print: a qualitative study into consumer perceptions, experiences and attitudes towards healthy and environmentally friendly food behaviours. *Appetite* 108: 117–131.

84 Lindh, H., Olsson, A., and Williams, H. (2016). Consumer perceptions of food packaging: contributing to or counteracting environmentally sustainable development? *Packaging Technology and Science* 29 (1): 3–23.

85 Young, S. (2008). Packaging and the environment: a cross-cultural perspective. *Design Management Review* 19 (4): 42–48.

86 Magnier, L., Schoormans, J., and Mugge, R. (2016). Judging a product by its cover: packaging sustainability and perceptions of quality in food products. *Food Quality and Preference* 53: 132–142.

87 Ericksen, P., Stewart, B., Dixon, J. et al. (2012). The value of a food system approach. In: *Food Security and Global Environmental Change*, 45–65. Routledge.

88 Robertson, G.L. (2013). Edible, biobased and biodegradable food packaging materials. In: *Food packaging principles and practice*, 3e, 49–90. Boca Raton, London, New York: CRC Press, Taylor and Francis Group.

89 Ahmed, A., Ahmed, N., and Salman, A. (2005). Critical issues in packaged food business. *British Food Journal*.

90 Silayoi, P. and Speece, M. (2004). Packaging and purchase decisions: An exploratory study on the impact of involvement level and time pressure. *British food journal*.

91 Dangelico, R.M. and Pujari, D. (2010). Mainstreaming green product innovation: Why and how companies integrate environmental sustainability. *Journal of business ethics* 95: 471–486.

92 Mai, R., Symmank, C., and Seeberg-Elverfeldt, B. (2016). Light and pale colors in food packaging: when does this package cue signal superior healthiness or inferior tastiness? *Journal of Retailing* 92 (4): 426–444.

93 Cinelli, P., Schmid, M., Bugnicourt, E. et al. (2014). Whey protein layer applied on biodegradable packaging film to improve barrier properties while maintaining biodegradability. *Polymer Degradation and Stability* 108: 151–157.

94 Davis, G. and Song, J.H. (2006). Biodegradable packaging based on raw materials from crops and their impact on waste management. *Industrial Crops and Products* 23 (2): 147–161.

95 Stewart, R. and Niero, M. (2018). Circular economy in corporate sustainability strategies: a review of corporate sustainability reports in the fast-moving consumer goods sector. *Business Strategy and the Environment* 27 (7): 1005–1022.

96 Ketelsen, M., Janssen, M., and Hamm, U. (2020). Consumers' response to environmentally-friendly food packaging-a systematic review. *Journal of Cleaner Production* 254: 120123.

Index

a

active nanocomposite films   264
active packaging   23, 24, 27, 30, 33, 36, 52, 65, 121, 169, 209, 219, 223, 226, 258, 261, 282, 286, 324, 325, 338, 339
active/smart carboxymethyl cellulose-polyvinyl alcohol composite films   39
agar   12, 40, 61, 92, 200, 205, 207–212, 218, 226, 243, 258, 263, 266, 338
agaran   207, 208
agarose   92, 207–209
aging   3, 12–14, 144
agro-waste based biopolymers
   cellulose   76–77
   edible coatings and films   80–82
   hemicellulose   77
   lignin   77–78
   pectin   79
   starch   78–79
alginate   26, 34–35, 37, 39, 40, 56, 79, 90, 92, 142, 200, 202–205, 216, 217, 222, 258, 285, 315, 316
antifungal activity of curcumin   167
aromatic compounds   314, 319–320
Aureobasidium pullulans   299

b

bacterial cellulose   279, 298–299
bacteriocins   35, 314, 320–321, 323, 325
bamboo   10, 335
bentonite-reinforced polymer nanocomposite   142–143
bio-based food packaging films   75, 336
bio-based polymers   1, 10, 12, 90, 265, 275, 279, 292, 295
bio-based polypropylene (Bio-PP)   296
biodegradable foams   89
biodegradable food packaging   2, 14, 34, 143, 224, 242–248, 265, 334–335
biodegradable nanocomposites (NCs)   122, 264, 335
biodegradable packaging   96–97, 257, 261, 267, 275, 300, 334, 335, 340, 344
biodegradable polyester PHA   6
biodegradable polymer blends and composites
   PCL and polyethylene blends   8–9
   PCL and polyvinyl chloride blends   9
   PLA and acrylobutadiene styrene (ABS) blends   8
   poly(butylene succinate) blends   10
   polylactic acid and polyethylene blends   8

Natural Materials for Food Packaging Application, First Edition.
Edited by Jyotishkumar Parameswaranpillai, Aswathy Jayakumar,
E. K. Radhakrishnan, Suchart Siengchin, and Sabarish Radoor.
© 2023 WILEY-VCH GmbH. Published 2023 by WILEY-VCH GmbH.

TPS and polypropylene blends 9
TPS/PE blends 9
biodegradable polymer-clay nanocomposites 138
biodegradable polymers 1, 2, 4–10, 12, 23, 24, 38, 90, 138, 187, 209, 257–259, 263, 275, 281, 284, 334–336, 343
biofilms 75–79, 81, 83, 106–109, 131, 167, 218, 286, 316
bioplastics 3, 243, 275, 335–337
biopolymers 2, 4, 7, 23, 26, 32, 34, 35, 38, 39, 41, 52, 55, 76–80, 83, 89, 90, 139, 141, 149, 180, 187, 188, 190, 193, 199–226, 242, 243, 245, 257, 258, 264, 266, 267, 275, 279, 283, 294, 300, 323, 333, 336–339, 345
burst strength 279, 283, 294, 295, 301
butyl-hydroxyanisole (BHA) 23
butyl-hydroxytoluene (BHT) 23

c

carboxymethylcellulose (CMC) 32, 34, 39, 40, 77, 92, 97, 98, 141, 248, 259–261, 263, 264
carrageenan 114, 168, 170, 200, 213, 214, 218, 222, 223, 258, 263, 266, 315, 324
casein 243, 286, 287
cello-biohydrolases 11
cellulose 4, 15, 25, 27, 32–34, 36, 39, 40, 55, 75–78, 89, 90, 109, 112, 113, 129, 143, 168–170, 187, 188, 220, 225, 226, 243, 247, 258, 259, 262, 263, 279–284, 289, 291, 294, 298–299, 335, 336, 338, 340, 343
cellulose and cellulosic derivatives-based food packaging films 32–34
cellulose ester-based films 55
cellulose nanofibers (CNFs) 79, 112, 113, 115, 141, 143, 225, 294
cellulosic reinforcements 188
chemical hydrogels 90
chitin 11, 27, 75, 77, 97, 220, 243, 244, 247, 249, 285, 315, 335

chitosan 4, 11, 15, 25, 27–30, 33, 36–38, 41, 55–58, 61, 64, 75, 77–79, 90, 97, 109, 112–114, 139, 144, 167–170, 200, 217, 220–226, 241–251, 258, 261–264, 279, 280, 282, 284–285, 289, 294, 295, 299, 315, 321, 324, 335, 338, 340
chitosan blends/composites 222–224
chitosan composites/plant extract-based food packaging films 36–38
chitosan (CS)/ hydroxypropyl methylcellulose blend 37
chitosan plant extract-based food packaging films 27–29
clay-based packaging 122
clay materials classification 127–128
CMC/ZnO nanocomposite films 264
Codium tomentosum seaweed extract (SE) 37
collagen 34, 76, 92, 222, 287–290, 320, 335, 340
collagen-based bioplastics 335
color agent effects 262–263
composites/plant extract-based food packaging films 35–41
consequence of storage time 12–14
containment 122, 256
curcumin 28
  active food packaging 169
  antimicrobial, antifungal and antioxidant properties of 166–167
  definition 165
  food packaging 168–169
  intelligent food packaging 170
  nanocomposite food packaging 169
  nanoencapsulation of 167–168
  structural characteristics of 165–166
cut jute fibres (CJF) 111, 122

d

diferuloylmethan e (1,7-bis(4-hydroxy-3-ethoxyphenyl)-1,6-heptadiene-3,5-dione) 165

different biobased packaging materials
  bioplastics   336
  biopolymers   336–338

e

eco friendly packaging   341
edible coatings   2, 3, 24, 80, 89, 245, 316, 320
edible coatings and films   80–82, 209, 220, 223
edible films in food packaging   339–340
elongation at break (EB)   36, 37, 55, 111, 113, 114, 131, 168, 181, 182, 185, 186, 188, 226, 245, 259, 277, 278
endo-cellulases   11
environmental impact of food packaging materials   14
environment and food packaging   341–342
essential oils (EO)   261, 316
  antibacterial properties of   61–64
  antioxidant properties of   58–60
  barrier properties   56
  challenges and future trends   65
  chemistry and classification of   52–54
  physical properties   56–58
  tensile properties   55–56
exo-cellulases   11
extended polystyrene (EPS)   89
extraction methods for plant extracts   24–25
extrusion coating and lamination   108

f

fiber reinforced biofilms, manufacturing of   107–109
fiber spinning   108
flame retardant micro-encapsulated ammonium polyphosphate   10
flax based films   113–114
flax cellulose nanocrystals (FCN) films   113
flaxseed gum (FG) films   114
flexible chitosan films   245

food packaging
  system   256
  thermal properties of   259–260
food packaging materials   275, 276
  characteristics of   242
  chitin and chitosan chemical structure   243
  chitin/chitosan derivatives   249
  chitosan   244
  chitosan and cell signalling pathways   250
  chitosan film   244–245
    chitosan-based films   245
    flexible packaging films   245
  crop protection   249–250
  films embedded with clays   246–247
  films embedded with nanomaterials   245–246
  films embedded with natural oils and extracts   247–248
  films embedded with polysaccharide particles, fibres and whiskers   247
  history of   241–242
  molecular imprinting technology   250
  need for biodegradable   242–243
  resorbable chitosan matrix   250
food preservation and storage
  aromatic compounds   319–320
  available solutions from the natural resources and combination with technology   314–315
  bacteriocins   320–321
  different extraction processes employed for natural materials   321–322
  effects of natural materials   323–325
  essential oils   316–318
  major objective of   313–314
  other animal-based antimicrobials   321
  phenolic compounds   318–319
  polysaccharides   315–316
  proteins   320
food quality attributes   256
fossil-based plastics   333
fucoidan   200, 216–218

g

gelatin   26, 34–35, 37, 39, 41, 56, 57, 76, 78, 79, 90, 131, 168, 170, 222, 223, 226, 259–262, 264–266, 279, 281, 282, 285, 288–290, 299, 324, 338, 340

gelatin and Alginate/plant extract-based food packaging films   34–35

gellan   300

"Generally Recognized As Safe" (GRAS)   134, 165, 285, 292, 299, 316

ginger EO (GEO)   57, 262

glass transition temperature   6, 37, 113, 180, 181, 186, 259, 261, 280, 293, 295

glucomannan   34, 61, 90, 261

gluten protein   291–292

glycerol   5, 9–12, 30, 79, 109, 112, 113, 131, 143, 180–182, 185, 186, 188, 219, 259, 260, 265, 266, 280, 281, 284–286, 288, 289, 295, 300

glycerol-containing TPS/PP polymer mixtures   9

green packaging materials   2, 335

Green Product innovation   334

h

heat-sealing process   278

hectorite-reinforced polymer nanocomposite   143–144

hemicellulose   34, 77, 78, 299

hemp based films   115

hot pressing method   183, 184

hydration-rich polymer gels   90

hydrogel-based 3D bioprinting   99

hydrogel-forming microneedles (HFMs)   100

hydrogels   90
  biodegradability   97–98
  biodegradable packaging   96–97
  classification according to network electrical charge   92
  classification according to polymeric composition   91
  classification based on configuration   91
  classification based on physical appearance   92
  classification based on the type of cross-linking   91
  classification of   93
  food packaging material   92–93
  functional properties   93
  hydrogel-based 3D bioprinting   99
  hydrogel-forming microneedles (HFMs)   100
  in vitro and food matrices   96
  latest development   98
  other potential applications in the food industry   98
  preparation of hydrogel film   92
  star-PEG-heparin hydrogel platform   100
  unravelling the antimicrobial activity of peptide hydrogel systems   99

hydroxypropyl methylcellulose (HPMC)   32, 34, 37, 40, 78, 284, 299

hydroxy propyl methyl cellulose (HPMC)-lignin   78

i

industrial agars   207

intelligent food packaging materials   170

interpenetrating polymer networks (IPN)   90, 91

j

jute based films   111–112

k

kaolinite   127, 128, 134, 145

kenaf based films   114–115

l

lactoferrin   321

laponite-reinforced polymer nanocomposite   141

levan   300

life cycle assessment (LCA)   2, 14

lifecycle of films   105

lignin   4, 25, 77–78, 115, 188–190, 299, 336, 338
lignin-cellulose nanocrystals (L-CNC)   115
linear low-density polyethylene (LLDPE) films   37, 39, 122, 133, 140, 262
lysozyme   97, 144, 321

m

main seaweed polysaccharides   201–202
marketing of natural materials in packaging   343–344
microencapsulation   30, 323
migration   4, 14, 25, 40, 52, 65, 81, 132, 145–149, 209, 315
modified chitosan
  chemical modifications of   220–222
  chitosan blends/composites   222–224
  nanomaterials   224–226
Modulus of Elasticity   277
molecular imprinting technology   250
montmorillonite (MMT)   77, 128, 131–135, 139–142, 144, 147–149, 262, 335
montmorillonite (MMT) based nanocomposite   139–141
multifunctional carboxymethyl cellulose/ agar-based smart films   40

n

nanocellulose   4–6, 77, 113, 127, 188, 225
nanoclay-based food packaging   123, 127, 146
nanoclay minerals   123
nanoencapsulation of curcumin   167–168
nanofibrillated cellulose (NFC) films   34
nano food packaging   338
nanotechnology   23, 122, 123, 146, 171, 224, 225, 263, 338, 344
native agars   207
natural agars   207
natural antibacterial agents   338, 339
natural biodegradable polymers
  polyanhydrides   7–8
  poly-caprolactone (PCL) based natural materials   5–6
  polycarbonate based natural materials   7
  polyglycolide-based natural materials   6–7
  poly-hydroxy alkanoate based natural materials   6
  poly-lactic acid-based natural materials   5
  polyurethanes   7
  soy-based bio-degradable polymers   7
  starch-based natural materials   4–5
natural/bioderived monomers   279, 292
natural clay-based food packaging films
  application of   135–145
  barrier property   132–133
  challenges of   145–149
  clay materials classification   127–128
  mechanical property   131
  oxygen and ethylene scavenging activity   133–134
  preparation of   128–130
  thermal stability   133
natural fibers   1, 105–116, 335
naturally occurring polymers   249, 279–292
natural materials
  drying methods and corresponding properties   323
  enhancement of packaging characteristics   323–324
  maintenance of physiochemical properties of raw and processed products   324–325
natural materials properties for food packaging
  barrier properties   10–11
  biodegradation properties   11–12
  consequence of storage time   12–14
natural polymer-based films   275
natural polymers   91, 92, 202, 264, 275, 279–300, 338, 339
nisin   40, 78, 320, 321, 325
non-degradable plastic packaging   333

p

package manufacturing   335
packaging   241
   active   23, 24, 27, 30, 33, 36, 52, 65, 121, 169, 209, 219, 223, 226, 258, 261, 282, 286, 324, 325, 338, 339
   biodegradable   96–97, 257, 261, 267, 275, 300, 334, 335, 340, 344
   biodegradable food   2, 14, 34, 143, 224, 242–248, 265, 334–335
   clay-based   122
   eco friendly   341
   edible films in food   339–340
   environmental impact of food   14
   environment and food   341–342
   green materials   2, 335
   intelligent food, materials   170
   marketing of natural materials in   343–344
   nanoclay-based food   123, 127, 146
   nano food   338
   non-degradable plastic   333
   sustainable   342–343
packaging films
   biodegradable polymers   275
   mechanical properties   276–279
   microorganisms   296–298
   natural polymers   279–296
packaging industry   41, 65, 92, 105, 121, 122, 191, 251, 257, 275, 333–335, 343–345
PCL and polyethylene blends   8–9
PCL and polyvinyl chloride blends   9
pectin   30, 75, 79, 90, 169, 170, 222, 226, 258, 285–286, 299, 316
pediocins   320, 321, 325
phenolic compounds   24, 25, 28, 30, 109, 201, 216, 222, 261, 314, 316, 318–319
photo-or thermo-oxidizable polymers   12
physical hydrogels   90
pineapple based films   112–113
pine needles   28
PLA and acrylobutadiene styrene (ABS) blends   8

plant based protein polymers   75
plant extract   23–41, 248, 257, 259–261, 264, 318, 323, 324, 335, 339
plant extracts-based food packaging films
   bibliographic research investigation   25–26
   cellulose and cellulosic derivatives-based food packaging films   32–34
   chitosan plant extract-based food packaging films   27–29
   composites/plant extract-based food packaging films   35–41
   extraction methods for plant extracts   24–25
   gelatin and alginate/plant extract-based food packaging films   34–35
   starch/extract-based food packaging films   30–31
plastic-based food packaging materials   1, 242
plasticization, starch   180–183
plasticizers   5, 9, 11, 13, 14, 30, 55, 80, 109, 111–113, 131, 180–188, 193, 214, 219, 244, 258, 259, 264–267, 280–289, 292, 293, 298–300, 301, 340
polyanhydrides   7–8
poly(butylene adipate-co-terephthalate) (PBAT)   39, 56, 187, 188, 190, 279, 281, 292, 294, 296
poly(butylene succinate) (PBS)   10, 122, 187, 279, 292, 295, 298
poly(butylene succinate) (PBS) blends   10, 122, 187, 279, 292, 295, 298
poly-caprolactone (PCL) based natural materials   5–6
polycarbonate based natural materials   7
polyethylene   6, 8, 9, 24, 34, 40, 41, 65, 115, 140, 145, 188, 245, 257, 279, 284, 286, 291, 296
polyethylene furanoate (PEF)   279, 292, 295

polyglycolide-based natural
    materials 6–7
poly-hydroxy alkanoate based natural
    materials 6
polyhydroxyalkanoates (PHAs) 6, 41,
    78, 279, 296–298, 338
poly(3-hydroxybutyrate) (PHB) 3, 6, 15,
    41, 78, 190, 297, 298
polyhydroxyl-3-butyrate-co-3-valerate
    (PHBV) 109–111, 297, 298
polylactic acid (PLA) 4–6, 8, 15, 26,
    39, 41, 65, 79, 111, 112, 122,
    129, 139–142, 187, 190, 191, 223,
    226, 279, 281, 292–296, 298,
    300, 335
polylactic acid and polyethylene
    blends 8
polylactic acid (PLA)-based natural
    materials 5
polymer-clay nanocomposites (PCNs)
    122, 123, 133, 135, 138, 140,
    147, 149
polymer nanoclay composites
    124–126
polymer nanocomposites (PNCs) 128,
    133, 141, 143, 146, 147, 150, 299
polyols plasticizers 265
polysaccharides 4, 23, 25, 30, 32, 34, 35,
    37, 55, 57, 58, 76, 80, 81, 92, 96, 97,
    109, 144, 168, 179, 199–219, 226,
    243, 245, 247, 250, 257, 258,
    264–266, 267, 279, 280, 285, 286,
    298–300, 314–316, 322, 335,
    338, 339
polysaccharides from seaweeds
    agar 205–212
    alginates 202–205
    carrageenan 213–216
    fucoidan 216–218
    main seaweed
        polysaccharides 201–202
    ulvan 218–219
polyurethanes 7, 122, 338
poly-vinyl-alcohol (PVA)-starch
    blend 38

proteins 4, 7, 23, 34, 55–57, 76, 80, 81,
    89, 90, 92, 113, 139, 166–168, 170,
    200, 201, 213, 243, 245, 248, 250,
    257, 258, 264–267, 279–282,
    286–292, 299, 300, 319–324, 335,
    339, 340
proteolytic enzymes 12
P3HB-co-3HV polymers 6
P3HB homogeneous polymers 6
pullulan 223, 299–300
puncture resistance 127, 276, 279, 286,
    287, 295

q
quercetin 248

r
rectorite-reinforced polymer
    nanocomposite 144
recycling 2, 122, 146–148, 263, 333,
    335, 341
rice straw based films 109, 110

s
seal strength 276, 278, 283, 285, 294
sepiolite-reinforced polymer
    nanocomposite 141–142
silver nanoparticles (AgNPs) 96, 219,
    246, 264
solution casting process 107
solution-induced intercalation 130
soy-based bio-degradable polymers 7
soy protein 90, 113, 279, 289–291, 299,
    324, 335, 340
Sphingomonas elodea 300
starch 78, 280, 283
    definition 179
    plasticization 180–183
    TPS, preparation of 180
starch-based natural materials 4–5
starch-based plastics 179–193
starch-chitosan-konjac glucomannan
    blend films (SCK) 61
starch composites/extract-based food
    packaging films 38–39

starch/extract-based food packaging films 30–31
star-PEG-heparin hydrogel platform 100
sulfated polysaccharides (SPs) 201
superabsorbent polymer (SAP) based hydrogel 90
sustainable food packaging 333, 343
sustainable packaging 220, 334, 336, 342–344
synthetic polymer-clay nanocomposites 135–137

t

tear resistance 187, 276, 278
tensile modulus 8, 142, 181, 182, 185, 186, 277
tensile properties 8, 10, 29, 55–56, 112–115, 188, 276, 277, 285, 301
tensile strength (TS) 8, 9, 29, 35–40, 55, 56, 80, 81, 90, 109, 111, 113–115, 131, 168, 181, 182, 185–189, 205, 245, 259, 276, 277
tert-butyl hydroxyquinone (TBHQ) 23
thermal stability of food packaging
    effects of color agent 262–263
    effects of nanomaterials 263–264
    emulsifiers/surfactants 266
    essential oils (EO) 261–262
    plant extract 260–261
    plasticizers effect 265–266
thermo gravimetric analysis (TGA) 186, 259, 261, 263
thermo-plasticization 180
thermoplastic starch (TPS)
    biopolymer blends 187–188
    biopolymer composites 188–191
    preparation of 180
    processing of 183–185

    properties of 185
        barrier properties 186–187
        mechanical properties 185–186
        thermal behaviour 186
tissue engineering (TE) 90, 99, 100, 205, 218, 250, 299
toxicity of nano clay 148–149
TPS and polypropylene blends 9
TPS/PE blends 9
turmeric 28, 165, 248

u

ulvan 200, 218–219

w

water-in-oil (W/O) nanoemulsion 35
water solubility (WS) 28, 29, 35, 37, 57, 77, 114, 115, 165, 168, 217, 259, 263, 284
water vapor permeability (WVP) 10, 11, 29, 36–40, 56, 89, 111, 113–115, 132, 141, 142, 168, 169, 181, 186–188, 217, 219, 223, 262, 324, 335
wheat straw based films 109, 111
wheat straw fibres (WSF) 109, 111
whey protein 90, 168, 287, 288, 290, 320, 323, 335, 340

x

xanthan 222, 299
Xanthomonas campestris 299

y

young's modulus 8, 9, 13, 55, 113–115, 131, 143, 277–278, 282, 284, 294, 298

z

zein protein 289–290, 323